Advanced Textbooks in Control and Signal Processing

Series Editors

Professor Michael J. Grimble, Professor of Industrial Systems and Director
Professor Michael A. Johnson, Professor of Control Systems and Deputy Director
Industrial Control Centre, Department of Electronic and Electrical Engineering,
University of Strathclyde, Graham Hills Building, 50 George Street, Glasgow G1 1QE,
U.K.

Other titles published in this series:

Genetic Algorithms: Concepts and Designs
K.F. Man, K.S. Tang and S. Kwong

Model Predictive Control
E. F. Camacho and C. Bordons

Introduction to Optimal Estimation
E.W. Kamen and J.K. Su

Discrete-time Signal Processing
D. Williamson

Neural Networks for Modelling and Control of Dynamic Systems
M. Nørgaard, O. Ravn, N.K. Poulsen and L.K. Hansen

Modelling and Control of Robot Manipulators (2nd Edition)
L. Sciavicco and B. Siciliano

Fault Detection and Diagnosis in Industrial Systems
L.H. Chiang, E.L. Russell and R.D. Braatz

Soft Computing
L. Fortuna, G. Rizzotto, M. Lavorgna, G. Nunnari, M.G. Xibilia and
R. Caponetto

Statistical Signal Processing
T. Chonavel – Translated by Janet Ormrod

Parallel Processing for Real-time Signal Processing and Control
M.O. Tokhi, M.A. Hossain and M.H. Shaheed
Publication due January 2003

T. Söderström

Discrete-time Stochastic Systems

Estimation and Control

 Springer

T. Söderström, PhD
Department of Systems and Control, Information Technology, Uppsala University, PO Box 337, SE 75105 Uppsala, Sweden

British Library Cataloguing in Publication Data
Soderstrom, Torsten
  Discrete-time stochastic systems : estimation and control.
  - 2nd ed. - (Advanced textbooks in control and signal
  processing)
  1.Stochastic systems 2.Discrete-time systems 3.Control
  theory
  I.Title
  003.8'5
  ISBN 1852336498

Library of Congress Cataloging-in-Publication Data
Söderström, Torsten.
  Discrete-time stochastic systems : estimation and control / T. Söderström.--2nd ed.
    p. cm. -- (Advanced textbooks in control and signal processing, ISSN 1439-2232)
  Includes bibliographical references and index.
  ISBN 1-85233-649-8 (alk. paper)
    1. Automatic control. 2. Control theory. 3. Discrete-time systems. I. Title. II. Series.
  TJ213 .S517 2002
  629.8--dc21   2002067022

ISSN 1439-2232
ISBN 1-85233-460-6 Springer-Verlag London Berlin Heidelberg
a member of BertelsmannSpringer Science+Business Media GmbH
http://www.springer.co.uk

1st edition published 1994, published by Prentice Hall, ISBN 0-13-309683-1

Typesetting: Electronic text files prepared by author
Printed and bound by Athenæum Press Ltd., Gateshead, Tyne & Wear
69/3830-543210 Printed on acid-free paper SPIN 10878170

To Marianne, Johanna, Andreas and David

Series Editors' Foreword

The topics of control engineering and signal processing continue to flourish and develop. In common with general scientific investigation, new ideas, concepts and interpretations emerge quite spontaneously and these are then discussed, used, discarded or subsumed into the prevailing subject paradigm. Sometimes these innovative concepts coalesce into a new sub-discipline within the broad subject tapestry of control and signal processing. This preliminary battle between old and new usually takes place at conferences, through the Internet and in the journals of the discipline. After a little more maturity has been acquired by the new concepts then archival publication as a scientific or engineering monograph may occur.

A new concept in control and signal processing is known to have arrived when sufficient material has evolved for the topic to be taught as a specialised tutorial workshop or as a course to undergraduate, graduates or industrial engineers. *Advanced Textbooks in Control and Signal Processing* are designed as a vehicle for the systematic presentation of course material for both popular and innovative topics in the discipline. It is hoped that prospective authors will welcome the opportunity to publish a structured and systematic presentation of some of the newer emerging control and signal processing technologies.

The essentials for any advanced course on control include a thorough understanding of state-space systems, both continuous- and discrete-time, the concepts of stochastic systems and insights into their optimal control systems. This textbook on *Discrete-time Stochastic Systems* by Torsten Söderström provides an invaluable introduction to these topics. It is a revised edition of an earlier Prentice Hall textbook which has benefited from a decade of classroom experience with Professor Söderström's course at Uppsala University, Sweden.

Apart from being used to support a full course, the text also has some interesting and useful features for the individual reader. The chapters are exceptionally well structured and can be used in a reference book fashion to instruct or review different technical topics, for example, spectral factorization. Unlike many linear stochastic control textbooks, Professor Söderström has given both time-domain and polynomial methods, proofs and techniques for topics like linear filtering and stochastic control systems. There are strong and fascinating links between these two approaches and it is invaluable to have them presented together in a single course textbook.

Every course lecturer likes to point their students to some topics which are a little more challenging and which might lead on to an interest deepening into research. Professor Söderström has included a chapter on nonlinear filtering which demonstrates how linear methods can be extended to deal with the more difficult nonlinear system problems. Each chapter is also accompanied by a bibliographical list of books and references to further reading for the interested reader.

The *Advanced Textbook in Control and Signal Processing* series seeks to create a set of books that are essential to a fundamental knowledge of the control and signal processing area. Professor Söderström's text is a welcome complement to other books in the series like Kamen and Su's *Introduction to Optimal Control* (ISBN 1-85233-133-X) and Williamson's *Discrete Time Signal Processing* (ISBN 1-85233-161-5) and we hope you will add *Discrete-time Stochastic Systems* to your library.

M.J. Grimble and M.A. Johnson
Industrial Control Centre
Glasgow, Scotland, U.K.
April, 2002

Preface

This book has been written for graduate courses in stochastic dynamic systems. It has emerged from various lecture notes (in Swedish and in English) that I have compiled and used in different courses at Uppsala University since 1978.

The current text is a second edition of a book originally published by Prentice Hall International in 1994. All chapters have been revised. A number of typographical and other errors have been corrected. Various new material and results, including further problems, have been added.

The reader is assumed to be somewhat familiar with dynamic systems and stochastic processes. In particular, it is assumed that the reader has a working knowledge of the following areas (or is prepared to re-examine this background elsewhere, should it be necessary):

- Fundamentals of linear discrete-time systems (such as state space models and their relationships with transfer function operators and weighting functions).
- Fundamentals of probability theory (including Gaussian distributed random vectors and conditional probabilities).
- Fundamentals of linear algebra and matrix calculations.
- Fundamentals of stochastic processes (such as the concepts of covariance function and spectral density, particularly in discrete time).

In compiling the manuscript, I have taken inspiration from various sources, including other books. Some parts reflect my own findings and derivations. The bibliographical notes at the end of each chapter give hints for further reading. These notes have intentionally been kept brief and there is no ambition to supply comprehensive lists of references. The cited references do contain, in many cases, extensive publication lists. Many books deal with the fundamentals of linear stochastic systems, using analysis of state space models leading to the celebrated Kalman filter. Treatments using polynomial methods, as presented here, seem much less frequent in the literature. The same comment applies to extensions to nonlinear cases and higher-order statistics.

Most of the chapters contain problems to be treated as exercises by the reader. Many of these are of pen-and-pencil type, while others require numerical computation using computer programs. Some problems are straightforward illustrations of the results in the text. Several problems, however, are designed to present extensions and to give further insight. To achieve knowledge and understanding of the estimation and control of stochastic systems, it is of great importance that the user gains experience by his or her own work. I hope that the problem sections will stimulate the reader to gain such experience. When using the text for graduate courses I have let the students work with some of the problems as an integral part of the course examination.

The structure of the book is as follows.

Chapter 2 begins by giving a short review of probability theory. Some useful properties of conditional probabilities and densities, and of Gaussian variables are stated. One section is devoted to complex-valued Gaussian variables.

Various model descriptions of stochastic dynamic systems are illustrated in Chapter 3. Some basic definitions are reviewed briefly. The important concepts of a Markov process and a state vector in a stochastic setting are discussed. Properties of covariance functions, spectra, and higher-order moments are also presented.

Chapter 4 covers the analysis of linear stochastic systems, with an emphasis on second-order properties such as the propagation of the covariance function and the spectral density in a system. Spectral factorization, which is a very fundamental topic, is also treated. It is concerned with how to proceed from a specified spectrum to a filter description of a signal, so constructed that optimal prediction and control can be derived easily from that filter model.

The topic of Chapter 5 is optimal estimation, where "optimal" refers to mean square optimality (*i.e.* the estimation error variance is minimized). Under certain conditions more general performance measures are also minimized. The general theory is given, showing that the optimal estimate can often be described by a conditional expectation.

The celebrated Kalman filter is derived in Chapter 6, using the results of the previous chapter. Optimal prediction and smoothing algorithms (assuming that future data are available) are presented for a general linear state space model.

Optimal prediction for processes given in filter or polynomial form is presented in Chapter 7. The basic relations for Wiener filtering are also derived. A general (single input, single output) estimation problem is solved by applying the Wiener filter technique and using a polynomial formalism. Where a time-invariant input–output perspective is sufficient, this gives a convenient and interesting alternative to the state space methodology based on the Kalman filter. The solution is naturally the same in the time-invariant case using either approach.

Chapter 8 is devoted to an example in which a detailed treatment using both the state space and the polynomial approaches is examined. The optimal filters, error variances, frequency characteristics, and so on, are examined, and it is shown how they are influenced by the measurement noise, *etc.* The calculations illustrate the close connections between the polynomial and the state space approaches.

Many systems are inherently nonlinear. In Chapter 9, some nonlinear filters and nonlinear effects such as quantization are dealt with. The greater part of the chapter deals with the extended Kalman filter and some variants thereof.

The topic of Chapter 10 is the control of stochastic systems. It is first shown that the way in which uncertainties are introduced can make a distinct difference to the way in which the system should be optimally controlled. Next, the general optimal control problem is handled by using the principle of dynamic programming. As this approach, while theoretically very interesting, requires extreme amounts of computation, some suboptimal schemes are also discussed.

Finally, optimal control for linear systems that are perturbed by process and measurement noise is the topic of Chapter 11. This problem, known as linear quadratic Gaussian control, has an elegant solution based on the so-called separation theorem, and this is also described. Also, the use of polynomial formalism to derive some simple optimal controllers and some comparisons with state space formalism are included.

It is worth stressing here that the Glossary (pp. xiii–xv) contains items that appear frequently in the book.

Several people have contributed directly or indirectly to these lecture notes. As mentioned above, I have been teaching stochastic systems regularly in Uppsala and elsewhere for more than a decade. The feedback I have received from the many students over the years has been very valuable in compiling the manuscript. For the second edition I thank all those who have pointed out unclear points and errors in the first version. Special thanks go to Fredrik Sandquist, who detected a tricky mistake, to Dr Erik G. Larsson who pointed out a flaw in a proof, and to Professor Torbjörn Wigren who gave many suggestions for improving the chapter on nonlinear estimation and also provided some further exercises. Dr Egil Sviestins and Dr Niclas Bergman have contributed with valuable comments on nonlinear estimation.

Several students who recently used the text in a graduate course have pointed out various typos or unclear points. I am grateful to Emad Abd-ElRady, Richard Abrahamsson, Bharath Bhikkaji, Hong Cui, Mats Ekman, Kjartan Halvorsen, Bengt Johansson, Erik K. Larsson, Kaushik Mahata, Hans Norlander and Erik Ohlander for numerous valuable comments. Needless to say, the responsibility for any remaining errors rests upon me.

Last, but not least, I also acknowledge the nice and smooth cooperation with Springer-Verlag, and the persons who in some way or another have been involved in the production of this book: Professor Michael Grimble, Professor Michael Johnson, Ms. Catherine Drury, Mr Frank Holzwarth, Mr Oliver Jackson and Mr Peter Lewis.

To all the above people I express my sincere thanks.

In control, feedback is an essential concept. This is also true for writing books. I welcome comments by the readers, and can be reached by the email address ts@syscon.uu.se.

Torsten Söderström
Department of Systems and Control
Uppsala University
P O Box 337, SE 751 05 Uppsala
Sweden
Spring 2002

Glossary

Notation

C	complex plane
CN	complex Gaussian distribution
E	expectation operator
$e(t)$	white noise (a sequence of independent random variables)
$G(q)$	transfer function operator
$H(q)$	transfer function operator, noise shaping filter
h	sampling interval
h_k	weighting function coefficient, $H(q) = \sum_{k=0}^{\infty} h_k q^{-k}$
I	identity matrix
I_n	$(n\|n)$ identity matrix
i	imaginary unit
K	optimal predictor gain (in Kalman filter)
K_f	optimal filter gain (in Kalman filter)
K_p	optimal predictor gain (in Kalman filter)
L	optimal feedback gain
$\mathcal{L}$	Laplace transform
$(m\|n)$	matrix is m by n
$\mathbf{N}(m, P)$	normal (Gaussian) distribution of mean value m and covariance matrix P
n	model order
P	probability
P	transition matrix for a Markov chain
$P(t)$	covariance matrix of state or state prediction error
p	differentiation operator
$p(x\|y)$	conditional probability of x given y
q	shift operator, $qx(t) = x(t+1)$
q^{-1}	backward shift operator, $q^{-1}x(t) = x(t-1)$
R	real axis
$r(t, s)$	covariance function
$r(\tau)$	covariance function
$r_{yu}(\tau)$	cross-covariance function between $y(t)$ and $u(t)$
$S(t)$	solution to Riccati equation for optimal control
A^T	transpose of the matrix A

t	time variable (integer-valued for discrete-time models)
tr	trace (of a matrix)
$u(t)$	input signal (possibly vector-valued)
V	loss function, performance index
$v(t)$	white noise (a sequence of independent random variables)
$x(t)$	state vector
Y^t	all available output measurements at time t, $Y^t = \{y(t), y(t-1)\ldots\}$
$y(t)$	output signal (possibly vector-valued)
$\hat{y}(t\vert t-1)$	optimal one-step predictor of $y(t)$
$\tilde{y}(t)$	prediction error, innovation
$\mathcal{Z}$	z transform
$\gamma(x; m, P)$	pdf of normal (Gaussian) distribution of mean value m and covariance matrix P
Δ	difference
$\delta_{t,s}$	Kronecker delta ($=1$ if $s = t$, else $= 0$)
$\delta(m, n)$	extended Kronecker delta ($=1$ if $m = n = 0$, else $= 0$)
$\delta(\tau)$	Dirac function
$\varepsilon(t)$	prediction error
Λ	covariance matrix of innovations
λ^2	variance of white noise
ρ	control weighting
ϕ	Euler matrix
$\phi_{cl}(z)$	closed loop characteristic polynomial
$\phi_{ol}(z)$	open loop characteristic polynomial
$\phi(z)$	spectrum
$\phi(\omega)$	spectral density
$\phi_u(\omega)$	spectral density of the signal $u(t)$
$\phi_{yu}(\omega)$	cross-spectral density between the signals $y(t)$ and $u(t)$
$\tilde{\phi}$	positive real part of spectrum
$\varphi(\omega)$	characteristic function
τ	time lag (in covariance function)
ω	angular frequency

Abbreviations

AR	autoregressive
AR(n)	AR of order n
ARE	algebraic Riccati equation
ARMA	autoregressive moving average
ARMA(n, m)	ARMA where AR and MA parts have orders n and m respectively
ARMAX	autoregressive moving average with exogenous input
cov	covariance matrix
deg	degree

FIR	finite impulse response
GMV	generalized minimum variance
GPC	generalized predictive control
iid	independent and identically distributed
IIR	infinite impulse response
IMM	interacting multi models
LLMS	linear least mean square
LQG	linear quadratic Gaussian
LTR	loop transfer recovery
MA	moving average
MA(n)	MA of order n
ML	maximum likelihood
pdf	probability density function
SISO	single input, single output
tr	trace (of a matrix)

Notational Conventions

$H^{-1}(q)$	$[H(q)]^{-1}$
$x^T(t)$	$[x(t)]^T$
A^{-T}	$[A^{-1}]^T$
$A \geq B$	the difference matrix $A - B$ nonnegative definite
$A > B$	the difference matrix $A - B$ positive definite
$\triangleq$	defined as
$\sim$	distributed as
$[\ ]_+$	causal part of a transfer function
$\overline{w}$	complex conjugate of w
w^T	transpose of w
w^*	conjugate transpose of w, $w^* = \overline{w}^T$
w^{-*}	$(w^{-1})^* = (w^*)^{-1}$

Conventions for Polynomials

$$A(z) = z^n + a_1 z^{n-1} + \ldots + a_n$$
$$A^*(z) = [A(z)]^* = z^{*n} + a_1^* z^{*(n-1)} + \ldots + a_n^*$$
$$A^*(z^{-*}) = z^{-n} + a_1^* z^{-(n-1)} + \ldots + a_n^*$$
$$A(z^{-1}) = z^{-n} + a_1 z^{-(n-1)} + \ldots + a_n$$

If $A(z)$ has real-valued coefficients

$$A^*(z^{-*}) = A(z^{-1})$$

On the unit circle

$$z^* = z^{-1}, \quad z^{-*} = z, \quad A(z) = A(z^{-*}), \quad A^*(z) = A^*(z^{-*})$$

Contents

1. Introduction

1.1 What is a Stochastic System?

A "stochastic system" is understood here as a *dynamic* system that has some kind of uncertainty. The type of uncertainty will be specified in a precise mathematical sense when dealing with methods of analysis and design. At this point, it is sufficient to say that the uncertainty will include disturbances acting on the system, sensor errors and other measurement errors, as well as partly unknown dynamics of the system. The uncertainties will be modelled in a probabilistic way using random variables and stochastic processes as important tools.

Theories of stochastic systems are very useful in many areas of systems science and information technology, such as controller design, filtering techniques, signal processing and communications. They give systematic techniques on how to model and handle random phenomena in dynamic systems.

Some typical illustrations of the usefulness of stochastic systems are given later in this chapter. They show that the concepts of stochastic dynamic systems can be useful for forecasting (Example 1.1), control under uncertainty (Example 1.2) and the design of filters (Example 1.3).

This book is aimed as an introduction to the properties of stochastic dynamic systems in discrete time. There are several reasons why the emphasis is on discrete-time systems only. One is that, today, processing equipment for filtering and control is very often based on digital hardware, so data are available only in discrete time. Another reason is that discrete-time stochastic processes are much easier to handle than their continuous-time counterparts, which have certain mathematical subtleties that are far from trivial to handle in a stringent way. Nevertheless, continuous-time processes will occasionally be discussed, especially as far as sampling is concerned.

Most of the material centres around the treatment of linear systems using variance criteria as measurements of performance. This is no doubt very useful in many areas of application. The combination of linear dynamics and quadratic performance criteria also leads to neat mathematical analysis. One should, however, remember that aspects other than low variance may sometimes be of importance. There can also be strong nonlinear effects to consider. Such aspects are only discussed briefly in the book, and the mathematics then

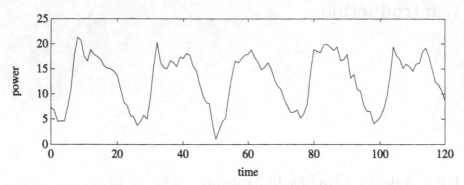

Fig. 1.1. Electric power consumption for a period of 120 h

no longer show the neat character of the linear quadratic case. Both input–output models and state space formalisms will be used extensively in the book. In the linear case, there are always close links between these two ways of treating dynamics, and it is fruitful to see how any concept appears in both types of model.

For illustration of the theories of stochastic systems that can be used, a few examples are in order.

Example 1.1 The consumption of electrical energy in an area varies considerably over time. A typical pattern is shown in Figure 1.1.

The energy consumption shows a regular variation through the day and decays to low values at night. There is also a random effect that adds to the regular effect. This random effect has several causes: effect of weather, special needs in industry, popular TV programs, *etc.* In order to generate the amount of power that is needed for every time instant, it is important to be able to forecast the demand a few hours ahead. The regular component of the consumption may be known, but there is a need to describe (*i.e.* model) the random contribution, and use that description to find good forecasts or *predictions* of its future value using currently available measurements.   □

Example 1.2 In the processing industry, there are many examples of production of paper, pulp, concrete, chemicals, *etc.*, where variations in raw material, temperature and several other effects produce random variations in the final product. For several reasons, the producer may want to reduce such variations. One reason could be the quality requirements of the customers. Another could be the need for more efficient saving of energy and raw material. A third could be that smaller variations allow a more economical setpoint. This is illustrated in Figure 1.2, which shows how a reduced variation can allow the setpoint to be chosen closer to a critical level.

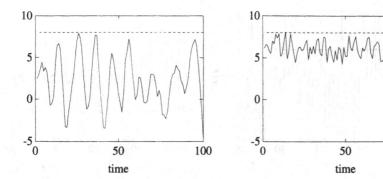

Fig. 1.2. Output variations (*solid lines*) around setpoints (*dotted lines*); critical values that presumably should not be passed (*dashed lines*); crude regulator (*left*) and a well-tuned regulator (*right*)

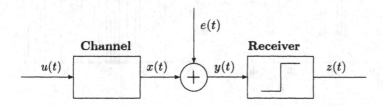

Fig. 1.3. Block diagram for a simple radio communication

To achieve efficient control of the process, it is often necessary to have a stochastic model of how the output is influenced both by the inputs (the control variables) and by disturbances. Such a model can then be the basis for the design of regulators, which seek to minimize the influence of the disturbances. □

Example 1.3 As yet another illustration, consider mobile radio communication, which in a very simplified form can be described as follows. The message to be transmitted is digitized. In this example it is represented as a binary signal, $u(t) = \pm 1$; see Figure 1.3.

The channel refers to the "system" or "filter" that describes how the signal is distorted before it arrives at the receiver. A typical reason for such distortions is that the signal propagates along several paths to the receiver. Signals that arrive after reflection travel a longer distance than direct signals and introduce a delay. There is often also noise, for example sensor noise in the receiver, $e(t)$, that adds to the signal, $x(t)$, so that the actual measurement is $y(t)$. A simple approach to reconstructing the transmitted signal is to take the

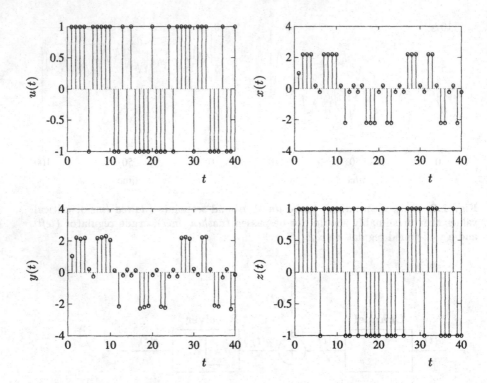

Fig. 1.4. Examples of signals for a digital radio communication

sign of $y(t)$ to form a binary signal $z(t)$. It should resemble the transmitted signal $x(t)$ for a good communication system. The procedure for determining $z(t)$ from the measurements $y(t)$ is called *equalizing*. The transmission causes a distortion of the transmitted signal, which is called *intersymbol interference*. A good equalizer will include a dynamic filter operating on $y(t)$ and not only the sign operator. To design such a filter, it is important to have a good description (*i.e.* a model) of the channel, and the statistical properties of the transmitted signal $x(t)$ and the disturbance $e(t)$. The "best" equalizer is a compromise between different objectives. Should there be no noise and the channel model be invertible, it is, of course, optimal to filter $y(t)$ by the inverse of the channel model. However, the inverse is often not stable, which makes the design more complicated. Another difficulty is how to take appropriate consideration of the noise. In the extreme case, when only the effect of the noise is considered, a filter giving zero as output would be ideal. In the general case, the filter must be a compromise between damping the noise and trying to "invert" the channel by a stable filter.

Figure 1.4 illustrates, by simulation for a simple case, what the signal $u(t)$, $x(t)$, $y(t)$ and $z(t)$ may look like.

This example can also illustrate the concept of *smoothing*. In order to reconstruct the transmitted signal $x(t)$ as efficiently as possible, it seems pertinent to allow the output $z(t)$ to depend not only on $y(s)$, $s \le t$, but also on future data, $y(s)$, $s \le t + \tau$. Such a principle would introduce a delay in the received message, so that ideally $z(t) = u(t - \tau)$. However, such a (small) delay can often be accepted, especially if it improves the quality of the outcome.   □

In various communication systems, such as radar, sonar and radio communications, it is convenient to describe the signals as being complex-valued. For example, in radar, the amplitude of the echo (response) is a measure of the effective size of the target, and the phase (due to Doppler shift in the carrier frequency) is a measure of the target's radial velocity towards the radar. In many parts of the book, complex-valued signals and processes are treated in order to make the treatment as general as possible. In other parts, though, the more traditional approach of considering only real-valued signals is employed.

The need for complex-valued signal models can be heuristically motivated in various ways.

- The signals are often of the narrow-band type, meaning that they have their energy concentrated in a small frequency region. The signals can therefore be (approximately) characterized as sinewaves. Interesting information is contained in the amplitude and the phase. To model amplitudes, phases and how they are affected by linear filtering, it is convenient to introduce complex-valued modelling of the signal.
- A radio communication signal contains a low-frequency message that is modulated using a carrier signal of high frequency. The transmitted signal then has a frequency content that is varied slightly around the carrier frequency. Distortion affects this frequency content. After demodulation, when retrieving the low-frequency message, it turns out that the frequency content is not symmetric. This can be viewed as a sign that a complex-valued description of the signal is needed.

Not only may the signal be complex-valued, but the dynamic system itself may also be complex-valued. Section 3.A gives a brief account of complex-valued models of narrow-band signals and the properties of linear dynamic complex-valued systems.

Bibliography

There is a huge literature on stochastic dynamic systems. For some alternative books on estimation and control, see, for example:

Anderson, B.D.O., Moore, J.B., 1979. *Optimal Filtering*. Prentice Hall, Englewood Cliffs, NJ.

Åström, K.J., 1970. *Introduction to Stochastic Control*. Academic Press, New York.

Borrie, J.A., 1992. *Stochastic Systems for Engineers*. Prentice Hall International, Hemel Hempstead, UK.

Brown, R.G., 1983. *Introduction to Random Signal Analysis and Kalman Filtering*. John Wiley & Sons, New York.

Grimble, M.J., Johnson, M.A., 1988. *Optimal Control and Stochastic Estimation*, vol. 2., John Wiley & Sons, Chichester.

Jazwinski, A.H., 1970. *Stochastic Processes and Filtering Theory*. Academic Press, New York.

Kailath, T., Sayed, A.H., Hassibi, B., 2000. *Linear Estimation*. Prentice Hall, Upper Saddle River, NJ.

Lewis, F.L., 1986. *Optimal Estimation*. John Wiley & Sons, New York.

Maybeck, P.S., 1979–1982. *Stochastic Models, Estimation and Control*, vols 1–3. Academic Press, New York.

Needless to say, there are also many books dedicated to stochastic processes in general.

For some collections of historical key papers on estimation of stochastic systems, see:

Kailath, T. (Ed.), 1977. *Linear Least-Squares Estimation*. Dowden, Hutchinson and Ross, Inc., Stroudsburg, PA.

Sorenson, H. (Ed.), 1985. *Kalman Filtering: Theory and Application*. IEEE Press, New York.

This book deals with analysis and design for given stochastic models. To build (or estimate) dynamic models from experimental data is called system identification. For some general texts on that subject, see:

Ljung, L., 1999. *Identification – Theory for the User*, second ed. Prentice Hall, Upper Saddle River, NJ.

Söderström, T., Stoica, P., 1989. *System Identification*. Prentice Hall International, Hemel Hempstead, UK.

2. Some Probability Theory

2.1 Introduction

Some fundamentals of probability theory are reviewed in this chapter. These concepts will be instrumental in describing random phenomena and properties of stochastic dynamic systems. The presentation will be intentionally brief, assuming that the reader already has some familiarity with the subject, and is intended mainly as a refresher chapter. The emphasis will be on multivariable random variables, Gaussian distributions, conditional distributions and complex-valued random variables.

2.2 Random Variables and Distributions

2.2.1 Basic Concepts

A random variable ξ is a function from an "event space" Ω to $\mathbf{R}$. Its "outcomes" (or realizations, or observed values) will be denoted by x. There is a probability measure associated with ξ so that subsets of Ω are assigned a probability $P(\Omega)$. There is a *distribution function* $F_\xi(x)$ and a *probability density function* (pdf) $p_\xi(x)$ defined as

$$F_\xi(x) = P(\xi \leq x) , \qquad (2.1)$$

$$p_\xi(x) = \frac{\mathrm{d}F_\xi(x)}{\mathrm{d}x} . \qquad (2.2)$$

The distribution function has the following properties:

- $F(x)$ is increasing.
- $\lim_{x\to\infty} F(x) = 1$.
- $\lim_{x\to-\infty} F(x) = 0$.

The pdf has the properties

- $p_\xi(x) \geq 0$.
- $\int_{-\infty}^{\infty} p_\xi(x)\,\mathrm{d}x = 1$.
- $P(a < \xi \leq b) = \int_a^b p_\xi(x)\,\mathrm{d}x$.

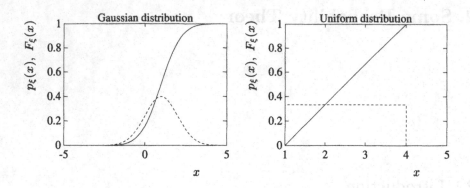

Fig. 2.1. Pdfs (*dashed lines*) and distribution functions (*solid lines*) for Gaussian and uniform distributions

Example 2.1 A *Gaussian*, or *normal*, *distribution* is characterized by

$$p_\xi(x) = \frac{1}{\sqrt{2\pi}\sigma} e^{-\frac{(x-m)^2}{2\sigma^2}} \qquad (\sigma > 0, \ x \in \mathbf{R}),$$

whereas a uniform distribution is given by

$$p_\xi(x) = \begin{cases} \frac{1}{b-a} & a < x \le b \ (b > a), \\ 0 & \text{elsewhere.} \end{cases}$$

These pdfs and associated distribution functions are illustrated in Figure 2.1.

□

Next consider random vectors. The previous development must be extended to multivariate ξ and x. Consider first the two-dimensional case, as generalization to an arbitrary dimension is straightforward.

Let the random variables ξ and η be characterized by a *joint distribution function*

$$F_{\xi,\eta}(x,y) = P(\xi \le x, \eta \le y) \tag{2.3}$$

and the associated (joint) pdf

$$p_{\xi,\eta}(x,y) = \frac{\partial^2 F_{\xi,\eta}(x,y)}{\partial x \partial y}. \tag{2.4}$$

These functions have the following properties:

- $F_{\xi,\eta}(x,y)$ is increasing in x and in y.
- $F_{\xi,\eta}(x,y) \to 1$ as $x \to \infty$ and $y \to \infty$.
- $F_{\xi,\eta}(x,y) \to 0$ as $x \to -\infty$ and $y \to -\infty$.
- $P((\xi,\eta) \in A) = \int \int_A p_{\xi,\eta}(x,y) \, dx \, dy$.
- $F_{\xi,\eta}(x,y) = \int_{-\infty}^{y} \int_{-\infty}^{x} p_{\xi,\eta}(x',y') \, dx' \, dy'$.

The relations to the marginal pdf $p_\xi(x)$ and the distribution function $F_\xi(x)$ are as follows:

- $F_\xi(x) = \lim_{y \to \infty} F_{\xi,\eta}(x,y)$.
- $p_\xi(x) = \int_{-\infty}^{\infty} p_{\xi,\eta}(x,y) \, dy$.

Example 2.2 Consider the joint pdf

$$p_{\xi,\eta}(x,y) = \frac{6}{7}(x+y)^2 , \qquad 0 \le x \le 1, \ 0 \le y \le 1 ,$$

for which the distribution function can be derived as follows:

$$F_{\xi,\eta}(x,y) = \int_0^x \int_0^y \frac{6}{7}(x'+y')^2 \, dx' \, dy'$$
$$= \frac{xy}{7}(2x^2 + 3xy + 2y^2) , \qquad 0 \le x \le 1, \ 0 \le y \le 1 .$$

The marginal pdf and the distribution functions of ξ and η are equal and are given by

$$F_\xi(x) = F_{\xi,\eta}(x,1) = \frac{x}{7}(2x^2 + 3x + 2) ,$$

$$p_\xi(x) = \frac{2}{7}(3x^2 + 3x + 1) .$$

These functions are illustrated in Figure 2.2. □

The expected value of a random variable, or its mean value, is

$$\mathbf{E}\,\xi = \int_{-\infty}^{\infty} x p_\xi(x) \, dx . \qquad (2.5)$$

More generally, the expected value of a function of ξ, say $g(\xi)$, is

$$\mathbf{E}\,g(\xi) = \int_{-\infty}^{\infty} g(x) p_\xi(x) \, dx . \qquad (2.6)$$

When $g(\xi)$ is a power of ξ (such as $g(\xi) = \xi, \xi^2$, etc.) the expected value is called a *moment* of ξ.

Let ξ be a random vector. Then its mean value is often denoted as

$$m = \mathbf{E}\,\xi = \int_{-\infty}^{\infty} x p_\xi(x) \, dx . \qquad (2.7)$$

The covariance matrix of ξ is defined as

$$\mathrm{cov}(\xi) = \mathbf{E}\,(\xi - m)(\xi - m)^T$$
$$= \mathbf{E}\,\xi\xi^T - mm^T . \qquad (2.8)$$

The following result will be used repeatedly in the book. It will be useful when evaluating various performance criteria.

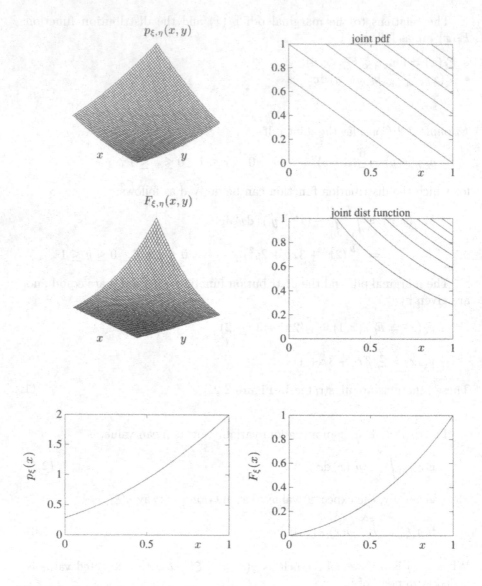

Fig. 2.2. Joint and marginal pdfs and distribution functions for Example 2.2. Joint pdfs and joint distributions given as mesh plots and with contour levels

Lemma 2.1 *Let x be a stochastic vector with mean m and covariance matrix P, and S a quadratic matrix. Then*

$$\mathbf{E}\, x^T S x = m^T S m + \operatorname{tr}(SP).$$

(2.9)

Proof Using straightforward calculations and properties of the trace operator:

$$\begin{aligned}
\mathbf{E}\, x^T S x &= \mathbf{E}\ \mathrm{tr}\,(S x x^T) = \mathrm{tr}\,(S\mathbf{E}\,[(x-m)(x-m)^T + m m^T]) \\
&= \mathrm{tr}\,(S[P + m m^T]) \\
&= \mathrm{tr}\,([m^T S m + S P]) \\
&= m^T S m + \mathrm{tr}\,(S P)\ .
\end{aligned}$$

$\blacksquare$

In subsequent chapters the same notation for random variables (such as ξ) and their outcomes (such as x) will often be used for simplicity.

2.2.2 Gaussian Distributions

Let m be an n-vector and P a symmetric positive definite ($P > 0$) $n|n$ matrix. The random vector is said to be normal or Gaussian distributed as

$$\xi \sim \mathbf{N}(m, P)\ ,$$

if its pdf is

$$\boxed{\begin{aligned}
p_\xi(x) &= \gamma(x; m, P) \\
&\triangleq \frac{1}{(2\pi)^{n/2}(\det P)^{1/2}} \exp\left(-\tfrac{1}{2}(x-m)^T P^{-1}(x-m)\right)\ .
\end{aligned}}
\qquad (2.10)$$

Note how this generalizes from the scalar case; see Example 2.1. The Gaussian pdf will be used repeatedly in the book. The specific notation $\gamma(x; m, P)$ is reserved for it.

There are several reasons why it is of particular interest to study the Gaussian distribution:

- Owing to the central limit theorem, the sum of many independent and equally distributed random variables can be well approximated by a Gaussian distribution. If disturbances are assumed to be due to the effects of many independent physical causes, it should therefore be relevant to model the total effect as Gaussian distributed random variables.
- Gaussian random variables have attractive mathematical properties; see below. In particular, linear transformations of Gaussian variables are still Gaussian distributed.
- For Gaussian distributed random variables, optimal estimates have a simple form. In contrast, in most other cases, it is very complicated to calculate optimal estimates explicitly.

The Gaussian distribution has the following properties:

- If $\xi \sim \mathbf{N}(m, P)$ then $\mathbf{E}\,\xi = m$.
- If $\xi \sim \mathbf{N}(m, P)$ then $\mathrm{cov}(\xi) = P$.
- If $\xi \sim \mathbf{N}(m, P)$ is a random n-vector, A an $r|n$ matrix of rank $r \le n$ and b an r vector, then $\eta = A\xi + b$ is also Gaussian distributed, $\eta \sim \mathbf{N}(Am + b, APA^T)$.

The third property mentioned above is very important. Gaussian distributions are preserved under linear transformations. As an implication, linear filtering of a Gaussian distributed input signal will produce a Gaussian distributed output.

One important implication of the first two properties is that the distribution is fully determined by the mean vector and the covariance matrix.

2.2.3 Correlation and Dependence

Let ξ and η be two random vectors with mean values m_ξ and m_η respectively. The vectors are said to be *uncorrelated* if

$$\mathbf{E}\,(\xi - m_\xi)(\eta - m_\eta)^T = 0 . \tag{2.11}$$

The vectors are said to be *independent* if

$$p_{\xi,\eta}(x,y) = p_\xi(x)p_\eta(y) . \tag{2.12}$$

The following results hold:

- ξ, η independent $\Longrightarrow$ ξ, η uncorrelated.
- ξ, η uncorrelated and Gaussian $\Longrightarrow$ ξ, η independent.

2.3 Conditional Distributions

Let ξ and η be two correlated random variables (possibly vector-valued). Assume that η is observed or measured to have an outcome y. This information is to be used to infer something about ξ. Given $\eta = y$, the pdf of ξ should take the form

$$p_{\xi|\eta=y}(x) = cp_{\xi,\eta}(x,y) ,$$

where c is a normalizing constant. Recall that $p_{\xi,\eta}(x,y)\,\mathrm{d}x\,\mathrm{d}y$ is the probability $P(x < \xi \le x + \mathrm{d}x, y < \eta \le y + \mathrm{d}y)$. As

$$1 = \int_{-\infty}^{\infty} p_{\xi|\eta=y}(x)\,\mathrm{d}x = c \int_{-\infty}^{\infty} p_{\xi,\eta}(x,y)\,\mathrm{d}x = cp_\eta(y)$$

it follows that $c = 1/p_\eta(y)$, and the *conditional pdf* is

$$p_{\xi|\eta=y}(x) = \frac{p_{\xi,\eta}(x,y)}{p_\eta(y)} . \tag{2.13}$$

In a more strict sense, (2.13) is often taken as the *definition* of the conditional pdf. One could possibly start with *Bayes' rule*

$$P(A|B) = \frac{P(AB)}{P(B)}$$

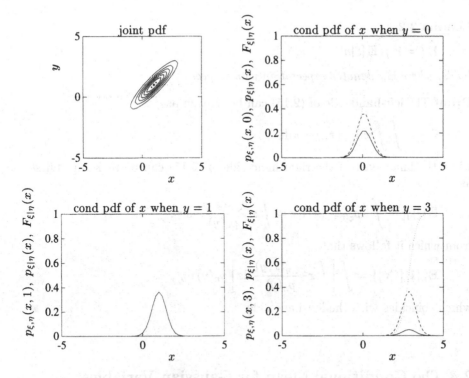

Fig. 2.3. Contour levels of the joint pdf, $p_{\xi,\eta}(x,y)$, the conditional pdf $p_{\xi|\eta=y}(x)$ and conditional distribution function $F_{\xi|\eta=y}(x)$ for $y = 0, 1$ and 3. (*Solid lines*) $p_{\xi,\eta}(x,y)$, (*dashed lines*) $p_{\xi|\eta=y}(x)$, (*dotted lines*) $F_{\xi|\eta=y}(x)$

for events A and B. There is, though, a difficulty with such an approach for "deriving" (2.13), as $P(\eta = y) = 0$ for a continuous random variable.

Using the relation between the joint and the marginal distributions, one finds

$$p_\xi(x) = \int p_{\xi,\eta}(x,y)\, \mathrm{d}y = \int p_{\xi|\eta=y}(x,y) p_\eta(y)\, \mathrm{d}y \,, \qquad (2.14)$$

which will be useful later.

Example 2.3 Let ξ and η be jointly Gaussian distributed:

$$\begin{pmatrix} \xi \\ \eta \end{pmatrix} \sim \mathbf{N}\left(\begin{pmatrix} 1 \\ 1 \end{pmatrix}, \begin{pmatrix} 1 & 0.9 \\ 0.9 & 1 \end{pmatrix} \right).$$

The joint pdf $p_{\xi,\eta}(x,y)$ and the conditional pdf $p_{\xi|\eta=y}(x)$ are illustrated in Figure 2.3. $\square$

The following lemma gives an important result concerning expectation and conditional expectation.

Lemma 2.2

$$\mathbf{E}\,\xi = \mathbf{E}_\eta\{\mathbf{E}\,[\xi|\eta]\} \tag{2.15}$$

holds, where $\mathbf{E}_\eta$ denotes expectation with respect to η.

Proof The left-hand side of (2.15) can be written as

$$\mathbf{E}\,\xi = \int\int x p_{\xi,\eta}(x,y)\,\mathrm{d}x\,\mathrm{d}y\;,$$

while the inner part of the right-hand side of (2.15) can be rearranged first as

$$\mathbf{E}\,[\xi|\eta] = \int x p_{\xi|\eta=y}(x)\,\mathrm{d}x = \int x\frac{p_{\xi,\eta}(x,y)}{p_\eta(y)}\,\mathrm{d}x\;,$$

from which it follows that

$$\mathbf{E}_\eta\{\mathbf{E}\,[\xi|\eta]\} = \int\left[\int x\frac{p_{\xi,\eta}(x,y)}{p_\eta(y)}\,\mathrm{d}x\right]p_\eta(y)\,\mathrm{d}y\;,$$

which coincides with the left-hand side. ∎

2.4 The Conditional Mean for Gaussian Variables

This section treats the case of Gaussian distributed random variables. It turns out that, for this case, the conditional pdf is Gaussian as well. The conditional mean is identical to the so-called linear least mean squares (LLMS) estimate, which will be derived in Section 5.3. The detailed result is given here as a lemma.

Lemma 2.3 *Let x and y be two real-valued, jointly Gaussian random variables*

$$\begin{pmatrix} x \\ y \end{pmatrix} \sim \mathbf{N}\left(\begin{pmatrix} m_x \\ m_y \end{pmatrix}, \begin{pmatrix} R_x & R_{xy} \\ R_{yx} & R_y \end{pmatrix}\right)\;. \tag{2.16}$$

Assume that the covariance matrix of the distribution is positive definite. Then the conditional pdf $p(x|y)$ is Gaussian as well:

$$p(x|y) \sim \mathbf{N}(m,P)\;, \tag{2.17}$$

with

$$m = \hat{x} = \mathbf{E}\,[x|y] = m_x + R_{xy}R_y^{-1}(y - m_y)\;,$$

$$P = \mathrm{cov}[x|y] = R_x - R_{xy}R_y^{-1}R_{yx}\;. \tag{2.18}$$

Proof Recall that the pdf of a Gaussian distribution $\mathbf{N}(m, P)$ is given by

$$p(x) = \frac{1}{(2\pi)^{n/2}(\det P)^{1/2}} \exp\left(-\frac{1}{2}(x-m)^T P^{-1}(x-m)\right),$$

where $n = \dim x$. Now set $n_x = \dim x$, $n_y = \dim y$ and $R = \begin{pmatrix} R_x & R_{xy} \\ R_{yx} & R_y \end{pmatrix}$.
Using Bayes' rule:

$$p(x|y) = \frac{p(x,y)}{p(y)} = \frac{(2\pi)^{n_y/2}(\det R_y)^{1/2}}{(2\pi)^{(n_x+n_y)/2}(\det R)^{1/2}}$$

$$\times \exp\left(-\frac{1}{2}(x^T - m_x^T \quad y^T - m_y^T)R^{-1}\begin{pmatrix} x - m_x \\ y - m_y \end{pmatrix}\right.$$

$$\left. + \frac{1}{2}(y - m_y)^T R_y^{-1}(y - m_y)\right). \tag{2.19}$$

Consider first the ratio of the determinants:

$$\det R = \det\begin{pmatrix} R_x & R_{xy} \\ R_{yx} & R_y \end{pmatrix}$$

$$= \det\left[\begin{pmatrix} I_{n_x} & -R_{xy}R_y^{-1} \\ 0 & I_{n_y} \end{pmatrix}\begin{pmatrix} R_x & R_{xy} \\ R_{yx} & R_y \end{pmatrix}\begin{pmatrix} I_{n_x} & 0 \\ -R_y^{-1}R_{yx} & I_{n_y} \end{pmatrix}\right]$$

$$= \det\begin{pmatrix} P & 0 \\ 0 & R_y \end{pmatrix} = \det P \det R_y$$

holds. Hence

$$\frac{(\det R_y)^{1/2}}{(\det R)^{1/2}} = \frac{1}{(\det P)^{1/2}}. \tag{2.20}$$

Proceed to examine the quadratic form in the exponent. It is first established that

$$R^{-1} = \begin{pmatrix} P^{-1} & -P^{-1}R_{xy}R_y^{-1} \\ -R_y^{-1}R_{yx}P^{-1} & R_y^{-1} + R_y^{-1}R_{yx}P^{-1}R_{xy}R_y^{-1} \end{pmatrix}.$$

This can be verified by direct multiplication (alternatively, use the hint of Exercise 3.4):

$$\begin{pmatrix} R_x & R_{xy} \\ R_{yx} & R_y \end{pmatrix}\begin{pmatrix} P^{-1} & -P^{-1}R_{xy}R_y^{-1} \\ -R_y^{-1}R_{yx}P^{-1} & R_y^{-1} + R_y^{-1}R_{yx}P^{-1}R_{xy}R_y^{-1} \end{pmatrix}$$

$$= \begin{pmatrix} PP^{-1} & -R_xP^{-1}R_{xy}R_y^{-1} + R_{xy}R_y^{-1} + R_{xy}R_y^{-1}R_{yx}P^{-1}R_{xy}R_y^{-1} \\ 0 & R_yR_y^{-1} \end{pmatrix}$$

$$= \begin{pmatrix} I_{n_x} & R_{xy}R_y^{-1} - PP^{-1}R_{xy}R_y^{-1} \\ 0 & I_{n_y} \end{pmatrix} = I_{n_x+n_y}.$$

Hence the quadratic form can be written as, skipping a factor $-\frac{1}{2}$:

$$(x^T - m_x^T \quad y^T - m_y^T) R^{-1} \begin{pmatrix} x - m_x \\ y - m_y \end{pmatrix} - (y - m_y)^T R_y^{-1}(y - m_y)$$

$$= (x - m_x)^T P^{-1}(x - m_x) - (x - m_x)^T P^{-1} R_{xy} R_y^{-1}(y - m_y)$$

$$- (y - m_y)^T R_y^{-1} R_{yx} P^{-1}(x - m_x)$$

$$+ (y - m_y)^T \{R_y^{-1} R_{yx} P^{-1} R_{xy} R_y^{-1}\}(y - m_y)$$

$$= (x - \hat{x})^T P^{-1}(x - \hat{x}) . \tag{2.21}$$

Inserting (2.20) and (2.21) into (2.19), it is found that

$$p(x|y) \sim N(m, P) ,$$

which is the stated result. ∎

Note that, for Gaussian variables, the conditional mean $\hat{x}$ becomes identical to the so-called LLMS estimate $\hat{x}_{\text{LLMS}}$. See Section 5.3.

In fact little more can be said about the optimal estimate $\hat{x}$ of a vector x when a correlated vector y is observed. It is easy to show (as in Section 5.3) that $x - \hat{x}$ *and y are uncorrelated.* It will also be shown that they are jointly Gaussian distributed and therefore *independent.* In fact:

$$\begin{pmatrix} x - \hat{x} \\ y \end{pmatrix} = \begin{pmatrix} I & -R_{xy} R_y^{-1} \\ 0 & I \end{pmatrix} \begin{pmatrix} x \\ y \end{pmatrix} + \begin{pmatrix} -m_x + R_{xy} R_y^{-1} m_y \\ 0 \end{pmatrix} .$$

This shows that $x - \hat{x}$ and y are jointly Gaussian. Further, as can also be predicted from the previous calculation:

$$\mathbf{E} \begin{pmatrix} x - \hat{x} \\ y \end{pmatrix} = \begin{pmatrix} 0 \\ m_y \end{pmatrix} ,$$

$$\mathbf{E} \begin{pmatrix} x - \hat{x} \\ y - m_y \end{pmatrix} ((x - \hat{x})^T \quad (y - m_y)^T)$$

$$= \begin{pmatrix} I & -R_{xy} R_y^{-1} \\ 0 & I \end{pmatrix} \begin{pmatrix} R_x & R_{xy} \\ R_{yx} & R_y \end{pmatrix} \begin{pmatrix} I & 0 \\ -R_y^{-1} R_{yx} & I \end{pmatrix} = \begin{pmatrix} P & 0 \\ 0 & R_y \end{pmatrix} ,$$

which demonstrates that the estimation error $x - \hat{x}$ is uncorrelated and independent of the measurement y.

There is another useful result dealing with the case when y can be split into two uncorrelated parts.

Lemma 2.4 *Let* $y = \begin{pmatrix} y_1 \\ y_2 \end{pmatrix}$ *with y_1 and y_2 uncorrelated and let* $\begin{pmatrix} x \\ y \end{pmatrix}$ *be jointly Gaussian distributed. Then*

$$\mathbf{E}[x|y] = \mathbf{E}[x|y_1] + \mathbf{E}[x|y_2] - \mathbf{E}[x] . \tag{2.22}$$

Proof Straightforward calculation, using Lemma 2.3

$$
\mathbf{E}\left[x|y\right] = m_x + R_{xy}R_y^{-1}(y - m_y)
$$

$$
= m_x + (R_{xy_1} \quad R_{xy_2}) \begin{pmatrix} R_{y_1} & 0 \\ 0 & R_{y_2} \end{pmatrix}^{-1} \begin{pmatrix} y_1 - m_{y_1} \\ y_2 - m_{y_2} \end{pmatrix}
$$

$$
= m_x + R_{xy_1}R_{y_1}^{-1}(y_1 - m_{y_1}) + R_{xy_2}R_{y_2}^{-1}(y_2 - m_{y_2})
$$

$$
= \mathbf{E}\left[x|y_1\right] + \mathbf{E}\left[x|y_2\right] - m_x .
$$

$\blacksquare$

2.5 Complex-Valued Gaussian Variables

Complex-valued Gaussian variables are treated in this section. The scalar case, which is relatively straightforward, is analyzed first. The multivariable case, which contains some technicalities, is treated next. Finally, the distributions of the norms of complex Gaussian variables are discussed.

2.5.1 The Scalar Case

Let x and y be zero mean Gaussian distributed random variables and set $w = x + iy$. Then define the variance of w as

$$
\mathbf{E}\,w\overline{w} = \mathbf{E}\,|w|^2 = \mathbf{E}\,(x + iy)(x - iy) = \mathbf{E}\,x^2 + \mathbf{E}\,y^2. \tag{2.23}
$$

Assume for simplicity that x and y are independent and *have equal variance*, say $\mathbf{E}\,x^2 = \mathbf{E}\,y^2 = \sigma^2/2$. Then $\mathbf{E}\,|w|^2 = \sigma^2$ and

$$
w \sim \mathbf{CN}(0, \sigma^2) \tag{2.24}
$$

is said to be complex Gaussian distributed with zero mean and variance σ^2. Note that in such a case

$$
\mathbf{E}\,w^2 = \mathbf{E}\,(x + iy)^2 = \mathbf{E}\,x^2 - \mathbf{E}\,y^2 + 2i\mathbf{E}\,xy = 0 . \tag{2.25}
$$

The pdf for w can be derived from

$$
p_w(w) = p\left(\begin{pmatrix} x \\ y \end{pmatrix}\right).
$$

Note that

$$
\begin{pmatrix} x \\ y \end{pmatrix} \sim \mathbf{N}\left(\begin{pmatrix} 0 \\ 0 \end{pmatrix}, \frac{\sigma^2}{2}\begin{pmatrix} 1 & 0 \\ 0 & 1 \end{pmatrix}\right)
$$

and therefore

$$p_w(w) = \frac{1}{(2\pi)^{2/2}\sqrt{\det(\sigma^2 I/2)}} \exp\left(-\tfrac{1}{2}(x\ \ y)(\tfrac{\sigma^2}{2}I)^{-1}\begin{pmatrix}x\\y\end{pmatrix}\right)$$

$$= \frac{1}{2\pi\tfrac{1}{2}\sigma^2} \exp\left(-(x\ \ y)\begin{pmatrix}x\\y\end{pmatrix}/\sigma^2\right)$$

$$= \frac{1}{\pi\sigma^2} e^{-(x^2+y^2)/\sigma^2} . \tag{2.26}$$

Next examine *linear transformations* of complex Gaussian variables. Let $\varphi = \varphi_R + i\varphi_I \in \mathbf{C}$, $w = x + iy \sim \mathbf{CN}(0,\sigma^2)$, and set $m = \varphi w$. Clearly:

$$\mathbf{E}\,m = 0 ,$$
$$\mathrm{var}(m) = \mathbf{E}\,\overline{m} = \mathbf{E}\,\varphi w\overline{w}\overline{\varphi} = \sigma^2|\varphi|^2 . \tag{2.27}$$

Moreover, the real and imaginary parts of m can be examined by rewriting it as

$$m = m_R + im_I = (\varphi_R + i\varphi_I)(x + iy) ,$$

giving

$$m_R = \varphi_R x - \varphi_I y ,$$
$$m_I = \varphi_R y + \varphi_I x ,$$

or

$$\begin{pmatrix}m_R\\m_I\end{pmatrix} = \begin{pmatrix}\varphi_R & -\varphi_I\\\varphi_I & \varphi_R\end{pmatrix}\begin{pmatrix}x\\y\end{pmatrix}. \tag{2.28}$$

This implies that

$$\mathbf{E}\begin{pmatrix}m_R\\m_I\end{pmatrix} = \begin{pmatrix}0\\0\end{pmatrix},$$

$$\mathbf{E}\begin{pmatrix}m_R\\m_I\end{pmatrix}(m_R\ \ m_I) = \begin{pmatrix}\varphi_R & -\varphi_I\\\varphi_I & \varphi_R\end{pmatrix}\mathbf{E}\begin{pmatrix}x\\y\end{pmatrix}(x\ \ y)\begin{pmatrix}\varphi_R & \varphi_I\\-\varphi_I & \varphi_R\end{pmatrix}$$

$$= \begin{pmatrix}\varphi_R & -\varphi_I\\\varphi_I & \varphi_R\end{pmatrix}\begin{pmatrix}\sigma^2/2 & 0\\0 & \sigma^2/2\end{pmatrix}\begin{pmatrix}\varphi_R & \varphi_I\\-\varphi_I & \varphi_R\end{pmatrix}$$

$$= \frac{\sigma^2}{2}|\varphi|^2 I .$$

As

$$\begin{pmatrix}m_R\\m_I\end{pmatrix} \sim \mathbf{N}\left(\begin{pmatrix}0\\0\end{pmatrix}, \frac{\sigma^2|\varphi|^2}{2}I\right)$$

obviously holds, one can conclude that

$$m \sim \mathbf{CN}(0, \sigma^2|\varphi|^2) . \tag{2.29}$$

2.5.2 The Multivariable Case

Before coping with the multivariable case, some technical results are needed, such as a convention for splitting up a complex-valued matrix, or vector, into real and imaginary parts.

In Section 2.5.1 it was shown how the scalar linear relationship

$$m = \varphi w \tag{2.30}$$

could be split up into real and imaginary parts as

$$\begin{pmatrix} m_\text{R} \\ m_\text{I} \end{pmatrix} = \begin{pmatrix} \varphi_\text{R} & -\varphi_\text{I} \\ \varphi_\text{I} & \varphi_\text{R} \end{pmatrix} \begin{pmatrix} w_\text{R} \\ w_\text{I} \end{pmatrix}. \tag{2.31}$$

This simple example gives some guidelines on how to link a complex-valued representation to a real-valued one.

In this section, let $f(w)$ denote the real-valued variable associated with w. In the above example, $f(w)$ and $f(\varphi)$ are given by

$$f(w) = \begin{pmatrix} w_\text{R} \\ w_\text{I} \end{pmatrix}, \quad f(\varphi) = \begin{pmatrix} \varphi_\text{R} & -\varphi_\text{I} \\ \varphi_\text{I} & \varphi_\text{R} \end{pmatrix}.$$

Next, this idea is formalized.

Convention 2.1 Let

$$w = (w_1 \ldots w_n)^T$$

be a complex-valued vector. Then define

$$f(w) \triangleq (w_{1\text{R}} \;\; w_{1\text{I}} \;\; w_{2\text{R}} \;\; w_{2\text{I}} \ldots w_{n\text{R}} \;\; w_{n\text{I}})^T,$$

where $w_{j\text{R}} = \text{Re}(w_j)$, $w_{j\text{I}} = \text{Im}(w_j)$. Similarly, if A is an $(m|n)$ matrix, introduce the matrix $f(A)$ of dimension $(2m|2n)$ as

$$f(A) = \begin{pmatrix} \tilde{a}_{11} & \ldots & \tilde{a}_{1n} \\ & \vdots & \\ \tilde{a}_{m1} & \ldots & \tilde{a}_{mn} \end{pmatrix},$$

where the block $\tilde{a}_{\mu\nu}$ is

$$\tilde{a}_{\mu\nu} = \begin{pmatrix} a_{\mu\nu\text{R}} & -a_{\mu\nu\text{I}} \\ a_{\mu\nu\text{I}} & a_{\mu\nu\text{R}} \end{pmatrix}$$

and where

$$a_{\mu\nu\text{R}} = \text{Re}(a_{\mu\nu}), \quad a_{\mu\nu\text{I}} = \text{Im}(a_{\mu\nu}). \qquad \square$$

Some results showing how transformations of A and w can be described equivalently by $f(A)$ and $f(w)$ will be needed later. A number of lemmas are now presented with this in mind.

Lemma 2.5 *The transformation $A \to f(A)$ is unique and invertible in the sense that*

$$A = B \Longleftrightarrow f(A) = f(B) .\tag{2.32}$$

Proof Obvious. ∎

Lemma 2.6 *Let A and B be two complex-valued matrices. Then*

$$f(AB) = f(A) \times f(B) .\tag{2.33}$$

Proof Evaluate the $\mu\nu$-block of each side:

$$(f(AB))_{\mu\nu} = f(\sum_k a_{\mu k} b_{k\nu}) = f(\sum_k (a_{\mu k\mathrm{R}} + ia_{\mu k\mathrm{I}})(b_{k\nu\mathrm{R}} + ib_{k\nu\mathrm{I}}))$$

$$= \sum_k f([a_{\mu k\mathrm{R}} b_{k\nu\mathrm{R}} - a_{\mu k\mathrm{I}} b_{k\nu\mathrm{I}}] + i[a_{\mu k\mathrm{R}} b_{k\nu\mathrm{I}} + a_{\mu k\mathrm{I}} b_{k\nu\mathrm{R}}])$$

$$= \sum_k \begin{pmatrix} a_{\mu k\mathrm{R}} b_{k\nu\mathrm{R}} - a_{\mu k\mathrm{I}} b_{k\nu\mathrm{I}} & -a_{\mu k\mathrm{R}} b_{k\nu\mathrm{I}} - a_{\mu k\mathrm{I}} b_{k\nu\mathrm{R}} \\ a_{\mu k\mathrm{R}} b_{k\nu\mathrm{I}} + a_{\mu k\mathrm{I}} b_{k\nu\mathrm{R}} & a_{\mu k\mathrm{R}} b_{k\nu\mathrm{R}} - a_{\mu k\mathrm{I}} b_{k\nu\mathrm{I}} \end{pmatrix} .$$

$$(f(A) \times f(B))_{\mu\nu} = \sum_k (f(A))_{\mu k} (f(B))_{k\nu}$$

$$= \sum_k \begin{pmatrix} a_{\mu k\mathrm{R}} & -a_{\mu k\mathrm{I}} \\ a_{\mu k\mathrm{I}} & a_{\mu k\mathrm{R}} \end{pmatrix} \begin{pmatrix} b_{k\nu\mathrm{R}} & -b_{k\nu\mathrm{I}} \\ b_{k\nu\mathrm{I}} & b_{k\nu\mathrm{R}} \end{pmatrix} .$$

Hence both sides of (2.33) are equal. ∎

Lemma 2.7 *Let A be invertible. Then*

$$f(A^{-1}) = (f(A))^{-1} .\tag{2.34}$$

Proof Apparently $f(I) = I$. (More exactly, $f(I_n) = I_{2n}$.) Using Lemma 2.6 with $B = A^{-1}$, one easily obtains

$$I = f(I) = f(A) \times f(A^{-1}) ,$$

which proves (2.34). ∎

Lemma 2.8

$$f(A^*) = (f(A))^T .\tag{2.35}$$

Proof By straightforward calculations,

$$(f(A^*))_{\mu\nu} = \begin{pmatrix} (A^*)_{\mu\nu R} & -(A^*)_{\mu\nu I} \\ (A^*)_{\mu\nu I} & (A^*)_{\mu\nu R} \end{pmatrix} = \begin{pmatrix} A_{\nu\mu R} & A_{\nu\mu I} \\ -A_{\nu\mu I} & A_{\nu\mu R} \end{pmatrix},$$

$$(f(A))_{\mu\nu}^T = \begin{pmatrix} A_{\nu\mu R} & -A_{\nu\mu I} \\ A_{\nu\mu I} & A_{\nu\mu R} \end{pmatrix}^T = \begin{pmatrix} A_{\nu\mu R} & A_{\nu\mu I} \\ -A_{\nu\mu I} & A_{\nu\mu R} \end{pmatrix}.$$

■

Lemma 2.9 *Let A be a quadratic matrix. Then*

$$A \text{ unitary} \iff f(A) \text{ orthogonal}. \tag{2.36}$$

Proof A unitary $\iff A^{-1} = A^* \iff f(A^{-1}) = f(A^*) \iff (f(A))^{-1} = (f(A))^T \iff f(A)$ orthogonal. ■

Lemma 2.10 *Let A be Hermitian with eigenvalues $\lambda_1, \ldots, \lambda_n$. The matrix $f(A)$ has eigenvalues $\lambda_1, \lambda_1, \lambda_2, \lambda_2, \ldots, \lambda_n, \lambda_n$.*

Proof As A is Hermitian, it can be diagonalized as $A = U^* D U$, with U unitary and D diagonal and real-valued elements. By Lemmas 2.6 and 2.8, $f(A) = f(U^*)f(D)f(U) = (f(U))^T f(D)f(U)$. Apparently, $f(D)$ is diagonal. The matrix A has the same eigenvalues as

$$D = \begin{pmatrix} \lambda_1 & & 0 \\ & \ddots & \\ 0 & & \lambda_n \end{pmatrix},$$

and $f(A)$ has the same eigenvalues as

$$f(D) = \begin{pmatrix} \lambda_1 & & & & 0 \\ & \lambda_1 & & & \\ & & \ddots & & \\ & & & \lambda_n & \\ 0 & & & & \lambda_n \end{pmatrix}.$$

■

Lemma 2.11 *Let A be Hermitian. Then*

$$\det[f(A)] = [\det A]^2. \tag{2.37}$$

Proof Following Lemma 2.10, $\det[f(A)] = \lambda_1^2 \lambda_2^2 \ldots \lambda_n^2 = (\lambda_1 \lambda_2 \ldots \lambda_n)^2 = [\det(A)]^2$. ■

Lemma 2.12 *Let A be a Hermitian matrix and w a vector. Then*

$$w^* A w = (f(w))^T f(A) f(w). \tag{2.38}$$

Proof As A is Hermitian, $a_{jkR} = a_{kjR}$, $a_{jkI} = -a_{kjI}$. Hence

$$w^*Aw = \sum_{j,k} \overline{w}_j a_{jk} w_k = \sum_k \overline{w}_k a_{kk} w_k$$

$$+ \sum_{j<k} \overline{w}_j a_{jk} w_k + \sum_{k<j} \overline{w}_j a_{jk} w_k$$

$$= \sum_k a_{kkR}(w_{kR}^2 + w_{kI}^2)$$

$$+ \sum_{j<k} \{(w_{jR} - iw_{jI})(a_{jkR} + ia_{jkI})(w_{kR} + iw_{kI})$$

$$+ (w_{kR} - iw_{kI})(a_{kjR} + ia_{kjI})(w_{jR} + iw_{jI})\}$$

$$= \sum_k a_{kkR}(w_{kR}^2 + w_{kI}^2)$$

$$+ \sum_{j<k} [(a_{jkR} + ia_{jkI})$$

$$\times \{(w_{jR}w_{kR} + w_{jI}w_{kI}) + i(w_{jR}w_{kI} - w_{jI}w_{kR})\}$$

$$+ (a_{jkR} - ia_{jkI})\{(w_{jR}w_{kR} + w_{jI}w_{kI}) + i(w_{kR}w_{jI} - w_{kI}w_{jR})\}]$$

$$= \sum_k a_{kkR}(w_{kR}^2 + w_{kI}^2)$$

$$+ 2\sum_{j<k} \{a_{jkR}(w_{jR}w_{kR} + w_{jI}w_{kI}) - a_{jkI}(w_{jR}w_{kI} - w_{jI}w_{kR})\} .$$

Moreover:

$$(f(w))^T f(A) f(w) = \sum_{j,k} (f(w))_j^T (f(A))_{jk} (f(w))_k$$

$$= \sum_{j,k} (w_{jR} \ \ w_{jI}) \begin{pmatrix} a_{jkR} & -a_{jkI} \\ a_{jkI} & a_{jkR} \end{pmatrix} \begin{pmatrix} w_{kR} \\ w_{kI} \end{pmatrix}$$

$$= \sum_{j,k} \{a_{jkR}(w_{jR}w_{kR} + w_{jI}w_{kI})$$

$$+ a_{jkI}(w_{jI}w_{kR} - w_{jR}w_{kI})\} ,$$

which proves the lemma. ∎

After all these lemmas, the scene is finally ready to cope with the multivariable complex Gaussian distribution. A very important special case will be treated. It can be viewed as a linear transformation of independent scalar $CN(0,1)$ variables. Compare this with filtering white, real-valued noise to obtain a general class of stationary stochastic processes.

Theorem 2.1 *Let*

$$w = (w_1 \ldots w_n)^T,$$

where $w_1, \ldots, w_n$ are independent and $\mathbf{CN}(0,1)$ and set

$$s = Aw ,\tag{2.39}$$

where A is a complex-valued $n|n$ matrix. Then

$$\mathbf{E}\, ss^* = AA^* \overset{\triangle}{=} P_s ,\tag{2.40}$$

and

$$s \sim \mathbf{CN}(0, P_s) ,$$

which means that

$$p_s(s) = \frac{1}{\pi^n \det P_s} \exp\left(-s^* P_s^{-1} s\right) .\tag{2.41}$$

Proof

$$f(s) = f(A)f(w), f(w) \sim \mathbf{N}(0, I/2)$$

clearly holds, and hence

$$f(s) \sim \mathbf{N}(0, f(A)(f(A))^T/2) .$$

Therefore:

$$
\begin{aligned}
p_s(s) &= \frac{1}{(\sqrt{2\pi})^{2n}\sqrt{\det[f(A)(f(A))^T/2]}} \\
&\quad \times \exp\left(-\frac{1}{2}(f(s))^T[f(A)(f(A))^T/2]^{-1}f(s)\right) \\
&= \frac{1}{\pi^n \sqrt{\det(f(A)(f(A))^T)}} \exp\left(-(f(s))^T[f(A)(f(A^T))]^{-1}f(s)\right) .
\end{aligned}\tag{2.42}
$$

Note, using Lemma 2.11 and (2.40):

$$
\begin{aligned}
\det[f(A)(f(A))^T] &= [\det f(A)]^2 = [\det(A)]^4 \\
&= [\det(A)\det(A^*)]^2 = [\det P_s]^2.
\end{aligned}
$$

From Lemmas 2.8 and 2.12, and (2.40), it further follows that

$$
\begin{aligned}
(f(s))^T[f(A)(f(A))^T]^{-1}f(s) &= (f(s))^T[f(A)f(A^*)]^{-1}f(s) \\
&= (f(s))^T[f(AA^*)]^{-1}f(s) \\
&= (f(s))^T[f(P_s)]^{-1}f(s) \\
&= (f(s))^T f(P_s^{-1})f(s) = s^* P_s^{-1} s .
\end{aligned}
$$

Inserting these expressions into (2.42) gives

$$p_s(s) = \frac{1}{\pi^n \det P_s} e^{-s^* P_s^{-1} s} ,$$

which is (2.41). ∎

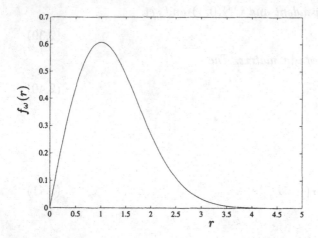

Fig. 2.4. The Rayleigh pdf

2.5.3 The Rayleigh Distribution

In the case where $w = x + iy$, where x and y are independent Gaussian random variables with zero mean and equal variance, the magnitude $|w|$ is said to have a Rayleigh distribution. This distribution is described and studied next.

Lemma 2.13 *Let x and y be independent and $\mathbf{N}(0, \sigma^2)$ distributed. Set $w = \sqrt{x^2 + y^2}$. Then the probability density function of w is*

$$f_w(r) = \frac{r}{\sigma^2} \, e^{-r^2/2\sigma^2}. \tag{2.43}$$

Proof First calculate the distribution function $F_w(r) = \int_0^r f_w(t) \, dt$ as follows:

$$
\begin{aligned}
F_w(r) &= P(w \leq r) = P(x^2 + y^2 \leq r^2) \\
&= \int\!\!\int_{x^2+y^2 \leq r^2} \frac{1}{2\pi\sigma^2} e^{-x^2/2\sigma^2} e^{-y^2/2\sigma^2} \, dx \, dy \\
&= \int_0^r 2\pi\rho \frac{1}{2\pi\sigma^2} e^{-\rho^2/2\sigma^2} \, d\rho = 1 - e^{-r^2/2\sigma^2}.
\end{aligned}
$$

The pdf is then easily found as

$$f_w(r) = \frac{d}{dr} F_w(r) = \frac{r}{\sigma^2} e^{-r^2/2\sigma^2}. \qquad \blacksquare$$

The pdf $f_w(r)$ is illustrated in Figure 2.4.

Remark The squared and normalized value, w^2/σ^2, is $\chi^2(2)$ distributed. A $\chi^2(n)$ distributed random variable is obtained as a sum of n independent squared Gaussian random variables of zero mean and unit variance. Its pdf is given by

$$p_\xi(x) = \frac{1}{2^{n/2}\Gamma(n/2)} x^{n/2-1} e^{-x/2} ,$$

where $\Gamma(\cdot)$ is the Γ function (for n an integer it holds that $\Gamma(n) = n!$). In particular, for $\chi^2(2)$ distributed random variables:

$$p_\xi(x) = \frac{1}{2}e^{-x/2} . \qquad \square$$

Some properties of the Rayleigh distribution are given next.

Lemma 2.14 *Let w be a Rayleigh distributed random variable, with the pdf given by (2.43). Then*

1. *$f_w(r)$ has its maximum for $r = \sigma$.*
2. *$\mathbf{E}\,w = \sqrt{\frac{\pi}{2}}\sigma \approx 1.25\sigma$.*
3. *$\mathrm{var}(w) = (2 - \frac{\pi}{2})\sigma^2 \approx 0.43\sigma^2$.*

Proof Setting $\frac{d}{dr}f_w(r) = 0$ easily gives

$$e^{-r^2/2\sigma^2} - r\frac{2r}{2\sigma^2}e^{-r^2/2\sigma^2} = 0 ,$$

from which it is concluded that the maximum of $f_w(r)$ occurs for $r = \sigma$.

The mean value is derived by the following calculation:

$$\mathbf{E}\,w = \int_0^\infty rf_w(r)\,dr = \int_0^\infty \frac{r^2}{\sigma^2} e^{-r^2/2\sigma^2}\,dr$$

$$= \left[-r\,e^{-r^2/2\sigma^2}\right]_0^\infty + \int_0^\infty e^{-r^2/2\sigma^2}\,dr = \frac{\sqrt{2\pi}}{2}\sigma .$$

Further:

$$\mathbf{E}\,w^2 = \int_0^\infty \frac{r^3}{\sigma^2} e^{-r^2/2\sigma^2}\,dr$$

$$= \left[-r^2\,e^{-r^2/2\sigma^2}\right]_0^\infty + \int_0^\infty 2r\,e^{-r^2/2\sigma^2}\,dr = 2\sigma^2 ,$$

and hence

$$\mathrm{var}(w) = \mathbf{E}\,w^2 - (\mathbf{E}\,w)^2 = \left(2 - \frac{\pi}{2}\right)\sigma^2 . \qquad \blacksquare$$

Exercises

Exercise 2.1 Consider a random variable ξ.

(a) Let ξ be uniformly distributed in the interval $(-a, a)$. Derive its moments $\mathbf{E}\,\xi^k$ for $k = 1, 2, 3, 4$.

(b) Let ξ be Gaussian distributed, $\xi \sim N(m, \sigma^2)$. Derive its moments $\mathbf{E}\,\xi^k$ for $k = 1, 2, 3, 4$.

Exercise 2.2 Consider the joint pdf of Example 2.2:

$$p_{\xi,\eta} = \frac{6}{7}(x+y)^2, \qquad 0 \le x \le 1, \qquad 0 \le y \le 1.$$

Derive the mean and the covariance matrix of the random vector $(\xi \ \eta)^T$.

Exercise 2.3 Let ξ and η be jointly Gaussian

$$\binom{\xi}{\eta} \sim \mathbf{N}\left(\binom{1}{2}, \begin{pmatrix} 1 & 2\rho \\ 2\rho & 4 \end{pmatrix}\right).$$

(a) Under what conditions on ρ is the covariance matrix positive definite?

(b) Sketch the contour levels of the joint pdf. Show that if $\rho \approx -1$, then (with high probability)

$$\eta \approx 4 - 2\xi.$$

(c) What is the conditional pdf $p_{\eta|\xi=x}(y)$? Compare this with the findings of (b).

Exercise 2.4 Let v and e be two random vectors of zero mean that are jointly Gaussian. Let e have a positive definite covariance matrix. Show that there exists a unique matrix B, such that

$$v = Be + w$$

with e and w being independent.

Exercise 2.5 Consider two random variables ξ and η.

(a) Assume that ξ and η are independent with pdfs $p_\xi(x)$ and $p_\eta(y)$ respectively. Set $\zeta = \xi + \eta$. Derive the pdf $p_\zeta(z)$.

(b) Assume in particular that ξ and η are both uniformly distributed over the interval (a, b). Give $p_\zeta(z)$ in an explicit form.

Exercise 2.6 Let ξ, η and ζ be random variables. Prove the relation

$$p_{\xi|\zeta=z}(x) = \int p_{\xi|\eta=y,\zeta=z}(x)p_{\eta|\zeta=z}(y)\,\mathrm{d}y.$$

Exercise 2.7 Consider a sinewave of the form

$$\xi(t) = A\cos(\omega t + \varphi) \,,$$

where the amplitude A is Gaussian, $A \sim N(m,\sigma)$ and the phase φ is uniformly distributed over $(0, 2\pi)$. The random variables A and φ are independent, ω is a deterministic angular frequency and t denotes time. Consider also a phase-shifted variant of the signal

$$\eta(t) = A\sin(\omega t + \varphi) \,.$$

(a) Show that $\mathbf{E}\,\xi(t) = 0$, $\mathbf{E}\,\eta(t) = 0$ and derive the covariance elements $\mathbf{E}\,\xi^2(t)$, $\mathbf{E}\,\eta^2(t)$, $\mathbf{E}\,\xi(t)\eta(t)$.

(b) Set $m = 0$. Introduce the random variable

$$\zeta = \xi + \mathrm{i}\eta \,.$$

Find the moments $\mathbf{E}\,\zeta$, $\mathbf{E}\,|\zeta|$, $\mathbf{E}\,\zeta^2$, $\mathbf{E}\,|\zeta|^2$.

(c) Set $m = 0$. Will ζ be complex Gaussian distributed?

Hint. Examine the ratio $(\mathbf{E}\,|\zeta|^2)/(\mathbf{E}\,|\zeta|)^2$.

Bibliography

For a general background on probability theory, many standard books can be consulted. A classical source for the topic of complex Gaussian random variables is

Goodman, N.R., 1963. Statistical analysis based on certain multivariate complex Gaussian distribution – (an introduction). *Annals of Mathematical Statistics*, 34, 152–176.

3. Models

3.1 Introduction

Some basic models and concepts are introduced in this chapter.

Stochastic processes are fundamental when describing random effects in dynamic systems and analyzing filtering. A very brief review of the fundamentals of stochastic processes is given in Section 3.2. Section 3.3 is devoted to a short discussion on Markov processes and how the important concept of a state vector appears for stochastic systems. Stationary stochastic processes are considered in Section 3.4, and some fundamental properties of the covariance function and the spectral density are established.

In some situations, moments of order higher than two (the covariance function) are useful for describing and analyzing stochastic processes. Section 3.5 gives a short introduction to bispectra, which constitute the simplest class of higher-order statistics.

Two appendices (Sections 3.A and 3.B) contain supplementary material on complex-valued models of signals and systems, and of Markov chains.

Note at this stage that noise models in terms of stochastic processes can be used in various ways.

- One possibility is to let the noise models describe *random disturbances* that are assumed to be present. Sensor noise and the effects of unmodelled sources are typical interpretations of such descriptions.
- Another aspect is to use the noise model as a means for expressing *model uncertainties*. If the model description relating input and output variables is uncertain, then the uncertainties can be incorporated in a noise model. In such a way one may, for example, design filters, predictors, or controllers, which have some robustness against modelling errors. Some ideas along this line will be presented in Section 7.5.
- A third possibility is to regard the noise model from a more pragmatic point of view. It may be interpreted as *tuning variables* by a user when constructing a filter. By changing the noise model parameters, the user can change the frequency properties of, say, an associated optimal filter. This idea will be illustrated in Section 6.7.

- Another illustration is given in Section 11.5.3, where a particular noise model is used as a means for achieving certain robustness properties of a feedback system designed by stochastic control theory.

3.2 Stochastic Processes

A very brief account on the fundamentals of stochastic processes is given in this section. The real-valued case is treated, and some extensions to complex-valued processes are relatively straightforward; see Sections 2.5 and 3.4.

Definition 3.1 *A stochastic process $x(t)$ is a family of (possibly vector-valued) random variables $\{x(t)\}$, where the time t belongs to an index set. (In discrete time, the index set is mostly $\mathbf{N}$, the natural numbers, or $\mathbf{Z}$, the integers.)*   □

In order to describe the properties of a stochastic process fully, its distribution function

$$P(x(t_1) \leq x_1, x(t_2) \leq x_2, \ldots, x(t_k) \leq x_k) \tag{3.1}$$

is needed for arbitrary k, $x_1, \ldots, x_k$, $t_1, \ldots, t_k$. In many cases, it is too cumbersome to use the whole distribution function and one resorts to using moments, often only first-order (the mean value) and second-order moments (the covariance function).

In the special case, where $x(t)$ is a Gaussian process, the distribution function is (multivariable) Gaussian. It will remain Gaussian after linear transformations and it will be completely characterized by its first- and second-order moments.

Next define the first- and second-order moments.

Definition 3.2 *Let $x(t)$ and $y(t)$ be vector-valued stochastic processes, not necessarily of the same dimension. The mean of $x(t)$ is defined as*

$$m(t) = m_x(t) \stackrel{\triangle}{=} \mathbf{E}\, x(t) . \tag{3.2}$$

The covariance function *of $x(t)$ (autocovariance function) is*

$$r(t, s) = r_x(t, s) \stackrel{\triangle}{=} \mathbf{E}\,[x(t) - m_x(t)][x(s) - m_x(s)]^T, \tag{3.3}$$

(which is a square matrix). The cross-covariance function *of $x(t)$ and $y(t)$ is*

$$r_{xy}(t,s) = \mathbf{E}\left[x(t) - m_x(t)\right]\left[y(s) - m_y(s)\right]^T . \tag{3.4}$$

□

Stationary processes are mostly used in the book. There are several concepts of stationarity.

Definition 3.3 *The stochastic process $x(t)$ is strictly stationary if its distribution is invariant to time shifts, that is:*

$$P(x(t_1 + \tau) \le x_1 , \ldots , x(t_k + \tau) \le x_k)$$

does not depend on τ. It is weakly stationary if the mean and the covariance functions are invariant to time shifts, that is:

$$\begin{aligned} m(t+\tau) &= m(t) \\ r(t+\tau , s+\tau) &= r(t,s) \end{aligned} \quad (\text{all } \tau) . \tag{3.5}$$

□

A strict stationary process is also weakly stationary, but the converse is not generally true. An important exception is Gaussian processes, for which the two concepts are equivalent. In what follows, it will not always be specified what type of stationarity is needed. In the case $x(t)$ is weakly stationary, the mean is apparently a constant (vector). Further, the covariance function depends only on the difference between its time arguments. It is therefore very common to denote the covariance function of stationary processes as

$$r(\tau) \overset{\triangle}{=} \mathbf{E}\left[x(t+\tau) - m\right]\left[x(t) - m\right]^T . \tag{3.6}$$

Some further concepts will be useful.

Definition 3.4 *A sequence of independent and identically distributed (iid) random variables is called* white noise. *Mostly, it is assumed that the mean value is zero.* □

Remark In many cases the random variables in a white noise sequence are not only independent but also identically distributed. In such a case one sometimes refer to iid random variables. □

Next, an important class of random processes, obtained by linear filtering of white noise, is introduced.

Definition 3.5 *Consider the stochastic process $y(t)$, given as the solution to the difference equation*

$$y(t)+a_1 y(t-1)+\ldots+a_n y(t-n) = e(t)+c_1 e(t-1)+\ldots+c_m e(t-m) \; , (3.7)$$

where $e(t)$ is white noise of zero mean. Such a process is called an autoregressive moving average (ARMA) process. To indicate the number of terms in (3.7), it is sometimes denoted as an ARMA(n,m) process. An ARMA$(n,0)$ process is called an autoregressive, AR or AR(n), process. An ARMA$(0,m)$ process is called a moving average, MA or MA(m), process. □

Introduce the polynomials

$$
\begin{aligned}
A(z) &= z^n + a_1 z^{n-1} + \ldots + a_n \; , \\
C(z) &= z^m + c_1 z^{m-1} + \ldots + c_m \; ,
\end{aligned}
\tag{3.8}
$$

and the shift operator q, $qx(t) = x(t+1)$. The ARMA model (3.7) can then be written compactly as

$$A(q)y(t-n) = C(q)e(t-m) \; . \tag{3.9}$$

As the white noise can be "relabelled" without changing the statistical properties of $y(t)$, (3.7) is much more frequently written in the form

$$A(q)y(t) = C(q)e'(t) \; . \tag{3.10}$$

where $e'(t) = e(t+n-m)$ is a "relabelled" white noise sequence.

Some illustrations of ARMA processes will be given in Example 3.2. ARMA processes will also be used in many subsequent examples.

Remark 1 As an alternative, one can work primarily with the *backward* shift operator q^{-1}. An ARMA model would then be written as

$$\overline{A}(q^{-1})y(t) = \overline{C}(q^{-1})e(t) \; ,$$

where the polynomials are

$$
\begin{aligned}
\overline{A}(q^{-1}) &= 1 + a_1 q^{-1} + \ldots + a_n q^{-n} \; , \\
\overline{C}(q^{-1}) &= 1 + c_1 q^{-1} + \ldots + c_m q^{-m} \; .
\end{aligned}
$$

The advantage of using the q-formalism is that stability corresponds to the "natural" condition $|z| < 1$. The advantage of the alternative q^{-1} formalism is that causality considerations become easier to handle. The q formalism is used in this book. □

Remark 2 It is usually assumed that the ARMA model is stationary. This means that the polynomial $A(z)$ has all zeros strictly inside the unit circle. In some situations, such as modelling a drifting disturbance, it is appropriate to allow $A(z)$ to have zeros on or even outside the unit circle. Then the process will not be stationary and should only be considered for a finite period of time. A drifting disturbance is obtained when $A(z)$ has a zero at $z = 1$. The special simple case $A(z) = z - 1$, $C(z) = z$ is known as a random walk. That particular model:

$$y(t) - y(t - 1) = e(t)$$

can easily be iterated to obtain

$$y(t) = y(0) + \sum_{s=1}^{t} e(s) \ .$$

The name "random walk" refers to a walk along a straight line where the increments $e(t)$ for each time step are independent.   □

Remark 3 One might add an input signal term to (3.8) in order to obtain an ARMA with eXogenous input (ARMAX) model:

$$A(q)y(t) = B(q)u(t) + C(q)e(t) \ . \tag{3.11}$$

The system must be *causal*, which means that $y(t)$ is not allowed to depend on *future* values of the input $u(t + \tau)$, $\tau > 0$. Hence it is required that $\deg A \geq \deg B$. Otherwise $y(t)$ would depend on *future* input values.   □

3.3 Markov Processes and the Concept of State

The concept of states is fundamental for deterministic dynamic systems. The state vector, usually denoted by $x(t)$, and its associated state equation, have the property that they contain all information about the history of the system that has any impact on its future behaviour. Phrased differently, to compute the system response at times $t_1 > t$, it is sufficient to know $x(t)$ and external input signals in the interval $[t, t_1]$.

This notation of state does not directly carry over to stochastic systems, as such systems contain random elements or signals. A Markov process is defined to begin with.

Definition 3.6 *Let $t_1 < t_2 < \ldots < t_k < t$. The process $x(t)$ is said to be a Markov process if the conditional pdf satisfies*

$$p(x(t)|x(t_1), \ldots x(t_k)) = p(x(t)|x(t_k)) \ . \tag{3.12}$$

□

Hence, if several old values of the process are known, only the most recent one has any influence on the future behaviour of the process.

If the process $x(t)$ only takes a finite number of values it is called a Markov chain. Markov chains are described and analyzed in Section 3.B.

The statistical properties of a Markov process can be calculated once the distribution for an initial time t_0 is known as well as the *transition pdfs* $p(x(t+1)|x(t))$. This is so as by Bayes' rule, for example:

$$p(x(t_0+2), x(t_0+1)) = p(x(t_0+2)|x(t_0+1))p(x(t_0+1))$$

$$= p(x(t_0+2)|x(t_0+1)) \int_{-\infty}^{\infty} p(x(t_0+1)|x(t_0))p(x(t_0)) \, dx(t_0) \ . (3.13)$$

Next, consider an important and general class of discrete-time processes that satisfy the Markov property. Let the sampling interval be one time unit. The conditional pdf, $p(x(t+1)|x(t))$, for how the process propagates from one time step to the next one, is important for describing the updating of $x(t)$. In particular, the new state, $x(t+1)$, will be modelled as the conditional mean

$$g(x(t),t) \triangleq \mathbf{E}\left[x(t+1)|x(t)\right]$$

$$= \int_{-\infty}^{\infty} x(t+1)p(x(t+1)|x(t)) \, dx(t+1) \ . \tag{3.14}$$

Thus consider processes of the form

$$x(t+1) = g(x(t),t) + v(x(t),t) \ , \tag{3.15}$$

where $v(t)$ is a random disturbance. Assume, further, that

- $\{v(x(t),t)\}$ is white noise, that is $v(x(t_1),t_1)$ and $v(x(t_2),t_2)$ are independent for $t_1 \neq t_2$.
- $g(x(t),t)$ does not depend on older values of $x(t-\tau)$.

Then the process is a Markov process. The vector $x(t)$ is called the *state vector*, as, owing to (3.13) and (3.14), $x(t)$ contains everything about the past that has any impact on the future behaviour of the process. The future behaviour will also be influenced by the *process noise* $v(\cdot)$. Note, though, that *future* values of $v(\cdot)$ will be independent of $x(t)$. Under the above assumptions, (3.15) is often called a *stochastic difference equation*. The conditional pdf $p(x(t+1)|x(t))$ is sometimes called the *hyperstate* of the process. In order to obtain a full description of the process, the hyperstate must be known. In the special case of linear systems with additive Gaussian process noise, the hyperstate turns out to be given by a Gaussian distribution. It can thus be characterized fully by the mean and the covariance.

In order to obtain a more complete model, (3.15) should be complemented by describing the outputs or measurements. These are often given in the form

$$y(t) = h(x(t), e(t)) \ , \tag{3.16}$$

where $e(t)$ is a random term called *measurement noise*. It is often assumed that $e(t)$ is white noise. (Then $e(t_1)$ and $e(t_2)$ are independent if $t_1 \neq t_2$.)

In most cases, one further assumes that $v(t_1)$ and $e(t_2)$ are independent for all t_1 and t_2.

Regarding systems equipped with a measurement equation, the notation of *state* sometimes requires that

$$p(x(t+1)|x(t), y(t)) = p(x(t+1)|x(t)) . \tag{3.17}$$

For this to hold for the model (3.15) and (3.16), one must generally require $v(t)$ and $e(t)$ to be independent. Otherwise, loosely speaking, $y(t)$ contains, through $e(t)$ and therefore through $v(t)$, additional information about $x(t+1)$.

It is often possible to transform a model where $v(t)$ and $e(t)$ are dependent into another one where the process noise and measurement noise are independent. The trick is to augment the state vector. The idea is illustrated by an example.

Example 3.1 Consider a process of the form

$$x(t+1) = g(x(t)) + g_2(x(t))v(t) ,$$
$$y(t) = h(x(t)) + e(t) ,$$

where $v(t)$ and $e(t)$ are mutually correlated, zero mean white noise sequences with covariance matrices

$$\mathbf{E}\, v(t)v^T(s) = R_1\delta_{t,s} ,$$
$$\mathbf{E}\, e(t)e^T(s) = R_2\delta_{t,s} ,$$
$$\mathbf{E}\, v(t)e^T(s) = R_{12}\delta_{t,s} ,$$

and $\delta_{t,s}$ is the Kronecker delta ($\delta_{t,s} = 1$ if $t = s$, and 0 elsewhere). Assume that one can find a matrix B so that

$$v(t) = Be(t) + w(t) ,$$

and where $e(t)$ and $w(t)$ are independent. See Exercise 2.4. Then set

$$\overline{x}(t) = \begin{pmatrix} x(t) \\ e(t) \end{pmatrix} , \qquad \overline{v}(t) = \begin{pmatrix} w(t) \\ e(t+1) \end{pmatrix} .$$

Clearly, $\overline{v}(t)$ will be a white noise sequence. Then it holds that

$$\overline{x}(t+1) = \begin{pmatrix} x(t+1) \\ e(t+1) \end{pmatrix} = \begin{pmatrix} g(x(t)) + g_2(x(t))(Be(t) + w(t)) \\ e(t+1) \end{pmatrix}$$
$$\stackrel{\triangle}{=} \overline{g}(\overline{x}(t)) + \overline{g}_2(\overline{x}(t))\overline{v}(t) ,$$
$$y(t) = h(x(t)) + e(t) \stackrel{\triangle}{=} \overline{h}(\overline{x}(t)) .$$

In particular, for the linear model

$$x(t+1) = Fx(t) + v(t) ,$$
$$y(t) = Hx(t) + e(t) ,$$

one obtains

$$\bar{x}(t+1) = \begin{pmatrix} Fx(t) + Be(t) + w(t) \\ e(t+1) \end{pmatrix} = \begin{pmatrix} F & B \\ 0 & 0 \end{pmatrix} \bar{x}(t) + \bar{v}(t) ,$$
$$y(t) = Hx(t) + e(t) = (H \quad I)\bar{x}(t) .$$

□

Remark One may question if it is reasonable to allow the process noise $v(t)$ and the measurement noise $e(t)$ to be correlated, that is $R_{12} \neq 0$. If $v(t)$ and $e(t)$ really correspond to a model where the states have physical interpretations, it is likely that $v(t)$ and $e(t)$ are uncorrelated. However, state space models may also be obtained in other ways. For example, assume that the second-order AR process

$$y(t) + a_1 y(t-1) + a_2 y(t-2) = \varepsilon(t) , \qquad \lambda^2 = \mathbf{E}\,\varepsilon^2(t) ,$$

is to be represented in state space form. There is no unique solution. One possibility is

$$x(t+1) = \begin{pmatrix} -a_1 & -a_2 \\ 1 & 0 \end{pmatrix} x(t) + v(t) ,$$
$$y(t) = (1 \quad 0)x(t) ,$$

with $R_1 = \begin{pmatrix} \lambda^2 & 0 \\ 0 & 0 \end{pmatrix}$. Note that $R_2 = 0$, $R_{12} = 0$ in this case. Another possibility is to choose

$$x(t+1) = \begin{pmatrix} -a_1 & 1 \\ -a_2 & 0 \end{pmatrix} x(t) + v(t) ,$$
$$y(t) = (1 \quad 0)x(t) + e(t) ,$$

with

$$R_1 = \lambda^2 \begin{pmatrix} a_1^2 & a_1 a_2 \\ a_1 a_2 & a_2^2 \end{pmatrix} , \qquad R_2 = \lambda^2 , \qquad R_{12} = \lambda^2 \begin{pmatrix} -a_1 \\ -a_2 \end{pmatrix} ,$$

for which apparently $R_{12} \neq 0$.

□

3.4 Covariance Function and Spectrum

Properties of the covariance function and the spectrum are examined in this section. In order to make the treatment general, the stochastic processes are allowed to be complex-valued.

Definition 3.7 *Let $x(t)$ be a stationary, zero mean, possibly complex-valued, stochastic process. Its covariance function is defined as*

$$r(\tau) \overset{\triangle}{=} \mathbf{E}\, x(t+\tau)x^*(t) . \tag{3.18}$$

The process has an associated covariance matrix

$$R(m) = \begin{pmatrix} r(0) & r(1) \ldots r(m-1) \\ r(-1) & r(0) \\ \vdots & \ddots & \vdots \\ r(-m+1) & \ldots & r(0) \end{pmatrix}. \tag{3.19}$$

□

Remark 1 To emphasize the connection to the process $x(t)$, the notations $r_x(\tau)$ and $R_x(m)$ will frequently be used. □

Remark 2 If $x(t)$ instead has a nonzero (constant) mean value m, the covariance function is

$$r(\tau) \overset{\triangle}{=} \mathbf{E}\left[x(t+\tau) - m\right][x(t) - m]^*. \tag{3.20}$$

□

Definition 3.8 *Let $x(t)$ and $y(t)$ be stationary processes with zero mean. Then the* cross-covariance function *is defined as*

$$r_{xy}(\tau) = \mathbf{E}\, x(t+\tau)y^*(t). \tag{3.21}$$

□

Next, the spectrum of a stationary process is defined.

Definition 3.9 *Let $x(t)$ be a (discrete-time) stationary process with covariance function $r(\tau)$. Then its* spectrum *is defined as*

$$\phi(z) \overset{\triangle}{=} \sum_{n=-\infty}^{\infty} r(n)z^{-n}. \tag{3.22}$$

□

Remark The series of (3.22) converges at least in a circle strip $R_1 < |z| < R_2$, $0 < R_1 < 1, 1 < R_2 < \infty$. □

Definition 3.10 *The spectrum considered at the unit circle, $\phi(e^{i\omega})$, is called the* spectral density. *It is a periodic function of ω, and is mostly studied for $|\omega| \leq \pi$.*

□

Remark 1 When a discrete-time model is sampled from a continuous-time model, it is often valuable to emphasize the influence of the sampling interval h. (In other situations, it is common practice to consider a normalized time scale and set $h = 1$.) In such a case, the definition of the spectrum is modified to

$$\phi(z) = h \sum_{n=-\infty}^{\infty} r(nh)z^{-n} \tag{3.23}$$

and the spectral density is $\phi(e^{i\omega h})$, $|\omega| \leq \pi/h$.   □

Remark 2 Many results for stationary processes do also apply for a wider class of signals. In a control context it is common to treat the input as a deterministic signal and model the effect of various disturbances as a stationary process with an underlying model of the form

$$y(t) = G(q)u(t) + v(t)$$

where $G(q)$ is a transfer function operator describing the input-output relation, and $v(t)$ is the total disturbance acting on the output.

The concept of quasi-stationary signals is often useful when characterizing deterministic and stochastic signals jointly. A signal $y(t)$ is said to be quasi-stationary if the limit

$$\lim_{N \to \infty} \frac{1}{N} \sum_{t=1}^{N} y(t + \tau)y^*(t) = R(\tau)$$

exists for all τ. This is the case for ergodic stochastic processes, where the right-hand side is precisely the covariance function. For other signals, one may denote the operation in the left-hand side by

$$\overline{\mathbf{E}}\, y(t + \tau)y^*(t)$$

and maintain the notation of "covariance function". Once a covariance function is introduced, one can continue to use the concept of spectrum as in Definition 3.9.

There are quite a number of different types of quasi-stationary signal. Transient phenomena in dynamic systems is one example. Another is periodic signals, which leads to covariance functions being periodic too, and to spectra being composed of a number of weighted delta functions.

As many of the results for stochastic processes described in this book are based on the representations in terms of covariance function and spectrum, they can potentially be extended to the wider class of quasi-stationary signals. Note in this context that by using spectral factorization, see Section 4.3, one can also derive time domain models of the signal itself.   □

Definition 3.11 *The* cross-spectrum *between two zero mean stationary processes* $x(t)$ *and* $y(t)$ *is defined as*

$$\phi_{xy}(z) = \sum_{n=-\infty}^{\infty} r_{xy}(n)z^{-n} \ . \tag{3.24}$$

□

It is easy to set up the inverse relation, which for a given spectrum, produces the covariance function. This is done as follows ($\oint$ denotes integration around the unit circle counterclockwise):

$$\frac{1}{2\pi \mathrm{i}} \oint \phi(z)z^m \frac{\mathrm{d}z}{z} = \frac{1}{2\pi \mathrm{i}} \oint \sum_{n=-\infty}^{\infty} r(n)z^{-n}z^m \frac{\mathrm{d}z}{z}$$

$$= \sum_{n=-\infty}^{\infty} r(n) \frac{1}{2\pi \mathrm{i}} \oint z^{-n+m} \frac{\mathrm{d}z}{z}$$

$$= \sum_{n=-\infty}^{\infty} r(n)\delta_{n,m} = r(m) \ . \tag{3.25}$$

Definition 3.12 *Let* $x(t)$ *be a stationary stochastic process with covariance function* $r(\tau)$. *The* positive real part of the spectrum *is defined as*

$$\tilde{\phi}(z) = \frac{1}{2}r(0) + \sum_{n=1}^{\infty} r(n)z^{-n} \ . \tag{3.26}$$

□

The term "positive real part" will be explained later (see Section 4.6). Note that $\tilde{\phi}(z)$ has a close connection to the z-transform of $\{r(n)\}_{n=0}^{\infty}$. In fact:

$$\tilde{\phi}(z) = R(z) - \frac{1}{2}r(0) \ , \tag{3.27}$$

where

$$R(z) = \sum_{n=0}^{\infty} r(n)z^{-n} = \mathcal{Z}(\{r(n)\}) \tag{3.28}$$

is the z-transform of the covariance function (for positive lags).

The connections between model representation, realization, covariance function and spectral density are illustrated in the following example.

Example 3.2 Consider an ARMA process

$$A(q)y(t) = C(q)e(t) .$$

The coefficients of the $A(q)$ and $C(q)$ polynomials determine the properties of the process. In particular, the zeros of $A(z)$, which are called the poles, determine the frequency contents of the process. The closer the poles are located towards the unit circle, the slower or more oscillating the process will be. Figure 3.1 illustrates the connections between the A and C coefficients, realizations of processes and their second-order moments as expressed by covariance function and spectral density.   □

Some useful properties of the covariance function $r(\tau)$ and the spectrum are presented in the following two lemmas.

Lemma 3.1 *Let $x(t)$ be a stationary process. Its covariance function has the following properties:*

(i) $r(\tau)$ *is generally complex-valued, $r(0)$ has real-valued and positive diagonal elements.*

(ii) $r(-\tau) = r^*(\tau) .$ (3.29)

(iii) $R(m)$ *is a Hermitian matrix (i.e. $R^*(m) = R(m)$).*

(iv) $\{r(\tau)\}$ *is a nonnegative definite sequence, that is, the matrix $R(m)$ is nonnegative definite for any m (i.e. $w^* R(m)w \geq 0$ for all w).*

Proof It is no restriction to assume that $x(t)$ has zero mean. Part (i) is obvious. Note that

$$r_{ii}(0) = \mathbf{E}\, x_i(t)x_i^*(t) = \mathbf{E}\, |x_i(t)|^2 \geq 0 .$$

Part (ii) follows from

$$r(-\tau) = \mathbf{E}\, x(t-\tau)x^*(t) = \mathbf{E}\, x(t)x^*(t+\tau)$$
$$= [\mathbf{E}\, x(t+\tau)x^*(t)]^* = r_x^*(\tau) .$$

Part (iii) is immediate from part (ii). Finally, consider part (iv). Introduce the vectors u and $\varphi(t)$ as

$$u^* = (u_1^*,\ u_2^* \ \ldots \ u_m^*) ,$$
$$\varphi(t) = (\, x^*(t-1)\, x^*(t-2) \ldots x^*(t-m)\,)^*$$

where u_1, u_2, etc., have the same dimensions as $x(t)$. Then

$$u^* Ru = (\, u_1^*\, u_2^* \ldots u_m^*\,)\, \mathbf{E}\, \varphi(t)\varphi^*(t) \begin{pmatrix} u_1 \\ u_2 \\ \vdots \\ u_m \end{pmatrix}$$

$$= \mathbf{E}\,[\sum_{k=1}^{m} u_k^* x(t-k)][\sum_{k=1}^{m} x^*(t-k)u_k] = \mathbf{E}\,|\sum_{k=1}^{m} u_k^* x(t-k)|^2 \geq 0 .$$

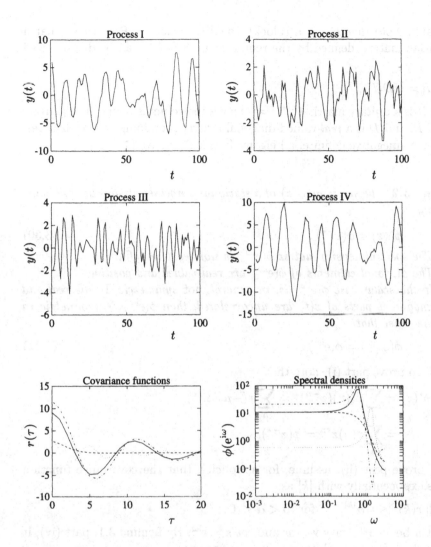

Fig. 3.1. Illustration of some ARMA processes.
Process I: $A(q) = q^2 - 1.5q + 0.8$, $C(q) = 1$.
Process II: $A(q) = q^2 - 1.0q + 0.3$, $C(q) = 1$.
Process III: $A(q) = q^2 - 0.5q + 0.8$, $C(q) = 1$.
Process IV: $A(q) = q^2 - 1.0q + 0.8$, $C(q) = q^2 - 0.5q + 0.9$.
First four plots show realizations of processes I–IV. Last two plots show covariance functions and spectral densities: Processes I (*solid lines*), II (*dashed lines*), III (*dotted lines*), IV (*dash–dotted lines*)

Since this holds for arbitrary vectors u and arbitrary m, it can be seen that part (iv) holds.   ∎

Remark. Note that R has a (block) Toeplitz structure. Recall also that a Hermitian matrix (defined by the requirement $R = R^*$) can be diagonalized as

$$R = UDU^*,$$

where U is a unitary matrix ($U^*U = I$) with the columns being the eigenvectors of R, and D is a *real*-valued diagonal matrix containing the eigenvalues. As R is nonnegative definite, all eigenvalues are positive (≥ 0).   □

Lemma 3.2 *The spectrum $\phi(z)$ of a stationary process satisfies the following relations:*

(i) $\phi^*(z) = \phi(z^{-*})$.   (3.30)

(ii) *The spectral density matrix $\phi(e^{i\omega})$ is nonnegative definite.*

(iii) *The diagonal elements of $\phi(e^{i\omega})$ are real-valued and positive.*

(iv) *In the scalar case $\phi(e^{i\omega})$ is, in general,* not *symmetric. If the real and imaginary parts of $x(t)$ are uncorrelated, then $\phi(e^{i\omega})$ is symmetric in the sense that*

$$\phi(e^{-i\omega}) = \phi(e^{i\omega}).$$ (3.31)

Proof To prove part (i), note that

$$\phi^*(z) = \sum_n r^*(n)(z^{-n})^* = \sum_n r(-n)(z^{-*})^n$$

$$= \sum_n r(n)z^{*n} = \phi(z^{-*}).$$

To prove part (ii), assume, for simplicity, that the covariance function decays exponentially with $|k|$ as

$$\| r(k) \| \leq C\alpha^{|k|}, \quad \text{for } 0 < \alpha < 1.$$

Let u be an arbitrary vector and set $z = e^{i\omega}$. By Lemma 3.1, part (iv), it follows that, N being arbitrary:

$$0 \leq \frac{1}{N} \left(u^* \ u^*z \ \ldots \ u^*z^{N-1} \right)$$

$$\times \begin{pmatrix} r(0) & r(1) \ldots r(N-1) \\ r(-1) & r(0) \\ \vdots & & \ddots \\ r(-N+1) & & r(0) \end{pmatrix} \begin{pmatrix} u \\ uz^{-1} \\ \\ uz^{-(N-1)} \end{pmatrix}$$

$$= \frac{1}{N}[Nu^*r(0)u + \sum_{k=1}^{N}(N - |k|)[u^*r(k)uz^{-k} + u^*r(-k)uz^k]]$$

$$= u^*(\sum_{k=-N}^{N} r(k)z^{-k})u - \frac{1}{N}\sum_{k=-N}^{N} |k|u^*r(k)uz^{-k}.$$ (3.32)

The second term in (3.32) can asymptotically be neglected as

$$\left| \frac{1}{N} \sum_{k=-N}^{N} |k| u^* r(k) u z^{-k} \right| \leq \frac{1}{N} \| u \|^2 \sum_{k=-N}^{N} |k| C \alpha^{|k|}$$

$$= \frac{2C}{N} \| u \|^2 \sum_{k=1}^{N} k \alpha^k \to 0 \quad \text{as} \quad N \to \infty .$$

By letting N tend to infinity in (3.32):

$$0 \leq u^* \left(\sum_{k=-\infty}^{\infty} r(k) z^{-k} \right) u = u^* \phi(z) u ,$$

which proves part (ii). Part (iii) follows directly from part (ii).

Finally, consider part (iv). Write the process as $w(t) = x(t) + iy(t)$ and let $x(t)$ and $y(t)$ be independent real-valued processes. Then

$$r_w(\tau) = \mathbf{E} \left[x(t+\tau) + iy(t+\tau) \right] [x(t) - iy(t)]$$
$$= [r_x(\tau) + r_y(\tau)] + i[-r_{xy}(\tau) + r_{yx}(\tau)]$$
$$= r_x(\tau) + r_y(\tau) .$$

Hence $\phi_w(\omega) = \phi_x(\omega) + \phi_y(\omega)$. However, for a *real*-valued process, such as $x(t)$, $r_x(\tau)$ is real-valued and

$$\phi_x(e^{-i\omega}) = \sum_{n=-\infty}^{\infty} r_x(n) \, e^{in\omega} = \sum_{n=-\infty}^{\infty} r_x(-n) \, e^{in\omega}$$

$$= \sum_{n=-\infty}^{\infty} r_x(n) \, e^{-in\omega} = \phi_x(e^{i\omega}) ,$$

from which the statement follows.   ∎

Remark It is a consequence of part (i) that the spectral density matrix $\phi(e^{i\omega})$ is Hermitian.   □

Corollary Let $x(t)$ and $y(t)$ be stationary processes with zero mean. Then

$$\phi_{xy}^*(z) = \phi_{yx}(z^{-*})$$ (3.33)

holds.

Proof Set

$$w(t) = \begin{pmatrix} x(t) \\ y(t) \end{pmatrix} .$$

This implies that

$$\phi_w(z) = \begin{pmatrix} \phi_x(z) & \phi_{xy}(z) \\ \phi_{yx}(z) & \phi_y(z) \end{pmatrix} .$$

Now use the relation in (3.30) to see that

$$\phi_w^*(z) = \begin{pmatrix} \phi_x^*(z) & \phi_{yx}^*(z) \\ \phi_{xy}^*(z) & \phi_y^*(z) \end{pmatrix} ,$$

$$\phi_w(z^{-*}) = \begin{pmatrix} \phi_x(z^{-*}) & \phi_{xy}(z^{-*}) \\ \phi_{yx}(z^{-*}) & \phi_y(z^{-*}) \end{pmatrix} .$$

Equating the lower left blocks gives the desired equality.   ■

The next lemma establishes a relation between $\phi(z)$ and $\tilde{\phi}(z)$.

Lemma 3.3

$$\phi(z) = \tilde{\phi}(z) + \tilde{\phi}^*(z^{-*}) \tag{3.34}$$

holds.

Proof By Definitions 3.9 and 3.12, and (3.29):

$$\phi(z) = \tilde{\phi}(z) + \frac{1}{2}r(0) + \sum_{n=-\infty}^{-1} r(n)z^{-n}$$

$$= \tilde{\phi}(z) + \frac{1}{2}r^*(0) + \sum_{n=1}^{\infty} r^*(n)z^n$$

$$= \tilde{\phi}(z) + \left[\frac{1}{2}r(0) + \sum_{n=1}^{\infty} r(n)(z^{-*})^{-n} \right]^* = \tilde{\phi}(z) + \tilde{\phi}^*(z^{-*}) .$$

■

Corollary

$$\phi(e^{i\omega}) = \tilde{\phi}(e^{i\omega}) + \tilde{\phi}^*(e^{i\omega}) , \tag{3.35}$$

and in the scalar case

$$\phi(e^{i\omega}) = 2\operatorname{Re}\tilde{\phi}(e^{i\omega}) \tag{3.36}$$

hold.

Proof Immediate by the substitution $z \to e^{i\omega}$.   ■

A specific feature of complex-valued processes is that the spectral density is no longer symmetric; see Lemma 3.2, part (iv). This is illustrated in the following example.

Example 3.3 Let $e(t)$ be a scalar, complex-valued white noise satisfying

$$\mathbf{E}\, e(t)e^*(s) = \delta_{t,s} ,$$

and set

$$x(t) = e(t) + \varphi e(t-1) \,,$$

$\varphi = a + ib$ being a complex constant. One easily obtains

$$r_x(0) = \mathbf{E}\, x(t)x^*(t) = 1 + \varphi\overline{\varphi} = 1 + |\varphi|^2 \,,$$
$$r_x(1) = \mathbf{E}\, x(t+1)x^*(t) = \varphi \,,$$
$$r_x(-1) = \mathbf{E}\, x(t)x^*(t+1) = \overline{\varphi} \,,$$
$$r_x(\tau) = 0, \qquad |\tau| > 1 \,,$$

and hence

$$\begin{aligned}\phi_x(e^{i\omega}) &= [1 + \varphi\overline{\varphi} + \varphi\, e^{-i\omega} + \overline{\varphi}\, e^{i\omega}]\\&= [1 + |\varphi|^2 + 2\,\mathrm{Re}(\varphi\, e^{-i\omega})]\\&= (1 + \varphi\, e^{-i\omega})(1 + \overline{\varphi}\, e^{i\omega}) = |1 + \varphi\, e^{-i\omega}|^2 \,,\end{aligned}$$

which illustrates that $\phi(e^{i\omega})$ is real-valued and positive. Further,

$$\begin{aligned}\phi_x(e^{i\omega}) &= 1 + a^2 + b^2 + (a+ib)\, e^{-i\omega} + (a-ib)\, e^{i\omega}\\&= 1 + a^2 + b^2 + 2a\cos\omega + 2b\sin\omega \,.\end{aligned}$$

The spectral density is not symmetric unless $b = 0$, which corresponds to φ being real-valued. $\qquad\square$

The next lemma gives a relation between the extrema of the spectral density $\phi(e^{i\omega})$ and the covariance matrix $R(m)$, (3.19).

Lemma 3.4 *The maximal and minimal eigenvalues of $R(m)$ satisfy*

$$\lambda_{\max}(R(m+1)) \geq \lambda_{\max}(R(m)) \,, \tag{3.37}$$
$$\lambda_{\min}(R(m+1)) \leq \lambda_{\min}(R(m)) \,, \tag{3.38}$$
$$\lim_{m\to\infty} \lambda_{\max}(R(m)) \geq \sup_{\omega}\sup_{u} \frac{u^*\phi(e^{i\omega})u}{u^*u} = \sup_{\omega} \lambda_{\max}[\phi(e^{i\omega})] \,, \tag{3.39}$$
$$\lim_{m\to\infty} \lambda_{\min}(R(m)) \leq \inf_{\omega}\inf_{u} \frac{u^*\phi(e^{i\omega})u}{u^*u} = \inf_{\omega} \lambda_{\min}[\phi(e^{i\omega})] \,. \tag{3.40}$$

Proof The maximal eigenvalue, $\lambda_{\max}(R(m))$, satisfies

$$\lambda_{\max}(R(m+1)) = \sup_{x} \frac{x^*R(m+1)x}{x^*x} \geq \frac{v^*R(m+1)v}{v^*v} \,, \qquad \text{for any } v \,.$$

Consider, in particular, vectors v of the form

$$v = \begin{pmatrix} \tilde{v} \\ 0 \end{pmatrix} \,.$$

Then

$$\lambda_{\max}(R(m+1)) \geq \frac{(\tilde{v}^*\ \ 0)R(m+1)\begin{pmatrix} \tilde{v} \\ 0 \end{pmatrix}}{\tilde{v}^*\tilde{v}} = \frac{\tilde{v}^*R(m)\tilde{v}}{\tilde{v}^*\tilde{v}} \,.$$

As this relation holds for any $\tilde{v}$, one can conclude that

$$\lambda_{\max}(R(m+1)) \geq \sup_{\tilde{v}} \frac{\tilde{v}^* R(m)\tilde{v}}{\tilde{v}^*\tilde{v}} = \lambda_{\max}(R(m)) \ .$$

The relation (3.38) is proved analogously.

Consider next a vector of the form

$$x^* = (u^* \ u^* z \ldots u^* z^{m-1}) \ ,$$

where $z = e^{i\omega}$. Then

$$\lambda_{\max}(R(m)) \geq \frac{x^* R(m)x}{x^*x}$$

$$= \frac{1}{mu^*u} \sum_{k=-m}^{m} (m - |k|)u^* r(k)u z^{-k}$$

$$= \frac{1}{u^*u}u^*\{ \sum_{k=-m}^{m} r(k)z^{-k}\}u - \frac{1}{mu^*u} \sum_{k=-m}^{m} |k|u^* r(k)u z^{-k} \ .$$

Now let m tend to infinity. The last sum will tend to zero, just as in the proof of Lemma 3.2. Thus

$$\lim_{m\to\infty} \lambda_{\max}(R(m)) \geq \frac{1}{u^*u}u^*\phi(e^{i\omega})u \ .$$

As this inequality holds for any u and ω, the inequality part of (3.39) follows. Then the equality follows, as for a general Hermitian matrix A

$$\sup_{x\neq 0} \frac{x^* Ax}{x^*x} = \lambda_{\max}(A)$$

where the right-hand side denotes the maximum eigenvalue of A. The relations in (3.40) are proved by similar arguments.   ∎

Remark Under weak stationarity assumptions, the inequalities in (3.39) and (3.40) can be replaced by equalities. It is, for example, sufficient to let $x(t)$ be filtered white noise.   □

3.5 Bispectrum

The bispectrum is a special case of higher-order spectra that generalizes the usual second-order spectrum. The bispectrum is the simplest form of higher-order spectra. Such spectra are useful tools for the following:

- extracting information due to deviations from a Gaussian distribution,
- estimating the phase of a non-Gaussian process,
- detecting and characterizing nonlinear mechanisms in time series.

It should be mentioned that bispectra are useful only for signals that do *not* have a pdf that is symmetric around its mean value. For signals *with* such a symmetry, spectra of higher order (such as fourth order) are needed.

In order to simplify the development here, it is generally assumed that signals have zero mean. This can also be stated as that only deviations of the signals from an operating point (given by the mean values) are considered.

Let $x(t)$ be a scalar stationary, real-valued process. Define its *third-order moment sequence*, $R(m, n)$, as

$$R(m, n) = \mathbf{E}\, x(t)x(t + m)x(t + n) , \tag{3.41}$$

and its *bispectrum* as

$$B(z_1, z_2) = \sum_{m=-\infty}^{\infty} \sum_{n=-\infty}^{\infty} R(m, n) z_1^{-m} z_2^{-n} . \tag{3.42}$$

The third-order moment sequence satisfies a number of symmetry relations.

Lemma 3.5 *For a stationary process*

$$
\begin{aligned}
R(m, n) &= R(n, m) \\
&= R(-n, m - n) \\
&= R(n - m, -m) \\
&= R(m - n, -n) \\
&= R(-m, n - m)
\end{aligned} \tag{3.43}
$$

holds.

Proof The equalities are easy to verify. In fact:

$$
\begin{aligned}
R(n, m) &= \mathbf{E}\, x(t)x(t + n)x(t + m) = R(m, n) . \\
R(-n, m - n) &= \mathbf{E}\, x(t)x(t - n)x(t + m - n) \\
&= \mathbf{E}\, x(t + n)x(t)x(t + m) = R(m, n) . \\
R(n - m, -m) &= \mathbf{E}\, x(t)x(t + n - m)x(t - m) \\
&= \mathbf{E}\, x(t + m)x(t + n)x(t) = R(m, n) . \\
R(m - n, -n) &= \mathbf{E}\, x(t)x(t + m - n)x(t - n) \\
&= \mathbf{E}\, x(t + n)x(t + m)x(t) = R(m, n) . \\
R(-m, n - m) &= \mathbf{E}\, x(t)x(t - m)x(t + n - m) \\
&= \mathbf{E}\, x(t + m)x(t)x(t + n) = R(m, n) . \quad \blacksquare
\end{aligned}
$$

Similarly, the bispectrum satisfies a number of symmetry relations and properties.

Lemma 3.6 *For a stationary process*

(i) $\qquad B(e^{i\omega_1}, e^{i\omega_2})$ is in general complex-valued. $\tag{3.44}$

(ii)

$$B(e^{i(\omega_1+2\pi)}, e^{i(\omega_2+2\pi)}) = B(e^{i(\omega_1+2\pi)}, e^{i\omega_2})$$
$$= B(e^{i\omega_1}, e^{i(\omega_2+2\pi)}) = B(e^{i\omega_1}, e^{i\omega_2}) \,. \quad (3.45)$$

(iii)

$$B(z_1, z_2) = B(z_2, z_1) = B^*(z_2^*, z_1^*) = B^*(z_1^*, z_2^*)$$
$$= B(z_1^{-1} z_2^{-1}, z_2) = B(z_2, z_1^{-1} z_2^{-1})$$
$$= B(z_1, z_1^{-1} z_2^{-1}) = B(z_1^{-1} z_2^{-1}, z_1) \,. \quad (3.46)$$

hold.

Proof Property (i) is trivial. Property (ii) is immediate from the definition in (3.42). It follows that

$$B(z_2, z_1) = \sum_m \sum_n R(m,n) z_2^{-m} z_1^{-n}$$
$$= \sum_m \sum_n R(n,m) z_1^{-n} z_2^{-m} = B(z_1, z_2) \,.$$
$$B^*(z_1^*, z_2^*) = (\sum_m \sum_n R(m,n) z_1^{*-m} z_2^{*-n})^*$$
$$= \sum_m \sum_n R(m,n) z_1^{-m} z_2^{-n} = B(z_1, z_2) \,.$$
$$B(z_1^{-1} z_2^{-1}, z_2) = \sum_m \sum_n R(m,n)(z_1^{-1} z_2^{-1})^{-m} z_2^{-n}$$
$$= \sum_m \sum_n R(-m, n-m)(z_1^m z_2^{-(n-m)})$$
$$= \sum_k \sum_\ell R(k,\ell) z_1^{-k} z_2^{-\ell} = B(z_1, z_2) \,.$$
$$B(z_1^{-1} z_2^{-1}, z_1) = \sum_m \sum_n R(m,n)(z_1^{-1} z_2^{-1})^{-m} z_1^{-n}$$
$$= \sum_m \sum_n R(n-m, -m) z_1^{m-n} z_2^m$$
$$= \sum_k \sum_\ell R(k,\ell) z_1^{-k} z_2^{-\ell} = B(z_1, z_2) \,. \qquad \blacksquare$$

3.A Appendix. Linear Complex-Valued Signals and Systems

3.A.1 Complex-Valued Model of a Narrow-Band Signal

A simple mathematical treatment may be as follows. Let the narrow-band signal be described as

$$s(t) = A(t)\cos(\omega_c t + \varphi(t)) \,. \qquad (3.47)$$

Here the amplitude $A(t)$ and the phase $\varphi(t)$ are slowly time-varying. The signal can then also be described by a complex-valued representation as

$$s(t) = \mathrm{Re}[\tilde{s}(t)] \,,$$
$$\tilde{s}(t) = A(t)\, e^{i(\omega_c t + \varphi(t))} \,. \qquad (3.48)$$

The amplitude $A(t)$ and the phase $\varphi(t)$ are easy to find in this representation.

Next, consider a time-delayed version of the signal. Assume that for the delay τ, the amplitude and phase remain unchanged ($A(t) = A(t-\tau)$, $\varphi(t) = \varphi(t-\tau)$). Then

$$\tilde{s}(t - \tau) = \tilde{s}(t) \, e^{-i\omega_c \tau} \,.$$

In particular, specialize to a delay of 90 degrees phase shift, that is $\tau = \pi/(2\omega_c)$, to obtain

$$s(t - \pi/2\omega_c) = A(t) \sin(\omega_c t + \varphi(t))$$

and hence the following connection between the real-valued and the complex-valued representation of narrow-band signals:

$$\tilde{s}(t) = s(t) + is(t - \pi/2\omega_c) \tag{3.49}$$

has been established.

The construction of $\tilde{s}(t)$ from discrete-time measurements is called quadrature sampling. The component $s(t) = \text{Re}[\tilde{s}(t)]$ is called the in-phase component, and $\text{Im}[\tilde{s}(t)] = s(t - \pi/(2\omega_c))$ is the quadrature component.

3.A.2 Linear Complex-Valued Systems

Most properties of linear systems continue to hold without change in the complex-valued case. The following standard representations:

state space form

$$\begin{aligned} x(t+1) &= Fx(t) + Gu(t) \,, \\ y(t) &= Hx(t) + Du(t) \,, \end{aligned} \tag{3.50}$$

transfer function operator

$$y(t) = H(q)u(t) = \frac{B(q)}{A(q)} u(t) \,, \tag{3.51}$$

weighting function

$$y(t) = \sum_{k=0}^{\infty} h_k u(t - k) \,, \tag{3.52}$$

can, for example, all be used. Note that these representations (the matrices F, G, H, D, the polynomial coefficients of $A(q)$ and $B(q)$, or the weighting function coefficients $\{h_k\}$) are now generally complex-valued. The usual relations between them apply, such as

$$H(q) = H(qI - F)^{-1}G + D = H(I - Fq^{-1})^{-1}Gq^{-1} + D \,,$$

$$H(q) = \sum_{k=0}^{\infty} h_k q^{-k} \,,$$

$$h(k) = HF^{k-1}G \,, \qquad k > 0; \ \ h(0) = D \,.$$

Further, the usual canonical forms of state space realizations apply, and can be used for transforming transfer function operators into state space forms.

Poles and zeros are defined in the usual way as the poles and zeros of the analytical function $H(z)$. More specifically, the poles are the solutions of $A(z) = 0$, and the zeros solve $B(z) = 0$. The poles can also be found as the eigenvalues of F.

Stability (or, in a strict sense, asymptotic stability) is defined in the usual way. The system is said to be asymptotically stable if, for zero input, the output converges to zero for arbitrary initial values. The system will be asymptotically

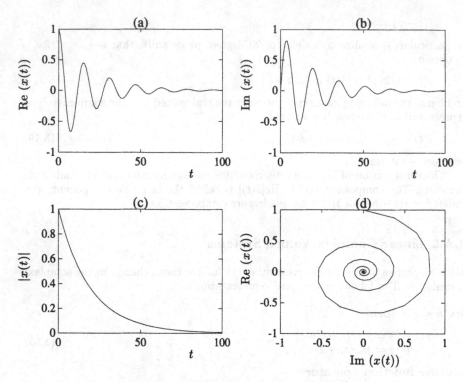

Fig. 3.2. The behaviour of $x(t)$, $x(t+1) = fx(t)$, $f = 0.9\,\mathrm{e}^{4\mathrm{i}}$, $x(0) = 1$.
(a) $\mathrm{Re}(x(t))$, **(b)** $\mathrm{Im}(x(t))$, **(c)** $|x(t)|$, **(d)** $\mathrm{Re}(x(t))$ versus $\mathrm{Im}(x(t))$

stable if and only if all poles lie strictly inside the unit circle. In such a case, the weighting function coefficients $\{h_k\}$ will decay exponentially to zero.

A system with all zeros inside the unit circle is called *minimum-phase*. Minimum-phase models will play a key role in many places; see Sections 4.3 and 7.2.

One *difference* from the real-valued case is that it is no longer necessary that the poles appear in complex conjugate pairs. As a matter of fact, oscillatory modes can be obtained with a single pole (in contrast to the real-valued case, where two poles are required), as illustrated in the following example.

Example 3.4 Consider a first-order system

$$x(t+1) = fx(t)\,, \quad x(0) = x_0\,.$$

In Figures 3.2 and 3.3, the behaviour of $x(t)$ for $f = 0.9\,\mathrm{e}^{4\mathrm{i}}$ and $f = 0.9\,\mathrm{e}^{-4\mathrm{i}}$, $x_0 = 1$ is shown.   □

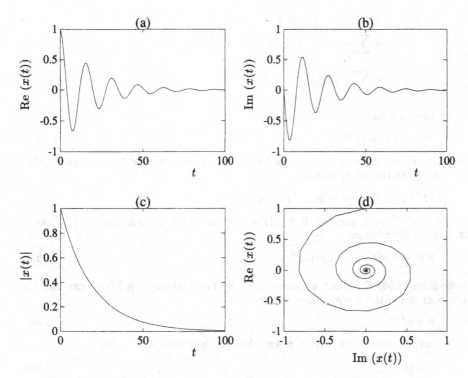

Fig. 3.3. The behaviour of $x(t)$, $x(t+1) = fx(t)$, $f = 0.9\,e^{-4i}$, $x(0) = 1$.
(a) $\mathrm{Re}(x(t))$, **(b)** $\mathrm{Im}(x(t))$, **(c)** $|x(t)|$, **(d)** $\mathrm{Re}(x(t))$ versus $\mathrm{Im}(x(t))$

3.B Appendix. Markov Chains

In this appendix, Markov chains are introduced and briefly analyzed. Such a process attains only discrete values and satisfies the Markov property in (3.12). Let the possible states of the chain be denoted by $\{S_k\}$. The Markov property then reads

$$P(x(t) = x_t | x(t-1) = x_{t-1}, \ldots x(0) = x_0)$$

$$= P(x(t) = x_t | x(t-1) = x_{t-1}). \tag{3.53}$$

In loose terms, only the most recent value of the state (i.e. $x(t-1)$) is of importance for the future behaviour of the process.

Introduce the probabilities

$$p_i(t) = P(x(t) \text{ in } S_i), \tag{3.54}$$

and the transition probabilities (which are assumed not to depend on t)

$$p_{ij} = P(x(t+1) \text{ in } S_j | x(t) \text{ in } S_i). \tag{3.55}$$

Using vector and matrix notation:

$$p(t) = (p_i(t)), \quad P = (p_{ij}),$$

where P will be called the transition matrix, Bayes' rule gives

$$p_j(t+1) = P(x(t+1) \text{ in } S_j)$$
$$= \sum_i P(x(t+1) \text{ in } S_j | x(t) \text{ in } S_i) P(x(t) \text{ in } S_i)$$
$$= \sum_i p_{ij} p_i(t) \,,$$

or, in compact form:

$$p(t+1) = p(t)P \,. \tag{3.56}$$

Transitions over a larger interval can be obtained by iteration of (3.56). Apparently, one can make the interpretation

$$(P^k)_{ij} = P(x(t+k) \text{ in } S_j | x(t) \text{ in } S_i) \,.$$

In what follows, assume that $x(t)$ is a *regular* Markov chain, in that it has a stationary distribution

$$p = \lim_{t\to\infty} p(t) = \lim_{t\to\infty} p(0)P^t \,. \tag{3.57}$$

Note from (3.56) that such a distribution can be obtained as a left eigenvector to P associated with an eigenvalue $\lambda = 1$:

$$p = pP \,. \tag{3.58}$$

It is, further, easy to see that P always has an eigenvalue $\lambda = 1$. Set

$$e = (1 \ldots 1)^T \,. \tag{3.59}$$

Then

$$(Pe)_i = \sum_j p_{ij} e_j = \sum_j p_{ij}$$
$$= \sum_j P(x(t+1) \text{ in } S_j | x(t) \text{ in } S_i)$$
$$= 1 \,.$$

As the elements of P are probabilities, and the elements of each row sum to 1, it is shown next that P cannot have any eigenvalue with magnitude larger than 1. The relation $Px = \lambda x$, with $\max |x_i| \overset{\triangle}{=} m$, implies that

$$|\lambda|m = \max_i |(\lambda x)_i| = \max_i |(Px)_i|$$
$$= \max_i \left| \sum_j p_{ij} x_j \right|$$
$$\leq \max_i \sum_j p_{ij} m$$
$$= m \,,$$

and hence that $|\lambda| \leq 1$.

Next, assume that all eigenvalues except $\lambda = 1$ lie strictly inside the unit circle, which will guarantee the existence of the limiting distribution in (3.57).

Example 3.5 Consider a binary Markov chain with

$$P = \begin{pmatrix} 1-\alpha & \alpha \\ \beta & 1-\beta \end{pmatrix}, \quad 0 \le \alpha \le 1, \quad 0 \le \beta \le 1.$$

The eigenvalues are easily found to be $\lambda_1 = 1$ and $\lambda_2 = 1 - \alpha - \beta$. The only cases when λ_2 will be located on the unit circle are the following.

I: $\alpha = \beta = 0$. Then $P = I$ and $p(t) = p(0)$. A stationary distribution exists, but it depends on the initial distribution.

II: $\alpha = \beta = 1$. Then $P = \begin{pmatrix} 0 & 1 \\ 1 & 0 \end{pmatrix}$ and no stationary distribution exists. The process will (for sure) jump between the two states at every time step. $\square$

In what follows, the mean value and the covariance function for a stationary Markov chain will be derived. Assume that the effect of the initial distribution has decayed and can be neglected. For this purpose, it will be convenient to introduce

$$S = \begin{pmatrix} s_1 & & 0 \\ & \ddots & \\ 0 & & s_n \end{pmatrix}, \tag{3.60}$$

where s_i is the value assigned to $x(t)$ when it is in stage S_i. Further, make an eigendecomposition of P as

$$PU = U\Lambda. \tag{3.61}$$

As it is known that e is an eigenvector with associated eigenvalue $\lambda = 1$, it is possible to write (3.61) in partitioned form as

$$P(e \; U_2) = (e \; U_2) \begin{pmatrix} 1 & 0 \\ 0 & F \end{pmatrix}, \tag{3.62}$$

where the matrix F has all eigenvalues strictly inside the unit circle. (If the eigenvalues are distinct, F can be chosen diagonal; otherwise it can be chosen in Jordan canonical form.) P has, similarly, a set of left eigenvectors

$$\begin{pmatrix} p \\ V_2 \end{pmatrix} P = \begin{pmatrix} 1 & 0 \\ 0 & F \end{pmatrix} \begin{pmatrix} p \\ V_2 \end{pmatrix}. \tag{3.63}$$

Note that the first row of (3.63) is nothing other than (3.58). The eigenvectors are assumed to be normalized in such a way that

$$\begin{pmatrix} p \\ V_2 \end{pmatrix} (e \; U_2) = I. \tag{3.64}$$

It is easy to obtain the (stationary) mean

$$m = \mathbf{E}\, x(t) = \sum_i s_i P(x(t) = s_i)$$

$$= pSe. \tag{3.65}$$

Let k be a nonnegative integer. The covariance function of $x(t)$ can be found as

$$r(k) \triangleq \mathbf{E}\, x(t+k)x(t) - m^2$$

$$= \sum_i \sum_j s_i s_j P(x(t+k) = s_i, x(t) = s_j) - m^2$$

$$= \sum_i \sum_j s_i s_j P(x(t+k) = s_i | x(t) = s_j) P(x(t) = s_j) - m^2$$

$$= \sum_i \sum_j p_j s_j (P^k)_{ji} s_i - m^2$$

$$= p S P^k S e - m^2 . \tag{3.66}$$

Using the eigendecomposition of (3.61) results in

$$P^k = (U\Lambda V)^k = U\Lambda^k V$$

$$= (e \;\; U_2) \begin{pmatrix} 1 & 0 \\ 0 & F^k \end{pmatrix} \begin{pmatrix} p \\ V_2 \end{pmatrix} , \tag{3.67}$$

which gives, with (3.66),

$$r(k) = pS(e \;\; U_2) \begin{pmatrix} 1 & 0 \\ 0 & F^k \end{pmatrix} \begin{pmatrix} p \\ V_2 \end{pmatrix} Se - m^2$$

$$= (pSe)^2 + pSU_2 F^k V_2 Se - m^2$$

$$= (pSU_2)F^k(V_2 Se) . \tag{3.68}$$

The square matrix F has dimension $n-1$. Now compare the expression in (3.68) with (4.21). It can be seen that

the covariance function of the Markov chain will have the same form as that of a linear dynamic system of order $n-1$. The associated spectrum is thus a rational function, just as is the spectrum of an ARMA process.

Example 3.6 Reconsider the binary Markov chain of Example 3.4 and assume that $0 < \alpha + \beta < 1$. The eigendecomposition of (3.61) turns out to be

$$F = 1 - \alpha - \beta ,$$

$$U = \begin{pmatrix} 1 & \alpha \\ 1 & -\beta \end{pmatrix} ,$$

$$V = \begin{pmatrix} \beta & \alpha \\ 1 & -1 \end{pmatrix} \frac{1}{\alpha + \beta} .$$

One obtains from (3.65)

$$m = \frac{1}{\alpha + \beta}(\beta s_1 + \alpha s_2) ,$$

and from (3.68)

$$r(k) = (1 - \alpha - \beta)^k \frac{\alpha\beta}{(\alpha + \beta)^2}(s_1 - s_2)^2 , \qquad k \geq 0 .$$

The associated spectrum can be found directly from Example 4.1:

$$\phi(z) = \frac{r(0)(1 - F^2)}{(z - F)(z^{-*} - F)} . \qquad\qquad \square$$

Exercises

Exercise 3.1 Consider a system given by

$$y(t) = \frac{q + 0.5}{q^2 - 1.5q + 0.7}u(t) + \frac{q + 0.7}{q - 0.8}\varepsilon(t) ,$$

where $\varepsilon(t)$ is white noise of zero mean and unit variance. Convert it into an ARMAX model. Also represent it in a state space model of the form

$$x(t + 1) = Fx(t) + Gu(t) + v(t) ,$$
$$y(t) = Hx(t) + Du(t) + e(t) ,$$

where $v(t)$ and $e(t)$ are white noise sequences with zero mean and covariance matrices

$$\mathbf{E} \begin{pmatrix} v(t) \\ e(t) \end{pmatrix} (v^T(t) \ \ e^T(t)) = \begin{pmatrix} R_1 & R_{12} \\ R_{21} & R_2 \end{pmatrix} .$$

Give R_1, R_{12} and R_2. Can the system be represented in such a state space form with $R_{12} = 0$?

Exercise 3.2 Consider a real-valued moving average process

$$y(t) = e(t) + c_1 e(t - 1) + \ldots + c_n e(t - n) , \quad \mathbf{E}\, e(t) = 0 , \quad \mathbf{E}\, e^2(t) = \lambda^2 .$$

Show that the covariance function $r(\tau)$ of $y(t)$ satisfies

$$r(0) = \lambda^2(1 + c_1^2 + \ldots + c_n^2) ,$$
$$r(\tau) = \lambda^2(c_\tau + c_1 c_{\tau+1} + \ldots c_{n-\tau} c_n) , \quad \tau = 1, \ldots, n - 1 ,$$
$$r(n) = \lambda^2 c_n ,$$
$$r(\tau) = 0 , \quad \tau > n ,$$
$$r(-\tau) = r(\tau) .$$

Exercise 3.3 Examine the bounds in (3.37), (3.38), (3.39) and (3.40) by some numerical examples. Are the bounds crude or sharp?

Exercise 3.4 Consider the system

$$x(t + 1) = Fx(t) + v(t) ,$$
$$y(t) = Hx(t) + e(t) ,$$

where $v(t)$ and $e(t)$ are jointly Gaussian white noise sequences with zero mean and covariance matrices

$$\mathbf{E}\, v(t)v^T(t) = R_1 ,$$
$$\mathbf{E}\, v(t)e^T(t) = R_{12} ,$$
$$\mathbf{E}\, e(t)e^T(t) = R_2 .$$

The initial value $x(t_0)$ is assumed to be Gaussian distributed, $x(t_0) \sim \mathbf{N}(m_0, R_0)$, and independent of the noise sources.

(a) Derive the conditional pdfs $p(x(t+1)|x(t))$ and $p(x(t+1)|x(t), y(t))$. When do they coincide?

Hint. If A, D and the matrix below are invertible:

$$\begin{pmatrix} A & B \\ C & D \end{pmatrix}^{-1} = \begin{pmatrix} A^{-1} & 0 \\ 0 & 0 \end{pmatrix} + \begin{pmatrix} -A^{-1}B \\ I \end{pmatrix} (D - CA^{-1}B)^{-1}$$
$$\times (-CA^{-1} \ \ I) \,.$$

holds.

(b) In case the two conditional pdfs differ, show that one can use

$$\overline{x}(t) = \begin{pmatrix} x(t) \\ y(t) \end{pmatrix}$$

as a state vector. Show that it satisfies

$$\overline{x}(t+1) = \overline{F}\overline{x}(t) + \overline{v}(t)$$

with $\overline{v}(t)$ white noise, and give explicit expressions for $\overline{F}$ and $\overline{R}_1 = \mathbf{E}\,\overline{v}(t)\overline{v}^T(t)$.

Exercise 3.5 Consider the following simplified model of amplitude modulation:

$$x(t) = a(t)\sin(\omega_0 t + \varphi)\,,$$

where $a(t)$ is a stationary stochastic process with zero mean, covariance function $r_a(\tau)$ and spectrum $\phi_a(z)$, and φ is a random variable that is uniformly distributed over $[0, 2\pi]$ and independent of $a(t)$.

Show that $x(t)$ is a stationary stochastic process with covariance function

$$r_x(\tau) = \frac{1}{2}r_a(\tau)\cos\omega_0\tau\,,$$

and spectrum

$$\phi_x(z) = \frac{1}{4}\phi_a(z\,e^{-i\omega_0}) + \frac{1}{4}\phi_a(z\,e^{i\omega_0})\,.$$

Remark The spectral density of $x(t)$ thus satisfies

$$\phi_x(e^{i\omega}) = \frac{1}{4}\phi_a(e^{i(\omega-\omega_0)}) + \frac{1}{4}\phi_a(e^{i(\omega+\omega_0)})\,.$$

If $a(t)$ has a lowpass character (*i.e.* $\phi_a(e^{i\omega})$ has most of its energy around $\omega = 0$), $x(t)$ will therefore have a bandpass character ($\phi_x(e^{i\omega})$ most of its energy around $\omega = \pm\omega_0$).

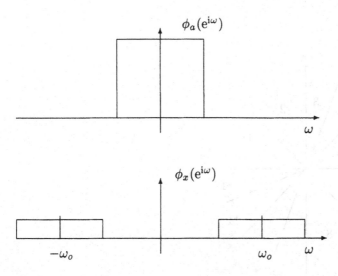

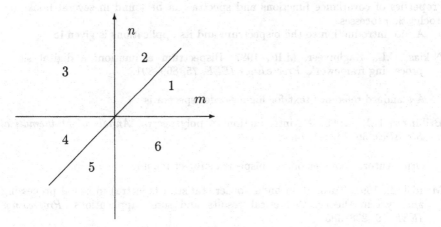

Exercise 3.6 Examine various "symmetries" of the third-order moment sequence and the bispectrum.

(a) Show that if a third order moment sequence is known in any of the six sectors below, it can easily be determined for all arguments:

(b) Show that if the bispectrum is known in any of the 12 sectors below, then it can easily be determined for all arguments:

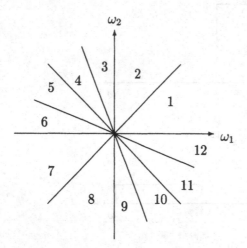

The bispectrum can be considered for arguments $z_k = e^{i\omega_k}$, $k = 1, 2$. This is sufficient since $B(z_1, z_2)$ is analytic. Note that the angles of the sectors are not 22.5 degrees.

Bibliography

Properties of covariance functions and spectra can be found in several books on stochastic processes.

A nice introduction to the bispectrum and its applications is given in

Nikias, C.L., Raghuveer, M.R., 1987. Bispectrum estimation: a digital signal processing framework. *Proceedings IEEE*, 75, 869–891.

A standard reference text for higher-order spectra is

Brillinger, D.R., 1965. An introduction to polyspectra. *Annals of Mathematical Statistics*, 36, 1351–1374.

Other tutorial sources on the bispectra subject include

Mendel, J., 1991. Tutorial on higher-order statistics (spectra) in signal processing and system theory: theoretical results and some applications. *Proceedings IEEE*, 79, 278–305.

Nikias, C.L., 1991. Higher-order spectra analysis. In Haykin, S. (Ed.), *Advances in Spectrum Analysis and Array Processing*. Prentice Hall, Englewood Cliffs, NJ.

Models for nonlinear time series and stochastic systems are outside the scope of this book. For a treatment, see

Tong, H., 1980. *Nonlinear Time Series: A Dynamical System Approach*. Clarendon Press, Oxford.

4. Analysis

4.1 Introduction

In this chapter stochastic models are analyzed in various ways. The following issues will be treated.

- *Linear filtering* and its effect on the spectrum. Complex-valued signals will be allowed (Section 4.2).
- *Spectral factorization* (Section 4.3), permitting complex-valued signals.
- A review of linear *continuous-time models* and their properties (Section 4.4), and *sampling* of such models (Section 4.5).
- Some aspects on the *positive real part of the spectrum* (Section 4.6).
- Properties of the *bispectrum* when stochastic signals are filtered (Section 4.7).
- *Algorithms for covariance calculations* based on finite-order models (Section 4.8).

4.2 Linear Filtering

In this section we examine how linear filtering influences spectral densities. The cases of transfer function models and state space models are treated in separate subsections.

4.2.1 Transfer Function Models

For transfer function models, the following result holds.

Lemma 4.1 *Let* $u(t)$ *be a stationary process with mean* m_u *and covariance function* $r_u(\tau)$ *and spectrum* $\phi_u(z)$. *Let*

$$y(t) = G(q)u(t) = \sum_{k=0}^{\infty} g_k u(t - k) , \qquad (4.1)$$

where $G(q)$ *is an asymptotically stable filter.*

Then $y(t)$ is a stationary process with mean $m_y = G(1)m_u$ and spectra

$$\phi_y(z) = G(z)\phi_u(z)G^*(z^{-*}) , \tag{4.2}$$

$$\phi_{yu}(z) = G(z)\phi_u(z) . \tag{4.3}$$

Proof Applying the expectation operator to both sides of (4.1), one obtains directly

$$\mathbf{E}\, y(t) = \mathbf{E}\sum_{k=0}^{\infty} g_k u(t-k) = \sum_{k=0}^{\infty} g_k m_u = G(1)m_u .$$

Next introduce the deviations $\tilde{y}(t)$, $\tilde{u}(t)$ from the mean values as

$$\tilde{y}(t) = y(t) - m_y , \qquad \tilde{u}(t) = u(t) - m_u .$$

It is easy to see that

$$\tilde{y}(t) = \sum_{k=0}^{\infty} g_k [\tilde{u}(t-k) + m_u] - m_y = \sum_{k=0}^{\infty} g_k \tilde{u}(t-k) .$$

Hence $\tilde{u}(t)$ propagates to $\tilde{y}(t)$ in the same way as $u(t)$ to $y(t)$. Therefore, one can assume, when coping with the covariance functions, that the signals have zero mean. (Alternatively, it may be said that the notation $\tilde{\ }$ is dropped in the following part of the proof.) By direct calculation:

$$r_y(t+\tau\,,t) = \mathbf{E}\, y(t+\tau)y^*(t) = \mathbf{E}\,[\sum_{k=0}^{\infty} g_k u(t+\tau-k)][\sum_{\ell=0}^{\infty} g_\ell u(t-\ell)]^*$$

$$= \sum_{k=0}^{\infty}\sum_{\ell=0}^{\infty} g_k \mathbf{E}\, u(t+\tau-k)u^*(t-\ell)g_\ell^*$$

$$= \sum_{k}\sum_{\ell} g_k r_u(\tau-k+\ell)g_\ell^* .$$

As this expression does not depend on t, it is concluded that $y(t)$ is a stationary stochastic process. Now proceed to derive the spectrum of $y(t)$:

$$\phi_y(z) = \sum_{n=-\infty}^{\infty} r_y(n)z^{-n}$$

$$= \sum_{n=-\infty}^{\infty}\sum_{k=0}^{\infty}\sum_{\ell=0}^{\infty} g_k r_u(n-k+\ell)g_\ell^* z^{-(n-k+\ell)}z^{-k}z^{\ell}$$

$$= [\sum_{k=0}^{\infty} g_k z^{-k}][\sum_{m=-\infty}^{\infty} r_u(m)z^{-m}][\sum_{\ell=0}^{\infty} g_\ell^* z^{\ell}]$$

$$= G(z)\phi_u(z)G^*(z^{-*}) ,$$

which is (4.2). Similarly, for the cross-covariance function and the cross-spectrum:

$$r_{yu}(\tau) = \mathbf{E}\, y(t+\tau)u^*(t) = \mathbf{E}\,[\sum_{k=0}^{\infty} g_k u(t+\tau-k)]u^*(t)$$

$$= \sum_{k=0}^{\infty} g_k r_u(\tau-k)\,,$$

$$\phi_{yu}(z) = \sum_{n=-\infty}^{\infty}\sum_{k=0}^{\infty} g_k r_u(n-k)z^{-(n-k)}z^{-k}$$

$$= [\sum_{k=0}^{\infty} g_k z^{-k}][\sum_{m=-\infty}^{\infty} r_u(m)z^{-m}] = G(z)\phi_u(z)\,. \qquad \blacksquare$$

The following corollary is a form of Parseval's relation. By specializing the matrix case to various elements several special cases can be derived.

Corollary Assume that $u(t)$ is an uncorrelated sequence with covariance matrix $\mathbf{E}\, u(t)u^*(t) = \Lambda$. Then

$$\boxed{\mathbf{E}\, y(t)y^*(t) = \sum_{k=0}^{\infty} g_k \Lambda g_k^* = \frac{1}{2\pi i}\oint G(z)\Lambda G^*(z)\frac{dz}{z}\,,} \qquad (4.4)$$

where the integration is counterclockwise around the unit circle.

Proof When $u(t)$ is an uncorrelated sequence

$$\mathbf{E}\, y(t)y^*(t) = r_y(0) = \sum_{k=0}^{\infty}\sum_{\ell=0}^{\infty} g_k \Lambda \delta_{-k+\ell,0} g_\ell^* = \sum_{k=0}^{\infty} g_k \Lambda g_k^*$$

$$= \frac{1}{2\pi i}\oint \phi_y(z)\frac{dz}{z} = \frac{1}{2\pi i}\oint G(z)\Lambda G^*(z)\frac{dz}{z}$$

holds. In the last equation, (4.2) and the fact that $z^{-*} = z$ on the unit circle have been used. $\qquad \blacksquare$

4.2.2 State Space Models

Next turn to the analysis of state space models.

Lemma 4.2 *Consider the system*

$$\begin{aligned} x(t+1) &= Fx(t) + v(t)\,, \\ y(t) &= Hx(t) + e(t)\,, \end{aligned} \qquad (4.5)$$

where $v(t)$ and $e(t)$ are complex-valued uncorrelated sequences with zero mean and

$$\mathbf{E} \begin{pmatrix} v(t) \\ e(t) \end{pmatrix} (v^*(s) \quad e^*(s)) = \begin{pmatrix} R_1 & R_{12} \\ R_{21} & R_2 \end{pmatrix} \delta_{t,s} . \tag{4.6}$$

Let the initial value $x(t_0)$ be a random vector with mean m_0 and covariance matrix R_0, and be independent of $v(t)$ and $e(t)$ for all $t \geq t_0$. Then the mean, $m_x(t)$, and the covariance function, $R_x(t,s)$, of the state vector satisfy

$$m_x(t+1) = F m_x(t) = F^{t+1-t_0} m_0 , \tag{4.7}$$

$$R_x(t+\tau, t) = F^\tau P(t) , \qquad \tau \geq 0 , \tag{4.8}$$

$$P(t+1) = F P(t) F^* + R_1 , \qquad P(t_0) = R_0 . \tag{4.9}$$

Proof Equation (4.7) is immediate from (4.5). Set $P(t) = R_x(t,t) = \mathbf{E}\,[x(t) - m_x(t)][x(t) - m_x(t)]^*$ and let $\tau > 0$. Then by straightforward calculation:

$$
\begin{aligned}
R_x(t+\tau, t) &= \mathbf{E}\,[x(t+\tau) - m_x(t+\tau)][x(t) - m_x(t)]^* \\
&= \mathbf{E}\,[F x(t+\tau-1) + v(t+\tau-1) - F m_x(t+\tau-1)] \\
&\quad \times [x(t) - m_x(t)]^* \\
&= F R_x(t+\tau-1, t) .
\end{aligned}
$$

Equation (4.8) is obtained by iteration. In order to obtain (4.9), proceed as follows (note that, by construction, $x(t)$ is uncorrelated with $v(s)$ for $t \leq s$):

$$
\begin{aligned}
P(t+1) &= \mathbf{E}\,[x(t+1) - m_x(t+1)][x(t+1) - m_x(t+1)]^* \\
&= \mathbf{E}\,[F x(t) + v(t) - F m_x(t)][F x(t) + v(t) - F m_x(t)]^* \\
&= \mathbf{E}\,F[x(t) - m_x(t)][x(t) - m_x(t)]^* F^* + \mathbf{E}\,v(t)v(t)^* \\
&= F P(t) F^* + R_1 .
\end{aligned}
$$
∎

Remark 1 Equation (4.9) is called a *Lyapunov equation*. The matrix $P(t)$ is apparently precisely the covariance matrix of the state vector $x(t)$:

$$P(t) = \mathrm{cov}[x(t)] = \mathbf{E}\,[x(t) - m_x(t)][x(t) - m_x(t)]^* . \tag{4.10}$$

□

Remark 2 If the system contains a deterministic input and reads

$$x(t+1) = Fx(t) + Gu(t) + v(t) ,$$
$$y(t) = Hx(t) + e(t) ,$$
(4.11)

instead of (4.5), the mean value $m_x(t)$ is modified to

$$m_x(t+1) = Fm_x(t) + Gu(t) ,$$
$$m_x(t_0) = m_0 ,$$
(4.12)

and the covariance properties are not changed.   □

Remark 3 The results extend easily to time-varying systems. In principle, F and H and the covariance matrices are used with the current time index instead of being treated as time-invariant factors. To be more specific, consider the system

$$x(t+1) = F(t)x(t) + v(t) ,$$
$$y(t) = H(t)x(t) + e(t) ,$$
(4.13)

$$\mathbf{E} \begin{pmatrix} v(t) \\ e(t) \end{pmatrix} \left(v^*(s) \ e^*(s) \right) = \begin{pmatrix} R_1(t) & R_{12}(t) \\ R_{21}(t) & R_2(t) \end{pmatrix} \delta_{t,s} .$$
(4.14)

Then, (4.7)–(4.9) are modified to (use the convention $\prod_{s=t}^{t-1} F(s) = I$)

$$m_x(t+1) = F(t)m_x(t) = \prod_{s=t_0}^{t} F(s)m_0 ,$$
(4.15)

$$R_x(t+\tau,t) = F(t+\tau-1)R_x(t+\tau-1,t)$$
(4.16)

$$= \prod_{s=t}^{t+\tau-1} F(s)P(t) , \qquad \tau \geq 0 ,$$
(4.17)

$$P(t+1) = F(t)P(t)F^*(t) + R_1(t) , \qquad P(t_0) = R_0 .$$
(4.18)

In what follows, it is mostly assumed that the system is time-invariant. Several results for optimal estimation and control can also be easily extended to the time-varying case, though.   □

Corollary 1 The output mean and covariance functions satisfy

$$m_y(t) = Hm_x(t) ,$$
(4.19)

$$R_y(t,t) = HP(t)H^* + R_2 ,$$
(4.20)

$$R_y(t+\tau,t) = HF^{\tau-1}[FP(t)H^* + R_{12}] , \qquad \tau > 0 .$$
(4.21)

Proof Equation (4.19) is immediate. So is (4.20), as

$$y(t) - m_y(t) = H[x(t) - m_x(t)] + e(t) .$$

Let $\tau > 0$. Then (recall that $y(t)$ is uncorrelated with $v(s)$ and $e(s)$, $t < s$)

$$
\begin{aligned}
R_y(t+\tau,t) &= \mathbf{E}\left[y(t+\tau) - m_y(t+\tau)][y(t) - m_y(t)\right]^* \\
&= \mathbf{E}\left[H\{x(t+\tau) - m_x(t+\tau)\} + e(t+\tau)][y(t) - m_y(t)\right]^* \\
&= H\mathbf{E}\left[F\{x(t+\tau-1) - m_x(t+\tau-1)\} + v(t+\tau-1)\right] \\
&\quad \times [y(t) - m_y(t)]^* \\
&= HF\mathbf{E}\left[x(t+\tau-1) - m_x(t+\tau-1)][y(t) - m_y(t)\right]^* \\
&= \ldots = HF^{\tau-1}\mathbf{E}\left[F\{x(t) - m_x(t)\} + v(t)][y(t) - m_y(t)\right]^* \\
&= HF^{\tau-1}\mathbf{E}\left[F\{x(t) - m_x(t)\}\{H[x(t) - m_x(t)]\}^*\right. \\
&\quad \left. + v(t)e(t)^*\right] \\
&= HF^{\tau-1}[FP(t)H^* + R_{12}] ,
\end{aligned}
$$

which is (4.21). ∎

Corollary 2 Assume, further, that the system is asymptotically stable. Then:

(i) The solution $P(t)$ to (4.9) converges to a solution of the Lyapunov equation

$$P = FPF^* + R_1 , \tag{4.22}$$

as $t - t_0 \to \infty$. The Lyapunov equation has a unique solution (for F having all eigenvalues strictly inside the unit circle). The solution is Hermitian and nonnegative definite.

(ii) The processes $x(t)$ and $y(t)$ are asymptotically stationary (as $t - t_0 \to \infty$).

Proof By straightforward iteration of (4.9):

$$
\begin{aligned}
P(t) &= F^{t-t_0} R_0 F^{*(t-t_0)} + \sum_{s=t_0}^{t-1} F^{t-1-s} R_1 F^{*(t-1-s)} \\
&= F^{t-t_0} R_0 F^{*(t-t_0)} + \sum_{j=0}^{t-t_0-1} F^j R_1 F^{*j} .
\end{aligned} \tag{4.23}
$$

When the matrix F has all eigenvalues strictly inside the unit circle, the above expression will converge, as $t - t_0 \to \infty$, to

$$P = \sum_{j=0}^{\infty} F^j R_1 F^{*j} , \tag{4.24}$$

which satisfies (4.22) by inspection.

The sum in (4.24) is convergent, which can be seen as follows. Assume first that F is diagonalizable, and write $F = SDS^{-1}$, where the diagonal matrix D contains the eigenvalues λ_j, $j = 1, \ldots, n$. Then each matrix element of (4.24) will consist of sums

$$\sum_{\mu,\nu=1}^{n} \sum_{j=0}^{\infty} \lambda_\mu^j \lambda_\nu^{*j} C_{\mu,\nu} .$$

Such sums are convergent as all eigenvalues of F lie strictly inside the unit circle for an asymptotically stable system. In the case there are multiple eigenvalues, the sums will rather take the form

$$\sum_{\mu,\nu} \sum_{j=0}^{\infty} p_{\mu,\nu}(j) \lambda_\mu^j \lambda_\nu^{*j}$$

where $p_{\mu,\nu}(j)$ is a polynomial, whose degree corresponds to the multiplicity of the repeated eigenvalue. The conclusion about convergence of the sum remains.

In order to show uniqueness of the solution, assume that P_1 and P_2 are solutions to (4.22) and set $X = P_1 - P_2$. It is then easy to derive the equality $X = FXF^*$. By successive iterations, one obtains $X = F^k X F^{*k}$ for any positive k. Considering the limiting case in particular gives $X = \lim_{k\to\infty} F^k X F^{*k} = 0$, which completes the proof of uniqueness.

From part (i) it follows easily that asymptotically (as $t - t_0 \to \infty$) $P(t)$ can be replaced by P and that $m_x(t)$ and $m_y(t)$ both tend to zero. Hence the covariance functions depend only on the difference between the time arguments, and the processes are stationary. ∎

Remark In the stationary case, the spectrum of $y(t)$ becomes

$$\phi_y(z) = H(zI - F)^{-1} R_1 (z^{-*}I - F)^{-*} H^* + R_2$$
$$+ H(zI - F)^{-1} R_{12} + R_{21}(z^{-*}I - F)^{-*} H^* . \qquad (4.25)$$

This can be proved by applying the definition of the spectrum to equations (4.19)–(4.21) for the covariance function. It can be derived more conveniently from (4.2) and (4.5). Set

$$u(t) = \begin{pmatrix} v(t) \\ e(t) \end{pmatrix} .$$

Then

$$y(t) = [H(qI - F)^{-1} \ I] u(t)$$

holds. By the identification

$$G(q) = [H(qI - F)^{-1} \ I] ,$$
$$\phi_u(z) = R_u = \begin{pmatrix} R_1 & R_{12} \\ R_{21} & R_2 \end{pmatrix} ,$$

and (4.2), (4.25) follows easily.   □

The results are illustrated by three examples.

Example 4.1 Consider the first-order real-valued process

$$x(t+1) = ax(t) + v(t) , \qquad |a| < 1 , \quad \mathbf{E}\, v^2(t) = \lambda^2 , \tag{4.26}$$

observed from $t = -\infty$. It is then stationary. Its variance P is the unique solution to the Lyapunov equation

$$P = a^2 P + \lambda^2 ,$$

giving $P = \lambda^2/(1 - a^2)$. Note that this is consistent with (4.24). The covariance function is, according to (4.21):

$$r_x(\tau) = a^\tau P = a^\tau \frac{\lambda^2}{1 - a^2} , \qquad \tau \geq 0 , \tag{4.27}$$

and the spectrum

$$\phi_x(z) = \frac{\lambda^2}{(z - a)(z^{-1} - a)} . \tag{4.28}$$

As a further illustration, note that expression in (4.28) for the spectrum follows directly by applying Lemma 4.1 to the model of (4.26). One may then apply (3.25) and residue calculus to find the covariance function. Let $\tau \geq 0$. Then

$$\begin{aligned}
r(\tau) &= \frac{1}{2\pi i} \oint \phi(z) z^\tau \frac{dz}{z} \\
&= \frac{1}{2\pi i} \oint \frac{\lambda^2}{(z - a)(z^{-1} - a)} z^\tau \frac{dz}{z} \\
&= \frac{1}{2\pi i} \oint \frac{\lambda^2 z^\tau}{(z - a)(1 - az)} \, dz \\
&= \frac{\lambda^2 a^\tau}{1 - a^2} .
\end{aligned}$$

In the above calculations $\oint$ denotes integration counterclockwise around the unit circle. The result is in full agreement with (4.27). The last equality follows as the pole $z = a$ is within the unit circle, whereas the other pole $z = 1/a$ is not.   □

Example 4.2 Consider the first-order ARMA process

$$y(t) + ay(t-1) = e(t) + ce(t-1) , \qquad |a| < 1 ,$$

where $e(t)$ is white noise of zero mean and variance λ^2. The stationary variance of $y(t)$ is sought. This variance can be calculated in many ways. Here it is chosen to derive it through a state space formalism. First, represent the system as

$$x(t+1) = \begin{pmatrix} -a & 1 \\ 0 & 0 \end{pmatrix} x(t) + \begin{pmatrix} 1 \\ c \end{pmatrix} v(t) ,$$

$$y(t) = (1 \ \ 0)x(t) .$$

In this case, $v(t) = e(t+1)$, again white noise of zero mean and variance λ^2. The "relabelling" of the time index in the noise sequence will not change its statistical properties. The steady state covariance matrix P can be found by solving the Lyapunov equation of (4.22). Set

$$P = \begin{pmatrix} p_{11} & p_{12} \\ p_{12} & p_{22} \end{pmatrix} ,$$

which gives

$$\begin{pmatrix} p_{11} & p_{12} \\ p_{12} & p_{22} \end{pmatrix} = \begin{pmatrix} -a & 1 \\ 0 & 0 \end{pmatrix} \begin{pmatrix} p_{11} & p_{12} \\ p_{12} & p_{22} \end{pmatrix} \begin{pmatrix} -a & 0 \\ 1 & 0 \end{pmatrix} + \lambda^2 \begin{pmatrix} 1 \\ c \end{pmatrix} (1 \ \ c) .$$

When reformulating this as a system of equations, note that, owing to symmetry, the 12 element and the 21 element of the equation will give the same information. Equating the elements gives

$$p_{11} = a^2 p_{11} - 2ap_{12} + p_{22} + \lambda^2 ,$$
$$p_{12} = \lambda^2 c ,$$
$$p_{22} = \lambda^2 c^2 .$$

The solution is now readily obtained:

$$\text{var}[y(t)] = (1 \ \ 0)P \begin{pmatrix} 1 \\ 0 \end{pmatrix} = p_{11} = \lambda^2 \frac{1 + c^2 - 2ac}{1 - a^2} . \qquad \Box$$

Example 4.3 As an illustration of the transient effects that can be modelled by the state space technique, consider a first-order model

$$x(t+1) = ax(t) + v(t) , \qquad v(t) \sim \mathbf{N}(0, \lambda^2) , \ \ x(t_0) \sim \mathbf{N}(m_o, R_0) ,$$

with Gaussian distributed noise and initial state $x(t_0)$. Figure 4.1 illustrates how the pdf propagates with time. $\qquad \Box$

4.2.3 Yule–Walker Equations

In this subsection, the so-called Yule–Walker equations for a real-valued ARMA process, and their possible role in determining a covariance function, are examined.

Consider first the case of an AR process

$$y(t) + a_1 y(t - 1) + \ldots + a_n y(t - n) = e(t) , \qquad \mathbf{E} e^2(t) = \lambda^2. \qquad (4.29)$$

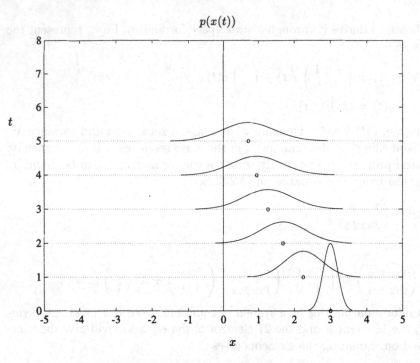

Fig. 4.1. Propagation of the pdf of $x(t)$ as a function of t, Example 4.3. Parameters: $a = 0.75$, $m_0 = 3$, $R_0 = 0.25$, $\lambda^2 = 0.04$. The values of $m(t)$ are denoted by circles

Note that $y(t)$ can be viewed as a linear combination of all old values of the noise, that is $\{e(s)\}_{s=-\infty}^{t}$. By multiplying $y(t)$ by a delayed value of the process, say $y(t - \tau)$, $\tau \geq 0$, and applying the expectation operator, one obtains

$$\mathbf{E}\, y(t - \tau)[y(t) + a_1 y(t - 1) + \ldots + a_n y(t - n)] = \mathbf{E}\, y(t - \tau)e(t) \ ,$$

or

$$r(\tau) + a_1 r(\tau - 1) + \ldots + a_n r(\tau - n) = \begin{cases} 0 \ , & \tau > 0 \ , \\ \lambda^2 \ , & \tau = 0 \ , \end{cases} \qquad (4.30)$$

which is called a Yule–Walker equation. Note in passing that it can be written as

$$A(q)r(t) = \begin{cases} 0 \ , & t > -n \ , \\ \lambda^2 \ , & t = -n \ . \end{cases}$$

By using (4.30) for $\tau = 0, \ldots, n$, one can construct the following system of equations for determining the covariance elements $r(0), r(1), \ldots, r(n)$:

$$\begin{pmatrix} 1 & a_1 & \cdots & a_n \\ a_1 & 1+a_2 & & a_n & 0 \\ \vdots & & \ddots & & \vdots \\ a_n & a_{n-1} & \cdots & & 1 \end{pmatrix} \begin{pmatrix} r(0) \\ \vdots \\ r(n) \end{pmatrix} = \begin{pmatrix} \lambda^2 \\ 0 \\ \vdots \\ 0 \end{pmatrix} . \tag{4.31}$$

The matrix appearing in (4.31) can be written as

$$\begin{pmatrix} 1 & & \cdots & 0 \\ a_1 & 1 \\ \vdots & & \ddots \\ a_n & a_{n-1} & & 1 \end{pmatrix} + \begin{pmatrix} 0 & a_1 & \cdots & & a_n \\ 0 & a_2 & & a_n & 0 \\ \vdots & & \cdot{\cdot}{\cdot} \\ & a_n & & & 0 \\ 0 & 0 \end{pmatrix} .$$

Once $r(0), \ldots, r(n)$ are known, (4.30) can be iterated (for $\tau = n + 1$, $n + 2, \ldots$) to find further covariance elements.

Example 4.4 Apply the Yule–Walker equations for computing the covariance function of a first-order AR process

$$y(t) - ay(t-1) = e(t) , \qquad \mathbf{E}\, e^2(t) = \lambda^2 .$$

The system of equations of (4.31) becomes

$$\begin{pmatrix} 1 & -a \\ -a & 1 \end{pmatrix} \begin{pmatrix} r(0) \\ r(1) \end{pmatrix} = \begin{pmatrix} \lambda^2 \\ 0 \end{pmatrix} ,$$

giving

$$r(0) = \frac{\lambda^2}{1 - a^2} , \qquad r(1) = \frac{a\lambda^2}{1 - a^2} .$$

Next, it is found from (4.30) that

$$r(\tau) - ar(\tau - 1) = 0 , \qquad \tau > 0 ,$$

which implies that

$$r(\tau) = a^\tau \frac{\lambda^2}{1 - a^2} , \qquad \tau \geq 0 .$$

This expression coincides with the previous findings in (4.27). $\square$

Consider next a full ARMA process

$$y(t) + a_1 y(t-1) + \ldots + a_n y(t-n)$$
$$= e(t) + c_1 e(t-1) + \ldots + c_m e(t-m) ,$$
$$\mathbf{E}\, e^2(t) = \lambda^2. \tag{4.32}$$

Now the cross-covariance function between $y(t)$ and $e(t)$ must be involved as an intermediate step. Multiplying (4.32) by $y(t-\tau)$, $\tau \geq 0$, and applying the expectation operator, gives

$$r_y(\tau) + a_1 r_y(\tau - 1) + \ldots + a_n r_y(\tau - n)$$
$$= r_{ey}(\tau) + c_1 r_{ey}(\tau - 1) + \ldots + c_m r_{ey}(\tau - m) \,. \tag{4.33}$$

In order to obtain the output covariance function $r_y(\tau)$, the cross-covariance function $r_{ey}(\tau)$ must first be found. This is done by multiplying (4.32) by $e(t - \tau)$, and applying the expectation operator, which leads to

$$r_{ey}(-\tau) + a_1 r_{ey}(-\tau + 1) + \ldots + a_n r_{ey}(-\tau + n)$$
$$= \lambda^2 [\delta_{\tau,0} + c_1 \delta_{\tau-1,0} + \ldots + c_m \delta_{\tau-m,0}] \,. \tag{4.34}$$

As $y(t)$ is a linear combination of $\{e(s)\}_{s=-\infty}^{t}$, it is found that $r_{ey}(\tau) = 0$ for $\tau > 0$. Hence (4.33) gives

$$\boxed{r_y(\tau) + a_1 r_y(\tau - 1) + \ldots + a_n r_y(\tau - n) = 0 \,, \qquad \tau > m \,.} \tag{4.35}$$

The use of (4.29)–(4.35) to derive the autocovariance function is illustrated next by applying them to a first-order ARMA process.

Example 4.5 Consider the ARMA process
$$y(t) + ay(t - 1) = e(t) + ce(t - 1) \,, \qquad \mathbf{E}\, e^2(t) = \lambda^2 \,.$$

In this case $n = 1$, $m = 1$. Equation (4.35) gives
$$r_y(\tau) + a r_y(\tau - 1) = 0 \,, \qquad \tau > 1 \,.$$

Using (4.33) for $\tau = 0$ and 1 gives
$$\begin{pmatrix} 1 & a \\ a & 1 \end{pmatrix} \begin{pmatrix} r_y(0) \\ r_y(1) \end{pmatrix} = \begin{pmatrix} 1 & c \\ c & 0 \end{pmatrix} \begin{pmatrix} r_{ey}(0) \\ r_{ey}(-1) \end{pmatrix} \,.$$

Now consider (4.34) for $\tau = 0$ and 1, which gives
$$\begin{pmatrix} 1 & 0 \\ a & 1 \end{pmatrix} \begin{pmatrix} r_{ey}(0) \\ r_{ey}(-1) \end{pmatrix} = \lambda^2 \begin{pmatrix} 1 \\ c \end{pmatrix} \,.$$

By straightforward calculations, it is found that
$$r_{ey}(0) = \lambda^2 \,,$$
$$r_{ey}(-1) = \lambda^2(c - a) \,,$$
$$r_y(0) = \frac{\lambda^2}{1 - a^2}(1 + c^2 - 2ac) \,,$$
$$r_y(1) = \frac{\lambda^2}{1 - a^2}(c - a)(1 - ac) \,,$$

and finally:
$$r_y(\tau) = \frac{\lambda^2}{1 - a^2}(c - a)(1 - ac)(-a)^{\tau-1} \,, \qquad \tau \geq 1 \,.$$

The expression above for $r_y(0)$ coincides with the former results for the variance in Example 4.2. □

4.3 Spectral Factorization

We now discuss how the concept of spectral factorization applies for complex-valued processes. Its role is instrumental in deriving (parametric) time-domain models from a spectrum representation.

4.3.1 Transfer Function Models

The key tool for the analysis will be Lemma 4.6 in Section 4.A. The lemma is applied to finite-order spectra in the following theorem.

Theorem 4.1 *Let $\phi(z)$ be a scalar spectrum that is rational in z, that is it can be written as*

$$\phi(z) = \frac{\sum_{|k| \le m} \beta_k z^k}{\sum_{|k| \le n} \alpha_k z^k} \,. \tag{4.36}$$

Then there are two polynomials:

$$\begin{aligned} A(z) &= z^n + a_1 z^{n-1} + \ldots + a_n \,, \\ C(z) &= z^m + c_1 z^{m-1} + \ldots + c_m \,, \end{aligned} \tag{4.37}$$

and a positive real number λ^2 so that

(i) $A(z)$ has all zeros inside the unit circle,
(ii) $C(z)$ has all zeros inside or on the unit circle,

(iii)

$$\phi(z) = \lambda^2 \, \frac{C(z)}{A(z)} \frac{C^*(z^{-*})}{A^*(z^{-*})} \,. \tag{4.38}$$

In the case where $\phi(e^{i\omega}) > 0$, $\forall \, \omega$, $C(z)$ will have no zeros on the circle.

Proof The statement is immediate from Lemma 4.6. ∎

Remark 1 Any continuous spectrum can be arbitrarily well approximated by a rational function in z as in (4.36), provided that m and n are appropriately chosen. Hence, the assumptions of the theorem are not restrictive, but the results are applicable, at least with a small approximation error, to a very wide class of stochastic processes. □

Remark 2 An important implication of Theorem 4.1 is that (as far as second-order moments are concerned) the underlying stochastic process can be regarded as generated by filtering white noise, that is as the ARMA process

$$A(q)y(t) = C(q)e(t) ,$$
$$\mathbf{E} \, e(t)e^*(t) = \lambda^2 . \tag{4.39}$$

This means, in particular, that for describing stochastic processes (as long as they have a rational spectral density), it is no restriction to assume the input signal to be white noise. It thus also gives a rationale for the state space model in (4.5) and the assumption that $v(t)$ and $e(t)$ are *white* noise sequences therein. Note that any linear state space model driven by correlated noise can be converted into the form of (4.5) by introducing additional state variables. In the representation in (4.39), the sequence $\{e(t)\}$ is called the *output innovations*. □

Remark 3 Spectral factorization can also be viewed as a form of aggregation of noise sources. Assume, for example, that an ARMA process

$$A(q)x(t) = C(q)v(t) \tag{4.40}$$

is observed in measurement noise

$$y(t) = x(t) + e(t) , \tag{4.41}$$

and that $v(t)$ and $e(t)$ are uncorrelated white noise sequences of zero mean and variances λ_v^2 and λ_e^2 respectively. As far as the second-order properties (such as the spectrum or the covariance function) are concerned, $y(t)$ can be viewed as generated from one single noise source as

$$A(q)y(t) = D(q)\varepsilon(t) . \tag{4.42}$$

The polynomial D and the noise variance λ_ε^2 are derived as follows. The spectrum is, according to (4.40) and (4.41):

$$\phi_y(z) = \phi_x(z) + \phi_e(z) = \lambda_v^2 \frac{C(z)C^*(z^{-*})}{A(z)A^*(z^{-*})} + \lambda_e^2 ,$$

while (4.42) gives

$$\phi_y(z) = \lambda_\varepsilon^2 \frac{D(z)D^*(z^{-*})}{A(z)A^*(z^{-*})} .$$

Equating these two expressions gives

$$\lambda_\varepsilon^2 D(z)D^*(z^{-*}) \equiv \lambda_v^2 C(z)C^*(z^{-*}) + \lambda_e^2 A(z)A^*(z^{-*}) . \tag{4.43}$$

The two representations given by (4.40) and (4.41), and by (4.42) of the processes $y(t)$, are displayed schematically in Figure 4.2. □

Remark 4 The theorem can be extended to the multivariable case. Assume that the spectral density $\phi(e^{i\omega})$ is rational and nonsingular. Then there is

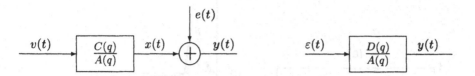

Fig. 4.2. Two representations of an ARMA process with noisy observations

a finite-dimensional filter $H(q)$ and a Hermitian positive definite matrix Λ, such that $H(q)$ and $H^{-1}(q)$ are asymptotically stable, $H(\infty) = I$, and

$$\phi(e^{i\omega}) = H(e^{i\omega})\Lambda H^*(e^{i\omega}) . \tag{4.44}$$

The condition $H(\infty) = I$ simply corresponds in the scalar case to the condition that $A(z)$ and $C(z)$ in (4.37) are *monic* polynomials, that is they have a leading coefficient equal to one. □

Example 4.6 As an illustration of the spectral factorization considered in Remark 3 above, consider an AR(2) process observed in some additional noise. In this case we take

$$A(q^{-1}) = 1 + a_1 q^{-1} + a_2 q^{-2} ,$$

with

$$a_1 = -1.5 , \qquad a_2 = 0.8 .$$

The measurements can be described as an ARMA(2,2) model; see (4.42):

$$y(t) = \frac{D(q^{-1})}{A(q^{-1})}\varepsilon(t) . \tag{4.45}$$

Now set

$$S = \lambda_e^2/\lambda_v^2 .$$

The value of S describes the amount of measurement noise. The zeros of the ARMA process in (4.45) varies from the origin when $S = 0$ to the positions of the zeros of $A(z)$ (in this case more precisely to $z = 0.75 \pm i0.487$) when S tends to infinity. How the zero locations vary with S are displayed in Figure 4.3.

The spectral density of $y(t)$ is, of course, also influenced by S. When S is close to zero, $y(t)$ is (almost) the AR process $1/A(q^{-1})v(t)$, which has a high peak in its spectrum. On the other hand, when S tends to infinity, $y(t)$ becomes more and more like white noise with a flat spectrum. The spectral

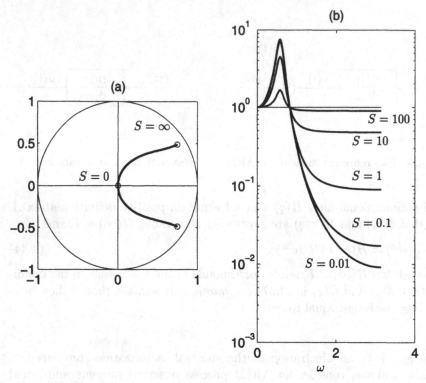

Fig. 4.3. (a) Zero locations for the ARMA process given by (4.45), as a function of the parameter S. (b) Normalized spectral densities, $\phi(e^{i\omega})/\phi(1)$, for some values of S

density is displayed in part Figure 4.3 (b) for a few S values. To facilitate comparisons, the normalized spectral density, $\phi(e^{i\omega})/\phi(1)$ is plotted versus the angular frequency ω. It is clear that the spectral density gradually becomes increasingly flat as the amount of measurement noise increases.

4.3.2 State Space Models

For state space models, the following result holds.

Theorem 4.2 *Consider the asymptotically stable system*

$$
\begin{aligned}
x(t+1) &= Fx(t) + v(t) , \\
y(t) &= Hx(t) + e(t) ,
\end{aligned}
\tag{4.46}
$$

where $\{v(t)\}$, $\{e(t)\}$ are zero mean white noise sequences with

$$
\mathbf{E} \begin{pmatrix} v(t) \\ e(t) \end{pmatrix} \begin{pmatrix} v^*(s) & e^*(s) \end{pmatrix} = \begin{pmatrix} R_1 & R_{12} \\ R_{21} & R_2 \end{pmatrix} \delta_{t,s} .
\tag{4.47}
$$

Assume that P is a Hermitian, positive definite solution to the algebraic Riccati equation *(ARE):*

$$P = FPF^* + R_1 - (FPH^* + R_{12}) \\ \times (HPH^* + R_2)^{-1}(HPF^* + R_{21}) \, . \tag{4.48}$$

Set

$$H(q) = I + H(qI - F)^{-1}K \, , \qquad \Lambda = HPH^* + R_2 \, , \tag{4.49}$$

where

$$K = (FPH^* + R_{12})(HPH^* + R_2)^{-1} \, . \tag{4.50}$$

Then

(i) $H(\infty) = I$ and $H(q)$ is asymptotically stable.
(ii) $H^{-1}(q) = I - H(qI - F + KH)^{-1}K \, . \tag{4.51}$
(iii) $H^{-1}(q)$ is stable.

(iv)
$$\phi_y(z) = H(z)\Lambda H^*(z^{-*}) \, . \tag{4.52}$$

Proof Part (i) is trivial by construction of $H(q)$. Equation (4.51) for $H^{-1}(q)$ is verified by direct multiplication:

$$H(q) \times [I - H(qI - F + KH)^{-1}K] \\ = I + H(qI - F)^{-1}K - H(qI - F + KH)^{-1}K \\ \quad - H(qI - F)^{-1}KH(qI - F + KH)^{-1}K \\ = I + H(qI - F)^{-1}[(qI - F + KH) - (qI - F) - (KH)] \\ \quad \times (qI - F + KH)^{-1}K \\ = I \, .$$

In order to prove (iii), it must be shown that $F - KH$ has all eigenvalues inside (or on) the unit circle. For this purpose, consider the system

$$z(t + 1) = (F - KH)^* z(t) \, , \tag{4.53}$$

and prove that it is stable by using the Lyapunov function $V(z) = z^* P z$. Obviously, $V(z) \geq 0$ with equality if and only if $z = 0$.

If it can be proved that

$$\Delta V(t) \overset{\Delta}{=} V(z(t + 1)) - V(z(t))$$

is nonpositive ($\Delta V(t) \leq 0$), then it follows that $V(t)$ is nonincreasing, and

$$V(z(t)) \leq V(z(0)) \, .$$

One can then conclude that

$$\| z(t) \| \leq C \| z(0) \|$$

for some constant C. [In fact, the tightest C is the condition number of P, $C = \lambda_{\max}(P)/\lambda_{\min}(P)$.] Hence the system in (4.51) is stable.

When $\Delta V(t) \leq 0$, it holds that $V(z(t))$ is monotonically decreasing and bounded from below. Hence $V(z(t))$ is then convergent. If it further can be shown that the limit is $V = 0$ (or equivalently, $\lim_{t \to \infty} z(t) = 0$), we have established *asymptotic* stability. (A sufficient condition would be that $\Delta V(t) = -z^*(t)Qz(t)$ with Q positive definite.)

Straightforward calculation gives

$$\Delta V(t) = z^*(t)[(F - KH)P(F - KH)^* - P]z(t) . \tag{4.54}$$

However, using (4.48):

$$
\begin{aligned}
(F - KH)&P(F - KH)^* - P = FPF^* - KHPF^* - FPH^*K^* \\
&+ KHPH^*K^* - P \\
&= -R_1 + K(HPH^* + R_2)K^* - KHPF^* - FPH^*K^* + KHPH^*K^* \\
&= -R_1 + K(2HPH^* + R_2)K^* - K[(HPH^* + R_2)K^* - R_{21}] \\
&\quad -[K(HPH^* + R_2) - R_{12}]K^* \\
&= -R_1 - KR_2K^* + KR_{21} + R_{12}K^* \\
&= -(I \;\; -K) \begin{pmatrix} R_1 & R_{12} \\ R_{21} & R_2 \end{pmatrix} \begin{pmatrix} I \\ -K^* \end{pmatrix} \leq 0 .
\end{aligned}
\tag{4.55}
$$

Hence $V(z(t))$ is nonincreasing with time, which proves stability.

Consider, finally, part (iv). Straightforward calculation gives,

$$
\begin{aligned}
\phi_y(z) &- H(z)\Lambda H^*(z^{-*}) \\
&= H(zI - F)^{-1}R_1(z^{-*}I - F)^{-*}H^* + R_2 + H(zI - F)^{-1}R_{12} \\
&\quad + R_{21}(z^{-*}I - F)^{-*}H^* - \Lambda - H(zI - F)^{-1}K\Lambda \\
&\quad - \Lambda K^*(z^{-*}I - F)^{-*}H^* - H(zI - F)^{-1}K\Lambda K^*(z^{-*}I - F)^{-*}H^* \\
&= H(zI - F)^{-1}[R_1 - (zI - F)P(z^{-*}I - F)^* - FP(z^{-*}I - F)^* \\
&\quad - (zI - F)PF^* - K\Lambda K^*](z^{-*}I - F)^{-*}H^* \\
&= H(zI - F)^{-1}[R_1 - P - FPF^* + FPF^* + FPF^* - K\Lambda K^*] \\
&\quad \times (z^{-*}I - F)^{-*}H^* \\
&= 0 . \quad\blacksquare
\end{aligned}
$$

Remark 1 The theorem implies that (as far as second order moments are concerned) the output can be represented equivalently as

$$y(t) = H(q)\tilde{y}(t) , \qquad \mathbf{E}\, \tilde{y}(t)\tilde{y}^*(s) = \Lambda\delta_{t,s} . \tag{4.56}$$

The white noise sequence $\{\tilde{y}(t)\}$ are the (*output*) *innovations*. They appear as the sequence $\{e(t)\}$ in (4.39). A state space representation of (4.56) is readily found:

$$x_0(t+1) = Fx_0(t) + K\tilde{y}(t) \, ,$$
$$y(t) = Hx_0(t) + \tilde{y}(t) \, . \tag{4.57}$$

Both forms, (4.56) and (4.57), are called the *innovations form* or innovations representation. The states $x_0(t)$ can be shown to be the optimal one-step prediction of the original states. Compare Section 6.4. □

Remark 2 It can be proved that the Riccati equation has at most one positive definite solution. See Exercise 6.11. Sometimes it has additional nonnegative definite solutions. If no positive definite solution exists, the innovations form will often be found by taking the "largest" nonnegative definite solution. More precisely, denote the solution to the Riccati equation that corresponds to the innovation form by P_*. Then, for any other symmetric solution P to the Riccati equation, it holds that

$$P_* - P \geq 0 \tag{4.58}$$

meaning that $P_* - P$ is nonnegative definite. See Exercise 4.25 for an illustration. □

Remark 3 Note that it is a consequence of (4.51) that the zeros of the innovations form are given by the eigenvalues of $F - KH$. □

Corollary Let the assumptions of the theorem be satisfied. Assume further that R_2 is positive definite, and introduce

$$\tilde{F} = F - R_{12}R_2^{-1}H \, ,$$
$$\tilde{R}_1 = R_1 - R_{12}R_2^{-1}R_{21} \, , \tag{4.59}$$

and factorize the matrix $\tilde{R}_1$ as

$$\tilde{R}_1 = BB^*. \tag{4.60}$$

Assume that the pair $(\tilde{F}, B)$ to be stabilizable. Then the filter $H^{-1}(q)$, (4.51), is asymptotically stable.

Proof The result follows if the following implication

$$\Delta V(t) = 0 \ \forall t \quad \Rightarrow \quad z(t) = 0 \ \forall t \, ,$$

can be proved. Here we get from (4.54) and (4.55):

$$\begin{aligned}
\Delta V(t) &= z^*(t)[(F - KH)P(F - KH)^* - P]z(t) \\
&= -z^*(t)[R_1 - KR_{21} - R_{12}K^* + KR_2K^*]z(t) \\
&= -z^*(t)[BB^* + R_{12}R_2^{-1}R_{21} - KR_{21} - R_{12}K^* + KR_2K^*]z(t) \\
&= - \parallel B^*z(t) \parallel - z^*(t)[K - R_{12}R_2^{-1}]R_2[K - R_{12}R_2^{-1}]^*z(t) \, .
\end{aligned}$$

Hence,

$$\Delta V(t) = 0 \quad \Rightarrow \quad \begin{cases} B^*z(t) = 0 \\ (K - R_{12}R_2^{-1})^*z(t) = 0 \, . \end{cases} \tag{4.61}$$

As $(\tilde{F}, B)$ is stabilizable, there exists some feedback L such that $\overline{F} \triangleq \tilde{F} - BL$ has all eigenvalues strictly inside the unit circle. The dynamics of the system in (4.53) become

$$
\begin{aligned}
z(t+1) &= (F - KH)^* z(t) \\
&= [(\tilde{F} + R_{12} R_2^{-1} H)^* - H^* K^*] z(t) \\
&= [\tilde{F}^* + H^* R_2^{-1} R_{21} - H^* R_2^{-1} R_{21}] z(t) \\
&= \tilde{F}^* z(t) \\
&= [\overline{F} + BL]^* z(t) \\
&= \overline{F}^* z(t) ,
\end{aligned}
$$

which is asymptotically stable. Note that in the above calculations we used (4.59) and (4.61). Hence the only possible limit point is $z = 0$, which proves asymptotic stability of (4.51) and (4.53).   ∎

Remark 1 As F was considered asymptotically stable in Theorem 4.2, in the case $R_{12} = 0$ it directly holds that $(\tilde{F}, B)$ is stabilizable.   □

Remark 2 The stability result of the corollary can be extended to unstable systems (F not necessarily asymptotically stable). The condition $(\tilde{F}, B)$ being stabilizable essentially means that all unstable modes of the system in (4.46) must be controllable from the "input" $v(t)$.   □

4.3.3 An Example

As an illustration, a simple spectral factorization problem will be solved using both the transfer function approach and the state space approach.

Reconsider Example 3.3, but now start with the spectral density (a, b being real-valued)

$$
\phi(e^{i\omega}) = 1 + a^2 + b^2 + 2a \cos \omega + 2b \sin \omega .
$$

Introducing $\varphi = a + ib$, it can also be written as

$$
\phi(e^{i\omega}) = [1 + \varphi \overline{\varphi} + \varphi e^{-i\omega} + \overline{\varphi} e^{i\omega}] .
$$

The transfer function approach is now used to find c and λ^2, such that

$$
1 + \varphi \overline{\varphi} + \varphi e^{-i\omega} + \overline{\varphi} e^{i\omega} = \lambda^2 (e^{i\omega} + c)(e^{-i\omega} + \overline{c}) ,
$$

subject to

$$
\begin{aligned}
&c \text{ complex, } |c| \le 1 , \\
&\lambda^2 \text{ real, } \quad \lambda^2 > 0 .
\end{aligned}
$$

Equating the powers of $e^{i\omega}$ gives the following two equations:

$$\begin{cases} 1 + \varphi\overline{\varphi} = \lambda^2(1 + c\overline{c}) \,, \\ \varphi = \lambda^2 c \,. \end{cases}$$

There are two solutions.

Solution 1 $c = \varphi, \lambda^2 = 1$. This solution is the one sought, provided that $|\varphi| \leq 1$.

Solution 2 $c = 1/\overline{\varphi}, \lambda^2 = \varphi\overline{\varphi}$. This solution is the one sought, if $|\varphi| \geq 1$.

Note that the two solutions correspond to reflected zero positions. The zero in the first case is $z_1 = -\varphi$, whereas in the second it is $z_2 = -1/\overline{\varphi} = 1/\overline{z}_1$.

Next, consider the state space approach and choose first to represent the system as

$$x(t+1) = \begin{pmatrix} 0 & 1 \\ 0 & 0 \end{pmatrix} x(t) + v(t) \,,$$

$$y(t) = (1 \ \ 0)x(t) \,,$$

$$\mathbf{E} \, v(t)v^*(s) = \begin{pmatrix} 1 & \overline{\varphi} \\ \varphi & \varphi\overline{\varphi} \end{pmatrix} \delta_{t,s} \,.$$

It is straightforward to verify that $y(t)$ will then have the given spectral density. Let the ARE have a solution

$$P = \begin{pmatrix} p_{11} & p_{12} \\ \overline{p_{12}} & p_{22} \end{pmatrix} \,.$$

(Note that p_{11} and p_{22} are real.) Writing the Riccati equation explicitly:

$$\begin{pmatrix} p_{11} & p_{12} \\ \overline{p_{12}} & p_{22} \end{pmatrix} = \begin{pmatrix} 0 & 1 \\ 0 & 0 \end{pmatrix} \begin{pmatrix} p_{11} & p_{12} \\ \overline{p_{12}} & p_{22} \end{pmatrix} \begin{pmatrix} 0 & 0 \\ 1 & 0 \end{pmatrix} + \begin{pmatrix} 1 & \overline{\varphi} \\ \varphi & \varphi\overline{\varphi} \end{pmatrix}$$

$$- \begin{pmatrix} 0 & 1 \\ 0 & 0 \end{pmatrix} \begin{pmatrix} p_{11} & p_{12} \\ \overline{p_{12}} & p_{22} \end{pmatrix} \begin{pmatrix} 1 \\ 0 \end{pmatrix} (p_{11})^{-1} (1 \ \ 0) \begin{pmatrix} p_{11} & p_{12} \\ \overline{p_{12}} & p_{22} \end{pmatrix} \begin{pmatrix} 0 & 0 \\ 1 & 0 \end{pmatrix} \,.$$

Evaluating the various matrix elements gives the following equations:

$$\begin{aligned} \text{Element 11:} \quad & p_{11} = p_{22} + 1 - p_{12}\overline{p_{12}}/p_{11} \,, \\ \text{Element 12:} \quad & p_{12} = \overline{\varphi} \,, \\ \text{Element 21:} \quad & \overline{p_{12}} = \varphi \,, \\ \text{Element 22:} \quad & p_{22} = \varphi\overline{\varphi} \,, \end{aligned}$$

which gives

$$p_{11} = 1 + \varphi\overline{\varphi} - \frac{\varphi\overline{\varphi}}{p_{11}} \,,$$

with the two solutions $p_{11} = 1$ and $p_{11} = \varphi\overline{\varphi}$. Thus it is found that the corresponding matrices are

$$P^{(1)} = \begin{pmatrix} 1 & \overline{\varphi} \\ \varphi & \varphi\overline{\varphi} \end{pmatrix} = \begin{pmatrix} 1 \\ \varphi \end{pmatrix} \begin{pmatrix} 1 & \overline{\varphi} \end{pmatrix},$$

$$P^{(2)} = \begin{pmatrix} \varphi\overline{\varphi} & \overline{\varphi} \\ \varphi & \varphi\overline{\varphi} \end{pmatrix}.$$

It is easy to verify that $P^{(1)}$ is positive semidefinite. The matrix $P^{(2)}$ is positive definite if $|\varphi| > 1$, indefinite if $|\varphi| < 1$ and coincides with $P^{(1)}$ if $|\varphi| = 1$. The solution sought is therefore given by $P^{(1)}$ for $|\varphi| \leq 1$ and by $P^{(2)}$ for $|\varphi| \geq 1$. Next, compute the corresponding values of K, $H(q)$ and Λ.

Case 1 $|\varphi| \leq 1$. Simple calculations give

$$K = \begin{pmatrix} \overline{p_{12}}/p_{11} \\ 0 \end{pmatrix} = \begin{pmatrix} \varphi \\ 0 \end{pmatrix},$$

$$H(q) = 1 + \begin{pmatrix} 1 & 0 \end{pmatrix} \begin{pmatrix} q & -1 \\ 0 & q \end{pmatrix}^{-1} \begin{pmatrix} \varphi \\ 0 \end{pmatrix} = 1 + \varphi q^{-1}, \quad \Lambda = p_{11} = 1.$$

Case 2 $|\varphi| \geq 1$. By similar means:

$$K = \begin{pmatrix} \overline{p_{12}}/p_{11} \\ 0 \end{pmatrix} = \begin{pmatrix} 1/\overline{\varphi} \\ 0 \end{pmatrix},$$

$$H(q) = 1 + \frac{1}{\overline{\varphi}} q^{-1},$$

$$\Lambda = p_{11} = \varphi\overline{\varphi}.$$

Note that the solution is, of course, the same as the one previously obtained using the transfer function approach. Also, note that for the case where $|\varphi| \leq 1$, no positive definite solution exists for the Riccati equation but that the (only) nonnegative definite solution leads to the correct innovations form.

4.4 Continuous-time Models

This section illustrates how some of the properties of discrete-time stochastic systems appear in analogue form for continuous-time models. However, some mathematical problems occur as well. White noise leads to considerable difficulties that must be solved in a mathematically rigorous way.

4.4.1 Covariance Function and Spectra

In continuous time, the covariance function of a process $y(t)$ is still defined as (compare (3.18)):

$$r(\tau) = \mathbf{E}\, y(t + \tau)y^*(t), \tag{4.62}$$

assuming for simplicity that $y(t)$ has zero mean. The spectrum will now be

$$\phi(s) = \int_{-\infty}^{\infty} r(\tau) \, e^{-s\tau} \, d\tau \, , \tag{4.63}$$

and the spectral density will be

$$\phi(i\omega) = \int_{-\infty}^{\infty} r(\tau) \, e^{-i\omega\tau} \, d\tau \, . \tag{4.64}$$

The inverse relation to (4.63) is

$$r(\tau) = \frac{1}{2\pi i} \int \phi(s) \, e^{s\tau} \, ds \, , \tag{4.65}$$

where integration is along the whole imaginary axis. One can therefore alternatively write

$$r(\tau) = \frac{1}{2\pi} \int_{-\infty}^{\infty} \phi(i\omega) \, e^{i\omega\tau} \, d\omega \, . \tag{4.66}$$

By setting $\tau = 0$ in (4.66), the usual interpretation is obtained: the spectral density describes how the variance or power is distributed over the frequency range.

The positive real part of the spectrum is defined as

$$\tilde{\phi}(s) = \int_{0}^{\infty} r(\tau) \, e^{-s\tau} \, d\tau \, . \tag{4.67}$$

It is thus precisely the ordinary Laplace transform of the covariance function.

The continuous-time variant of the results of Lemma 3.2 is as follows:

- $\phi^*(s) = \phi(-\bar{s})$.
- $\phi(i\omega)$ is nonnegative definite.
- The diagonal elements of $\phi(i\omega)$ are real-valued and positive.
- In the scalar case $\phi(i\omega) = \phi(-i\omega)$ if and only if the real and imaginary parts of the process are uncorrelated.
- $\phi_{xy}^*(s) = \phi_{yx}(-\bar{s})$.
- $\phi(s) = \tilde{\phi}(s) + \tilde{\phi}^*(-\bar{s})$.

4.4.2 Spectral Factorization

Consider a stationary stochastic process described by a spectral density $\phi(i\omega)$ that is a rational function of $i\omega$. For simplicity, assume that the process is scalar and real-valued. Then, by pure analogy with the discrete-time case (Theorem 4.1), it is found that

$$\phi(i\omega) = \frac{B(i\omega)B(-i\omega)}{A(i\omega)A(-i\omega)} \, , \tag{4.68}$$

where the polynomials

$$A(p) = p^n + a_1 p^{n-1} + \ldots + a_n ,$$
$$B(p) = b_1 p^{n-1} + \ldots + b_n ,$$

(4.69)

have all zeros in the left half plane (*i.e.* in the stability area). Here p is an arbitrary polynomial argument, but it will subsequently denote the differentiation operator ($py(t) = \dot{y}(t)$).

The effect of filtering a stationary process, say $u(t)$, with an asymptotically stable filter, say $H(p)$, can be easily phrased using the spectra. Let the filtering be described by

$$y(t) = H(p)u(t) .$$

(4.70)

Then

$$\phi_y(s) = H(s)\phi_u(s)H^*(-\bar{s}) ,$$

(4.71)

again paralleling the discrete-time case (Lemma 4.1). As a consequence, one can interpret any process with a rational spectral density given by (4.68) as having been generated by filtering as in (4.70) by using

$$H(p) = \frac{B(p)}{A(p)} .$$

(4.72)

The signal $u(t)$ would then have a *constant* spectral density, $\phi_u(i\omega) \equiv 1$. As for the discrete-time case, such a process is called *white noise*.

4.4.3 White Noise

From (4.66) it can immediately be seen that white noise, if meaningful at all, leads to mathematical difficulties. For such a case, it must follow that

$$r(0) = \frac{1}{2\pi} \int_{-\infty}^{\infty} \phi(i\omega) \times 1 \, d\omega = \frac{1}{2\pi} \int_{-\infty}^{\infty} 1 \, d\omega = \infty .$$

(4.73)

Continuous-time white noise thus has infinite variance. Generalizing the discrete-time case, one would also expect white noise to be uncorrelated for different time arguments, which would require

$$r(\tau) = 0 , \qquad \tau \neq 0 .$$

(4.74)

In attempting to comply with the requirements of (4.73) and (4.74), it is necessary to step outside the space of integrable functions and use distributions (generalized functions) instead. As an attempt, let

$$r(\tau) = r_0 \delta(\tau) ,$$

(4.75)

where $\delta(\tau)$ is Dirac's δ-function. Then (4.63) gives

$$\phi(s) = \int_{-\infty}^{\infty} r_0 \delta(\tau) \, e^{-s\tau} \, d\tau$$

$$= r_0 ,$$

(4.76)

which has the desired form (*i.e.* it is constant as a function of s).

For practical purposes, with some care in the calculations, one can use the concept of continuous-time white noise and proceed as one would expect. To be more stringent, though, note that it is *far from trivial* that differential equations such as

$$y(t) = \frac{B(p)}{A(p)} e(t) , \tag{4.77}$$

that is

$$A(p)y(t) = B(p)e(t) , \tag{4.78}$$

where $e(t)$ is white noise, have reasonable meaning or well-defined solutions. Those who are mathematically inclined often choose to cope with this problem by using Wiener processes, which can be regarded as integrated white noise.

4.4.4 Wiener Processes

Introduce, in a *heuristic* way, an integrated white noise process for $t > 0$:

$$w(t) = \int_0^t e(s) \, ds , \tag{4.79}$$

where $e(t)$ is assumed to have zero mean and covariance function $r_0 \delta(\tau)$.

Remark Note that the discrete-time correspondence to a Wiener process is a random walk process $y(t)$ given by $y(0) = 0$, and

$$y(t) = y(t-1) + v(t) ,$$
$$= \sum_{s=1}^{t} v(s) , \tag{4.80}$$

where $\{v(s)\}$ is discrete-time white noise. □

Some properties of Wiener processes are now analyzed. In a stringent analysis it is necessary to establish that the integral is well defined and that integration and expectation can commute. A more heuristic analysis is performed here.

First note that, by its construction, *increments of a Wiener process over disjoint intervals are uncorrelated*. If $t_1 < t_2 < t_3 < t_4$ it is obvious that

$$w(t_4) - w(t_3) = \int_{t_3}^{t_4} e(s) \, ds$$

and

$$w(t_2) - w(t_1) = \int_{t_1}^{t_2} e(s) \, ds$$

are uncorrelated.

Next consider the covariance function of $w(t)$. The Wiener process is not a stationary process. Let $\tau > 0, t > 0$. Then

$$
\begin{aligned}
r(t+\tau,t) &= \mathbf{E}\, w(t+\tau)w^*(t) \\
&= \mathbf{E}\,[w(t+\tau) - w(t) + w(t)]w^*(t) \\
&= \mathbf{E}\, w(t)w^*(t) \\
&= \mathbf{E}\, \int_0^t e(s')\,ds' \int_0^t e^*(s'')\,ds'' \\
&= \int_0^t \int_0^t \mathbf{E}\, e(s')e^*(s'')\,ds'\,ds'' \\
&= \int_0^t \int_0^t r_0\delta(s' - s'')\,ds'\,ds'' \\
&= \int_0^t r_0\,ds \\
&= r_0 t \,.
\end{aligned} \tag{4.81}
$$

In particular:

$$
\begin{aligned}
&\mathbf{E}\,[w(t+\Delta t) - w(t)][w(t+\Delta t) - w(t)]^* \\
&= \mathbf{E}\, w(t+\Delta t)w^*(t+\Delta t) - \mathbf{E}\, w(t+\Delta t)w^*(t) \\
&\quad -\mathbf{E}\, w(t)w^*(t+\Delta t) + \mathbf{E}\, w(t)w^*(t) \\
&= r_0(t+\Delta t) - r_0 t - r_0 t + r_0 t = r_0\Delta t \,.
\end{aligned} \tag{4.82}
$$

In the limiting case, this can be written as

$$
\mathbf{E}\, dw\, dw^* = r_0\, dt \,. \tag{4.83}
$$

It can be stated informally that white noise is the derivative of a Wiener process:

$$
\frac{dw(t)}{dt} = e \,, \tag{4.84}
$$

cf. (4.79).

4.4.5 State Space Models

Differential equation models with white noise e as input can be written in state space form as

$$
\dot{x} = Ax + Be \,. \tag{4.85}
$$

In a more stringent analysis, (4.85) is "multiplied by" dt and written in the form

$$dx = Ax \, dt + B \, dw ,\tag{4.86}$$

which is called a *stochastic differential equation*. Here $dw = e \, dt$ is the increment of a Wiener process $w(t)$.

For practical use, the more intuitive form, (4.85), can be used. Forms where the white noise term appears under an integral sign are "safer" to use than others.

Next, the covariance function of $x(t)$ in (4.85) is examined in order to find the continuous-time correspondence to Lemma 4.2. The solution to the differential equation in (4.85) can be written in the usual way:

$$x(t) = e^{A(t-t_0)}x(t_0) + \int_{t_0}^{t} e^{A(t-s)} Be(s) \, ds .\tag{4.87}$$

Assume that e has covariance function $R_0\delta(\tau)$. Assume, further, that the initial state $x(t_0)$ has zero mean and covariance matrix P_0 and is independent of $e(s)$. It is easily found that $x(t)$ has zero mean:

$$\mathbf{E}\, x(t) = e^{A(t-t_0)}\, \mathbf{E}\, x(t_0) + \int_{t_0}^{t} e^{A(t-s)} BE\, e(s) \, ds = 0 .\tag{4.88}$$

The covariance matrix of $x(t)$, say $P(t)$, can be evaluated as follows. The two terms in (4.87) are apparently uncorrelated. Therefore,

$$\begin{aligned}
P(t) = \mathbf{E}\, x(t)x^*(t) &= e^{A(t-t_0)} P(t_0) e^{A^*(t-t_0)}\\
&\quad + \mathbf{E}\,[\int_{t_0}^{t} e^{A(t-s')} Be(s') \, ds'][\int_{t_0}^{t} e^{A(t-s'')} Be(s'') \, ds'']^*\\
&= e^{A(t-t_0)} P(t_0) e^{A^*(t-t_0)}\\
&\quad + \int_{t_0}^{t}\int_{t_0}^{t} e^{A(t-s')} BR_0\delta(s' - s'')B^* e^{A^*(t-s'')} \, ds' \, ds''\\
&= e^{A(t-t_0)} P(t_0) e^{A^*(t-t_0)}\\
&\quad + \int_{t_0}^{t} e^{A(t-s)} BR_0B^* e^{A^*(t-s)} \, ds .
\end{aligned}\tag{4.89}$$

By straightforward differentiation of (4.89), it is found that $P(t)$ satisfies the differential equation (of Lyapunov type)

$$\dot{P}(t) = AP(t) + P(t)A^* + BR_0B^* .\tag{4.90}$$

Assume that A is asymptotically stable. It follows that, in the limit as $t - t_0 \to \infty$, the covariance matrix P is the unique (and nonnegative definite) solution to

$$\boxed{0 = AP + PA^* + BR_0B^* .}\tag{4.91}$$

This is a continuous-time *Lyapunov equation*.

Note from (4.89) that P can also be represented in the integral form, *cf.*
(4.24):

$$P = \int_{-\infty}^{t} e^{A(t-s)} BR_0 B^* e^{A^*(t-s)} \, ds$$

$$= \int_{0}^{\infty} e^{At} BR_0 B^* e^{A^* t} \, dt \,. \tag{4.92}$$

It is also possible to evaluate the covariance function of $x(t)$. Let $\tau > 0$. Then

$$R_x(t+\tau, t) = \mathbf{E}\, x(t+\tau) x^*(t)$$

$$= \mathbf{E}\left[e^{A\tau} x(t) + \int_{t}^{t+\tau} e^{A(t+\tau-s)} Be(s) \, ds \right] x^*(t)$$

$$= e^{A\tau} P(t) \,. \tag{4.93}$$

Hence $x(t)$ is a stationary process when $t_0 \to -\infty$. Its spectral density can
be found by applying the definition to the equation (4.93) for $R(\tau)$. It can
also be derived more simply by noting that

$$x(t) = [pI - A]^{-1} Be(t) \,, \tag{4.94}$$

giving, with (4.71):

$$\phi_x(s) = [sI - A]^{-1} BR_0 B^* [-\bar{s}I - A]^{-*} \,. \tag{4.95}$$

Finally, note that (in the stationary case) the covariance function $R(\tau)$ satisfies the *Yule–Walker equation*, that is $R(\tau)$ satisfies the following differential
equation:

$$(p^n + a_1 p^{n-1} + \ldots + a_n) R(\tau) = 0 \,, \qquad \tau > 0 \,, \tag{4.96}$$

where

$$p^n + a_1 p^{n-1} + \ldots + a_n = \det(pI - A) = 0 \tag{4.97}$$

is the characteristic equation of the matrix A.

This follows from the Cayley–Hamilton theorem (which states that every
square matrix satisfies its own characteristic equation):

$$(p^n + a_1 p^{n-1} + \ldots + a_n) R(\tau) = (p^n + a_1 p^{n-1} + \ldots + a_n) e^{A\tau} P$$

$$= (A^n + a_1 A^{n-1} + \ldots + a_n I) e^{A\tau} P$$

$$= 0 \,. \tag{4.98}$$

4.5 Sampling Stochastic Models

4.5.1 Introduction

There are many interesting features in sampling a stochastic process. Note
that, in this form of sampling, only the output values at the sampling events

are used. The output values have, of course, the same statistical properties before and after sampling. Hence, the output covariance function must be retained. This form of sampling differs from sampling a *deterministic* control system, in which case the sampling is characterized by the input being kept constant over the sampling intervals. As a side comment, recall that in filter design still another form of sampling is used, namely *approximation* of the continuous-time frequency function. There is no *unique* way in which such an approximation must be performed. Also note that some further results on sampling will appear in Section 4.8.4.

4.5.2 State Space Models

Consider a continuous-time model

$$\dot{x} = Ax + v \, , \tag{4.99}$$

where v is white noise with covariance function $r_v(\tau) = R_c \delta(\tau)$. Now sample this process by considering it for times $t = 0, h, 2h, \ldots$. Then (k being an integer)

$$x(kh + h) = e^{Ah} x(kh) + \int_{kh}^{kh+h} e^{A(kh+h-s)} v(s) \, ds$$

$$\stackrel{\triangle}{=} Fx(kh) + e(kh) \, , \tag{4.100}$$

where

$$F = e^{Ah} \, ,$$
$$e(kh) = \int_{kh}^{kh+h} e^{A(kh+h-s)} v(s) \, ds \, . \tag{4.101}$$

By its construction, it follows that $e(kh)$ is (discrete-time) white noise. Note that (4.100) is therefore a standard discrete-time stochastic state space model. When dealing exclusively with discrete-time models, the time scale is usually normalized by using the sampling interval as the time unit. This would imply that, in such a case, (4.100) can be rewritten in the simplified form

$$x(k + 1) = Fx(k) + e(k) \, . \tag{4.102}$$

In order to emphasize the relation to continuous time, it may be useful to keep the notation h explicitly in the time argument.

The covariance matrix of the white noise sequence $e(kh)$ in (4.101) can be evaluated as follows:

$$R_d \stackrel{\triangle}{=} \mathbf{E} \, e(kh)e^*(kh)$$

$$= \mathbf{E} \int_{kh}^{kh+h} \int_{kh}^{kh+h} e^{A(kh+h-s')} v(s')v^*(s'') e^{A^*(kh+h-s'')} \, ds' \, ds''$$

$$= \int_{kh}^{kh+h} \int_{kh}^{kh+h} e^{A(kh+h-s')} R_c \delta(s' - s'') e^{A^*(kh+h-s'')} \, ds' \, ds''$$

$$= \int_{kh}^{kh+h} e^{A(kh+h-s)} R_c e^{A^*(kh+h-s)} \, ds$$

$$= \int_{0}^{h} e^{As} R_c e^{A^*s} \, ds \, . \tag{4.103}$$

Remark. Note that even if R_c is singular (say of rank 1), the covariance matrix R_d of the sampled noise process will mostly (say, generically) be non-singular (and hence positive definite). □

4.5.3 Aliasing

In this section, the *aliasing* or *folding effect* when sampling a continuous-time stochastic process is investigated.

Consider a continuous-time process with covariance function $r(\tau)$ and spectral density

$$\phi_c(i\omega) = \int_{-\infty}^{\infty} r(\tau) e^{-i\tau\omega} \, d\tau \, . \tag{4.104}$$

Then

$$r(\tau) = \frac{1}{2\pi} \int_{-\infty}^{\infty} \phi_c(i\omega) e^{i\tau\omega} \, d\omega \, . \tag{4.105}$$

Now sample this signal. The covariance sequence is then

$$r_k = r(kh) \, , \qquad k = 0, \pm 1, \ldots$$

and the associated discrete-time spectral density is

$$\phi_d(e^{i\omega h}) = h \sum_{k=-\infty}^{\infty} r_k e^{-ik\omega h} \, , \tag{4.106}$$

which is defined for $|\omega| \le \pi/h$.

Next derive a relation between ϕ_c and ϕ_d. The key tool is Lemma 4.7, given in Section 4.A.

This gives

$$\phi_d(e^{i\omega h}) = h \sum_{k=-\infty}^{\infty} r(kh) e^{-ik\omega h}$$

$$= \frac{h}{2\pi} \sum_{k=-\infty}^{\infty} e^{-ik\omega h} \int_{-\infty}^{\infty} \phi_c(i\omega') e^{ikh\omega'} \, d\omega'$$

$$= \frac{h}{2\pi} \int_{-\infty}^{\infty} \phi_c(i\omega') \left[\sum_{k=-\infty}^{\infty} e^{-ik(\omega-\omega')h} \right] d\omega'$$

$$= \frac{h}{2\pi} \int_{-\infty}^{\infty} \phi_c(i\omega') 2\pi \sum_{j=-\infty}^{\infty} \delta(\omega'h - \omega h - 2\pi j) \, d\omega'$$

$$= \int_{-\infty}^{\infty} \phi_c(i\omega') \sum_{j=-\infty}^{\infty} \delta\left(\omega' - \omega - \frac{2\pi}{h}j\right) d\omega' \, ,$$

and, in final form:

$$\phi_d(e^{i\omega h}) = \sum_{j=-\infty}^{\infty} \phi_c\left(i(\omega + j\frac{2\pi}{h})\right) \, . \tag{4.107}$$

The geometric interpretation of this result is that the spectral density is folded at $\omega = \pm\frac{2\pi}{h}, \pm\frac{4\pi}{h}, \ldots$ into the interval $(-\pi/h, \pi/h)$ and that all contributions are added.

Remark Equation (4.107) also illustrates the sampling theorem. Assume that the signal has no energy above the Nyquist frequency $\omega_N = \pi/h$. Then $\phi_c(i\omega)$ will vanish as soon as $|\omega| > \omega_N$, and (4.107) simplifies to

$$\phi_d(e^{i\omega h}) = \phi_c(i\omega) \, . \tag{4.108}$$

Under such circumstances, it seems plausible that there is "no loss of information" due to the sampling. In fact, one can prove (and *this* is what Shannon's sampling theorem is about) that for such a band-limited signal one can reconstruct its value at any time from the discrete-time measurement $\{y(kh)\}_{k=-\infty}^{\infty}$. The reconstruction happens to be given by

$$f(t) = \sum_{k=-\infty}^{\infty} f(kh) \frac{\sin \pi \frac{t-kh}{h}}{\pi \frac{t-kh}{h}} \, . \tag{4.109}$$

□

Example 4.7 Consider a first-order model given by

$$(p + a)y(t) = av(t) \, ,$$

where $v(t)$ is white noise with covariance function $\delta(t)$. Its continuous-time spectral density is readily obtained from (4.71):

$$\phi_c(i\omega) = \frac{a^2}{\omega^2 + a^2} \, .$$

The sampled form of $y(t)$ is readily obtained from (4.100) and (4.101) as

$$y(kh + h) - e^{-ah}y(kh) = e(kh) \, ,$$

where the discrete-time white noise $e(kh)$ has variance

$$\mathbf{E}\, e^2(kh) = \lambda^2 = \frac{a}{2}(1 - e^{-2ah}) \, .$$

The discrete-time spectral density is

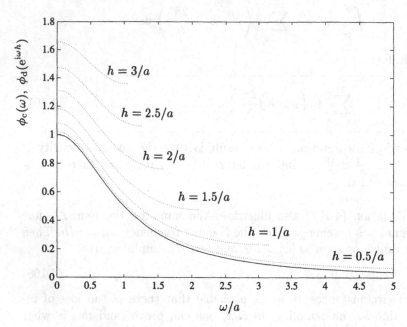

Fig. 4.4. Frequency folding when sampling a stochastic process with spectral density $\phi_c(\omega) = a^2/(\omega^2 + a^2)$. The *solid line* shows $\phi_c(\omega)$; the *dotted lines* show $\phi_d(e^{i\omega h})$ for some different sampling intervals h

$$\phi_d(e^{i\omega h}) = \frac{\lambda^2 h}{|e^{i\omega h} - e^{-ah}|^2} = \frac{ah}{2} \frac{1 - e^{-2ah}}{1 + e^{-2ah} - 2e^{-ah}\cos\omega h} \,, \quad |\omega| \leq \frac{\pi}{h}\,.$$

By straightforward calculation, it can be shown that for fixed ω, $\phi_d(e^{i\omega h})$ decreases monotonously to $\phi_c(i\omega)$ as h tends to zero. The relation between the continuous-time and discrete-time spectral densities is illustrated in Figure 4.4.   □

4.6 The Positive Real Part of the Spectrum

In this section, it will be shown how the positive real part of the spectrum can be conveniently found for finite-order models. The relations derived can be used for algorithmic purposes, as explained in Section 4.8. The case of real-valued processes is considered.

4.6.1 ARMA Processes

A detailed analysis of $\tilde{\phi}(z)$ for ARMA processes will be useful. The following lemma turns out to be useful.

Lemma 4.3 *Consider the identity*

$$\lambda^2 C(z)C(z^{-1}) \equiv D(z)A(z^{-1}) + A(z)D(z^{-1}) \,, \qquad (4.110)$$

where

$$\begin{aligned}
A(z) &= z^n + a_1 z^{n-1} + \ldots + a_n \,, \\
C(z) &= z^n + c_1 z^{n-1} + \ldots + c_n \,, \qquad (4.111) \\
D(z) &= d_0 z^n + d_1 z^{n-1} + \ldots + d_n \,.
\end{aligned}$$

Assume that the polynomial $A(z)$ is known and has all zeros strictly inside the unit circle.

(i) *Assume that λ^2 and $C(z)$ are given. Then there exists a unique $D(z)$ satisfying (4.110).*

(ii) *Assume that $D(z)$ is given. Then there are λ^2 and $C(z)$ satisfying (4.110), if and only if $D(z)/A(z)$ is positive real, that is:*

$$\mathrm{Re}\, \frac{D(e^{i\omega})}{A(e^{i\omega})} \geq 0 \quad \forall \omega \,. \qquad (4.112)$$

If (4.112) is satisfied, $C(z)$ can be chosen with all zeros inside or on the unit circle. If strict inequality holds in (4.112), then $C(z)$ can be chosen with all zeros strictly inside the unit circle.

Remark Equation (4.110) is an example of a Diophantine equation (in polynomials). Diophantine equations are generally of the form

$$A(z)X(z) + B(z)Y(z) = C(z) \,, \qquad (4.113)$$

where $A(z)$, $B(z)$, $C(z)$ are given polynomials of z, and $X(z)$ and $Y(z)$ are unknown polynomials. In order to obtain a solution, one must require that $A(z)$ and $B(z)$ are coprime (or that a common factor also appears in $C(z)$; in such a case, it can be cancelled) and that $X(z)$ and $Y(z)$ are allowed to have appropriate degrees. Diophantine equations of such a form also appear very naturally in polynomial pole placement of control systems. They will appear repeatedly in simplified form in this book in the treatment of prediction problems. □

Proof Consider first part (i). Note that (4.110) is a Diophantine equation, but with some symmetry constraints. The key property to use is that $A(z)$ and $A(z^{-1})$ are coprime. More specifically, the identity can easily be reformulated as a system of linear equations in the unknowns $d_0, \ldots, d_n$. Hence a unique solution exists if and only if

$$D(z)A(z^{-1}) + A(z)D(z^{-1}) \equiv 0 \qquad (4.114)$$

implies $D(z) = 0$. However, (4.114) easily gives

$$D(z)A(z^{-1}) \equiv -A(z)D(z^{-1}) .$$

Noting that both sides must have the same zeros and that $A(z)$ and $A(z^{-1})$ do not have any common zero, it follows that $D(z)$ must be a multiple of $A(z)$. However, inserting $D(z) = \alpha A(z)$ into (4.114) gives $2\alpha A(z)A(z^{-1}) \equiv 0$, from which it is concluded that $\alpha = 0$ and $D(z) = 0$.

The proof of part (ii) relies on spectral factorization. Clearly, solutions with respect to λ^2 and $C(z)$ exist if and only if

$$D(e^{i\omega})A(e^{-i\omega}) + A(e^{i\omega})D(e^{-i\omega}) \geq 0, \quad \forall \omega .$$

After division by $A(e^{i\omega})A(e^{-i\omega})$, this is written equivalently as

$$0 \leq \frac{D(e^{i\omega})}{A(e^{i\omega})} + \frac{D(e^{-i\omega})}{A(e^{-i\omega})} = 2 \operatorname{Re} \frac{D(e^{i\omega})}{A(e^{i\omega})} .$$

Hence, it has been proved that (4.112) is a necessary and sufficient condition for the existence of a solution. That $C(z)$ can be chosen with all zeros inside or on the unit circle (strictly inside if all inequalities are strict) follows from the properties of spectral factorization. See Section 4.3.   ∎

Remark Note that the identity in (4.110) can be written as

$$\lambda^2 \frac{C(z)C(z^{-1})}{A(z)A(z^{-1})} \equiv \frac{D(z)}{A(z)} + \frac{D(z^{-1})}{A(z^{-1})} . \tag{4.115}$$

The situation dealt with in part (i) of the lemma can thus be interpreted as a *partial fraction decomposition*. To be more precise, let $A(z)$ have zeros $\{p_i\}$, and assume them to be distinct and nonzero for simplicity. A partial fraction decomposition of the left-hand side of (4.115) can then be written, using the symmetry in the problem, as

$$\lambda^2 \frac{C(z)C(z^{-1})}{A(z)A(z^{-1})} = \beta_0 + \sum_{i=1}^{n} \left(\frac{\beta_i}{z - p_i} + \frac{\beta_i}{z^{-1} - p_i} \right) .$$

It is possible to merge $\beta_0/2$ and the fractions with poles in $\{p_i\}$ as

$$\frac{\beta_0}{2} + \sum_{i=1}^{n} \frac{\beta_i}{z - p_i} = \frac{D(z)}{A(z)} . \qquad \square$$

The function $\tilde{\phi}(z)$ can now be characterized. The result is formulated as a lemma.

Lemma 4.4 *Consider the ARMA process*

$$A(q)y(t) = C(q)e(t) . \tag{4.116}$$

Then $\tilde{\phi}(z)$, the positive real part of the spectrum, as defined by (3.26), satisfies

$$\tilde{\phi}(z) = \frac{D(z)}{A(z)} ,$$

$$(4.117)$$

where $D(z)$ is the solution of (4.110).

Proof The covariance function $r(k)$ satisfies the *Yule–Walker equations*, see Section 4.2.3:

$$r(k) + a_1 r(k-1) + \ldots + a_n r(k-n) = 0 , \qquad k > n .$$

$$(4.118)$$

Using the properties of z-transforms, this implies that

$$A(z)R(z) = F_0(z) ,$$

where

$$R(z) = \sum_{k=0}^{\infty} r(k) z^{-k}$$

is the z-transform of the covariance function and $F_0(z)$ is a polynomial of degree n that depends on the initial conditions of (4.118). Now, using (3.27):

$$A(z)\tilde{\phi}(z) = A(z) \left[R(z) - \frac{1}{2} r(0) \right] = F_0(z) - \frac{1}{2} r(0) A(z)$$

$$= F(z) ,$$

where $F(z)$ is a polynomial of degree n. Using the relation in (3.35):

$$\phi(e^{i\omega}) = \tilde{\phi}(e^{i\omega}) + \tilde{\phi}^*(e^{i\omega}) = 2 \operatorname{Re} \tilde{\phi}(e^{i\omega}) ,$$

which gives

$$\lambda^2 \frac{C(e^{i\omega})C(e^{-i\omega})}{A(e^{i\omega})A(e^{-i\omega})} = \frac{F(e^{i\omega})}{A(e^{i\omega})} + \frac{F(e^{-i\omega})}{A(e^{-i\omega})} ,$$

or

$$\lambda^2 C(z)C(z^{-1}) = F(z)A(z^{-1}) + F(z^{-1})A(z) .$$

Hence, $F(z)$ is nothing other than $D(z)$, *cf.* (4.110). This proves (4.117). ∎

Remark 1 One can alternatively proceed as follows to prove Lemma 4.4. Invoking Lemma 4.3, the spectrum of $y(t)$ can be written as

$$\phi(z) = \lambda^2 \frac{C(z)C(z^{-1})}{A(z)A(z^{-1})} = \frac{D(z)}{A(z)} + \frac{D(z^{-1})}{A(z^{-1})} .$$

$$(4.119)$$

The series above can be viewed as a Laurent series. It will converge in a strip including the unit circle.

However, the spectrum can also be written as (3.22)

$$\phi(z) = \sum_{n=-\infty}^{\infty} r(n)z^{-n} = \left[\frac{1}{2}r(0) + \sum_{n=1}^{\infty} r(n)z^{-n} \right]$$
$$+ \left[\frac{1}{2}r(0) + \sum_{n=-\infty}^{-1} r(n)z^{-n} \right]$$
$$\stackrel{\triangle}{=} \tilde{\phi}(z) + \tilde{\phi}(z^{-1}) .$$

which also converges on the unit circle. As the Laurent series expansion is unique, it can be concluded that (4.117) is true. This idea is closely related to the concepts of splitting a general filter into a causal and an anticausal part, which will be discussed in Section 7.3.3.   □

Remark 2 $D(z)/A(z)$ is *positive real*. This is easy to see, since for any ω

$$\text{Re} \, \frac{D(e^{i\omega})}{A(e^{i\omega})} = \text{Re} \, \tilde{\phi}(e^{i\omega}) = \frac{1}{2}\phi(e^{i\omega}) \geq 0 .$$ □

Example 4.8 For illustration, compute $\tilde{\phi}(z)$ for a first-order ARMA process

$$y(t) + ay(t-1) = e(t) + ce(t-1) , \quad \mathbf{E} \, e^2(t) = \lambda^2 . \tag{4.120}$$

First, to recapitulate from Example 4.5:

$$r(0) = \frac{\lambda^2}{1-a^2}(1 + c^2 - 2ac) ,$$
$$r(\tau) = \frac{\lambda^2}{1-a^2}(c-a)(1-ac)(-a)^{\tau-1} , \quad \tau \geq 1 .$$

Direct use of Definition 3.12 now gives

$$\tilde{\phi}(z) = 0.5r(0) + \sum_{n=1}^{\infty} r(n)z^{-n}$$
$$= \frac{1}{2}\frac{\lambda^2}{1-a^2}(1 + c^2 - 2ac)$$
$$+ \frac{\lambda^2}{1-a^2}(c-a)(1-ac) \sum_{n=1}^{\infty}(-a)^{n-1}z^{-n}$$
$$= \frac{\lambda^2}{1-a^2}\left[\frac{1}{2}(1 + c^2 - 2ac) + (c-a)(1-ac)\frac{z^{-1}}{1+az^{-1}} \right]$$
$$= \frac{\lambda^2}{1-a^2}\frac{1}{z+a}\left[\frac{1}{2}(z+a)(1 + c^2 - 2ac) + (c-a)(1-ac) \right] . \tag{4.121}$$

Next consider the identity in (4.110), which gives

$$\lambda^2(z+c)(z^{-1}+c) \equiv (d_0 z + d_1)(z^{-1}+a) + (z+a)(d_0 z^{-1} + d_1) ,$$

or, after equating the powers of z:

$$\begin{pmatrix} 2 & 2a \\ a & 1 \end{pmatrix} \begin{pmatrix} d_0 \\ d_1 \end{pmatrix} = \begin{pmatrix} \lambda^2(1+c^2) \\ \lambda^2 c \end{pmatrix} ,$$

with the solution

$$d_0 = \frac{1}{2} \frac{\lambda^2}{1-a^2}(1+c^2-2ac) ,$$

$$d_1 = \frac{1}{2} \frac{\lambda^2}{1-a^2}(-a-ac^2+2c) .$$

By Lemma 4.4:

$$\tilde{\phi}(z) = \frac{d_0 z + d_1}{z+a} ,$$

which coincides with the direct calculations above. $\square$

4.6.2 State Space Models

In this section, a standard state space model

$$\begin{aligned} x(t+1) &= Fx(t) + v(t) , \\ y(t) &= Hx(t) + e(t) , \end{aligned} \tag{4.122}$$

is considered, where $v(t)$ and $e(t)$ are mutually independent, white noise sequences of zero mean and covariances R_1 and R_2 respectively. A state space representation of $\tilde{\phi}(z)$, the positive real part of the spectrum, is sought. The solution is given in the following lemma.

Lemma 4.5 *Consider the system in (4.122). Then*

$$\tilde{\phi}(z) = J + H(zI - F)^{-1} FPH^* , \tag{4.123}$$

where

$$J = (HPH^* + R_2)/2 , \tag{4.124}$$

$$P = FPF^* + R_1. \tag{4.125}$$

Proof The proof is by direct verification. From (4.123), it is clear that $\tilde{\phi}(z)$ can be interpreted as a weighted sum of negative powers of z. By straightforward calculation

$$\begin{aligned} &\tilde{\phi}(z) + \tilde{\phi}^*(z^{-*}) \\ &= J + H(zI-F)^{-1}FPH^* + J^* + HPF^*(z^{-1}I - F^*)^{-1}H^* \\ &= J + J^* + H(zI-F)^{-1}[FP(z^{-1}I - F^*) + (zI - F)PF^*] \\ &\quad \times (z^{-1}I - F^*)^{-1}H^* \end{aligned}$$

$$= R_2 + H(zI - F)^{-1}[(zI - F)P(z^{-1}I - F^*) + FP(z^{-1}I - F^*)$$
$$+ (zI - F)PF^*](z^{-1}I - F^*)^{-1}H^*$$
$$= R_2 + H(zI - F)^{-1}[P - FPF^*](z^{-1}I - F^*)^{-1}H^*$$
$$= R_2 + H(zI - F)^{-1}R_1(z^{-1}I - F^*)^{-1}H^* = \phi(z) .$$

Referring to Lemma 3.3, the proof is thus completed.   ∎

Remark 1 Note that (4.125) is the standard Lyapunov equation for determining the state covariance matrix P. Furthermore, it can be seen that $2J = r_y(0)$.   □

Remark 2 Making a series expansion of (4.123):

$$\tilde{\phi}(z) = J + H(zI - F)^{-1}FPH^*$$
$$= J + z^{-1}H(I - z^{-1}F)^{-1}FPH^*$$
$$= J + z^{-1}H \sum_{j=0}^{\infty} z^{-j}F^j FPH^*$$
$$= J + H \sum_{k=1}^{\infty} z^{-k}F^k PH^* ,$$

and, comparison with Definition 3.12:

$$\tilde{\phi}(z) = \frac{1}{2}r(0) + \sum_{n=1}^{\infty} r(n)z^{-n} ,$$

gives by identifying coefficients

$$r(0) = 2J = HPH^* + R_2 ,$$
$$r(k) = HF^k PH^*, \quad k \geq 1 ,$$

which is perfectly in agreement with (4.20) and (4.21).   □

Example 4.9 As an illustration, reconsider Example 4.8, but use the state space formalism. First, represent the ARMA model

$$y(t) + ay(t - 1) = e(t) + ce(t - 1)$$

in state space form as

$$x(t + 1) = \begin{pmatrix} -a & 1 \\ 0 & 0 \end{pmatrix} x(t) + \begin{pmatrix} 1 \\ c \end{pmatrix} v(t) ,$$
$$y(t) = (1 \quad 0)x(t) .$$

See Example 4.2. From that example, recall that

$$P = \lambda^2 \begin{pmatrix} \frac{1+c^2-2ac}{1-a^2} & c \\ c & c^2 \end{pmatrix} .$$

Now, (4.124) implies

$$J = \frac{1}{2}(p_{11} + 0) = \frac{\lambda^2}{2} \frac{1 + c^2 - 2ac}{1 - a^2} \, ,$$

and (4.123) finally gives

$$\tilde{\phi}(z) = \frac{p_{11}}{2} + (1 \ \ 0) \begin{pmatrix} z + a & -1 \\ 0 & z \end{pmatrix}^{-1} \begin{pmatrix} -a & 1 \\ 0 & 0 \end{pmatrix}$$

$$\times \begin{pmatrix} p_{11} & p_{12} \\ p_{12} & p_{22} \end{pmatrix} \begin{pmatrix} 1 \\ 0 \end{pmatrix}$$

$$= \frac{p_{11}}{2} + \frac{1}{z + a}(-ap_{11} + p_{12})$$

$$= \frac{(p_{11}/2)z + (-ap_{11}/2 + p_{12})}{z + a} \, .$$

As

$$-ap_{11}/2 + p_{12} = \frac{\lambda^2}{2(1 - a^2)}[-a(1 + c^2 - 2ac) + 2c(1 - a^2)]$$

$$= \frac{\lambda^2}{2(1 - a^2)}[2c - a - ac^2] \, ,$$

it is finally seen that

$$\tilde{\phi}(z) = \frac{\lambda^2}{2(1 - a^2)} \frac{(1 + c^2 - 2ac)z + (2c - a - ac^2)}{z + a} \, ,$$

which is in agreement with the result of Example 4.8.   □

A related problem concerns the conditions under which a "state space representation"

$$\tilde{\phi}(z) = J + H(zI - F)^{-1}G \tag{4.126}$$

is positive real. In terms of stochastic systems, this would mean that $\tilde{\phi}(z)$ is the positive real part of a spectrum. The answer to the above question is given by the Kalman–Yakubovich lemma, also called the positive real lemma. It states that $\tilde{\phi}(z)$ is positive real if and only if there exists a positive definite Hermitian matrix P and matrices L and W such that

$$\boxed{\begin{aligned} P &= F^*PF + LL^* \, , \\ F^*PG &= H^* - LW \, , \\ W^*W &= J + J^* - G^*PG \, . \end{aligned}} \tag{4.127}$$

It is straightforward to demonstrate sufficiency (the "if part"). Using (4.126):

$$\tilde{\phi}(z) + \tilde{\phi}^*(z^{-*}) = J + H(zI - F)^{-1}G + J^* + G^*(z^{-1}I - F^*)^{-1}H^*$$
$$= W^*W + G^*PG + (W^*L^* + G^*PF)(zI - F)^{-1}G + G^*(z^{-1}I - F^*)^{-1}$$
$$\times (F^*PG + LW)$$
$$= [W^* + G^*(z^{-1}I - F^*)^{-1}L][W + L^*(zI - F)^{-1}G]$$
$$+ G^*(z^{-1}I - F^*)^{-1}[(z^{-1}I - F^*)P(zI - F) + (z^{-1}I - F^*)PF$$
$$+ F^*P(zI - F) - LL^*](zI - F)^{-1}G$$
$$= [W^* + G^*(z^{-1}I - F^*)^{-1}L][W + L^*(zI - F)^{-1}G]$$
$$+ G^*(z^{-1}I - F^*)^{-1}[P - F^*PF - LL^*](zI - F)^{-1}G$$
$$= [W + L^*(z^{-*}I - F)^{-1}G]^*[W + L^*(zI - F)^{-1}G],$$

which has the form of a spectrum in factorized form. For proofs of the necessity of (4.127), see the literature, cited in the Bibliography section.

The condition in (4.127) can also be formulated as

$$\begin{pmatrix} P - F^*PF & H^* - F^*PG \\ H - G^*PF & J + J^* - G^*PG \end{pmatrix} = \begin{pmatrix} L \\ W^* \end{pmatrix} (L^* \ W) \geq 0. \tag{4.128}$$

The left-hand side of (4.128) must thus be positive semidefinite. As an equation in P, this condition must have a positive definite solution.

4.6.3 Continuous-time Processes

Some results analogous to those in Section 4.6.1, but for (asymptotically stable) continuous-time processes, will now be developed. Consider, therefore, processes of the form

$$y(t) = \frac{B(p)}{F(p)}e(t),$$
$$B(p) = b_1 p^{n-1} + \ldots + b_n, \tag{4.129}$$
$$F(p) = p^n + f_1 p^{n-1} + \ldots + f_n,$$

with $e(t)$ being continuous-time white noise, $\mathbf{E}\,e(t)e(s) = \delta(t - s)$.

Let $r(\tau)$ be the covariance function of $y(t)$ and $R(s) = \tilde{\phi}(s)$ its Laplace transform; see (4.67). Noting that $r(\tau)$ satisfies the Yule–Walker equation (see (4.96))

$$(p^n + f_1 p^{n-1} + \ldots + f_n)r(\tau) = 0, \tag{4.130}$$

it is found that

$$\tilde{\phi}(s) = R(s) = \frac{D(s)}{F(s)} \tag{4.131}$$

for some polynomial

$$D(s) = d_1 s^{n-1} + \ldots + d_n, \tag{4.132}$$

which accounts for the initial values of (4.130). Note that $d_0 = 0$ due to the variance $r(0)$ being finite. As explained in Section 4.4:

$$\phi(i\omega) = \tilde{\phi}(i\omega) + \tilde{\phi}(-i\omega) = 2 \text{ Re } \tilde{\phi}(i\omega) \tag{4.133}$$

holds, which implies

$$\frac{B(i\omega)B(-i\omega)}{F(i\omega)F(-i\omega)} = \frac{D(i\omega)}{F(i\omega)} + \frac{D(-i\omega)}{F(-i\omega)} \,,$$

and finally:

$$B(s)B(-s) = D(s)F(-s) + D(-s)F(s) \,. \tag{4.134}$$

It is immediate from (4.133) that $D(i\omega)/F(i\omega)$ is positive real.

4.7 Effect of Linear Filtering on the Bispectrum

This section deals with how the bispectrum is affected by linear filtering. Some preliminary results are first established.

Gaussian processes

Let $x(t)$ be zero mean Gaussian. Then

$$R(m,n) = 0 \,, \qquad B(z_1, z_2) = 0 \,. \tag{4.135}$$

Non-Gaussian white noise

Let $x(t)$ be non-Gaussian *white* noise, so that $x(t)$ and $x(s)$ are independent for $t \neq s$, $\mathbf{E}\, x(t) = 0$, $\mathbf{E}\, x^2(t) = \sigma^2$, $\mathbf{E}\, x^3(t) = \beta$. Then

$$R(m,n) = \mathbf{E}\, x(t)x(t+m)x(t+n) = \beta\delta(m,n) \,. \tag{4.136}$$

Here the usual Kronecker delta notation has been extended, in the sense that

$$\delta(m,n) = \begin{cases} 1, & m = n = 0 \,, \\ 0, & \text{elsewhere} \,. \end{cases} \tag{4.137}$$

The bispectrum becomes a constant:

$$B(z_1, z_2) = \beta \,. \tag{4.138}$$

Effect of linear filtering

After these preludes, consider now a linear, asymptotically stable filter of the form

$$y(t) = \sum_{k=0}^{\infty} h_k u(t-k) \,, \tag{4.139}$$

and set

$$H(z) = \sum_{k=0}^{\infty} h_k z^{-k} \,. \tag{4.140}$$

Both $u(t)$ and $y(t)$ are assumed to be scalar signals. Let $u(t)$ be a stationary process with zero mean, third-moment sequence $R_u(m, n)$ and bispectrum $B_u(z_1, z_2)$. The corresponding quantities $R_y(m, n)$ and $B_y(z_1, z_2)$ for the output $y(t)$ are sought. By straightforward calculations

$$R_y(m, n) = \mathbf{E}\, y(t)y(t+m)y(t+n)$$

$$= \sum_{i=0}^{\infty}\sum_{j=0}^{\infty}\sum_{k=0}^{\infty} h_i h_j h_k \mathbf{E}\, u(t-i)u(t+m-j)u(t+n-k)$$

$$= \sum_{i=0}^{\infty}\sum_{j=0}^{\infty}\sum_{k=0}^{\infty} h_i h_j h_k R_u(m+i-j, n+i-k)\,,$$

$$B_y(z_1, z_2) = \sum_{m=-\infty}^{\infty}\sum_{n=-\infty}^{\infty} R_y(m, n) z_1^{-m} z_2^{-n}$$

$$= \sum_{m=-\infty}^{\infty}\sum_{n=-\infty}^{\infty}\sum_{i=0}^{\infty}\sum_{j=0}^{\infty}\sum_{k=0}^{\infty} h_i h_j h_k$$

$$\times R_u(m+i-j, n+i-k) z_1^{-m} z_2^{-n}$$

$$= \sum_{i=0}^{\infty}\sum_{j=0}^{\infty}\sum_{k=0}^{\infty} h_i h_j h_k \sum_{m=-\infty}^{\infty}\sum_{n=-\infty}^{\infty} R_u(m+i-j, n+i-k)$$

$$\times z_1^{-(m+i-j)} z_2^{-(n+i-k)} z_1^{i-j} z_2^{i-k}$$

$$= \sum_{i=0}^{\infty}\sum_{j=0}^{\infty}\sum_{k=0}^{\infty} h_i h_j h_k z_1^{i-j} z_2^{i-k} B_u(z_1, z_2)$$

$$= H(z_1^{-1} z_2^{-1}) H(z_1) H(z_2) B_u(z_1, z_2)\,. \tag{4.141}$$

The result in (4.141) is a generalization of (4.2), which applies for the usual (second-order) spectrum. It is well known that the spectral density (the power spectrum) does not carry information about the phase properties of a filter. In contrast to this, phase properties can be recovered from the bispectrum when it exists. This point is illustrated by a simple example.

Example 4.10 Let $e(t)$ be non-Gaussian white noise with

$$\mathbf{E}\, e(t) = 0\,, \qquad \mathbf{E}\, e^2(t) = 1\,, \qquad \mathbf{E}\, e^3(t) = \beta \neq 0\,. \tag{4.142}$$

Let a and b be two real-valued parameters in the interval $(-1, 1)$ and consider the following processes:

$$\begin{aligned} y_1(t) &= (1 + aq^{-1})(1 + bq^{-1})e(t) \\ &= e(t) + (a+b)e(t-1) + abe(t-2)\,. \end{aligned} \tag{4.143}$$

$$y_2(t) = (1 + aq^{-1})(b + q^{-1})e(t)$$

$$= be(t) + (ab + 1)e(t - 1) + ae(t - 2) . \tag{4.144}$$

$$y_3(t) = (a + q^{-1})(b + q^{-1})e(t)$$
$$= abe(t) + (a + b)e(t - 1) + e(t - 2) . \tag{4.145}$$

Note that $y_1(t)$ is obtained with a minimum-phase filter, $y_3(t)$ with a maximum-phase filter and $y_2(t)$ with a mixed-phase filter.

The power spectra for the three processes are identical, since they have identical autocovariance functions. In fact:

$$r_{y_1}(0) = 1 + (a + b)^2 + a^2 b^2 = 1 + a^2 + b^2 + 2ab + a^2 b^2 ,$$
$$r_{y_1}(1) = (a + b)(1 + ab) ,$$
$$r_{y_1}(2) = ab ,$$
$$r_{y_1}(\tau) = 0 , \qquad \tau > 2 .$$

$$r_{y_2}(0) = b^2 + (ab + 1)^2 + a^2 = r_{y_1}(0) ,$$
$$r_{y_2}(1) = (ab + 1)(a + b) = r_{y_1}(1) ,$$
$$r_{y_2}(2) = ab = r_{y_1}(2) ,$$
$$r_{y_2}(\tau) = 0 , \qquad \tau > 2 .$$

$$r_{y_3}(0) = a^2 b^2 + (a + b)^2 + 1 = r_{y_1}(0) ,$$
$$r_{y_3}(1) = (ab + 1)(a + b) = r_{y_1}(1) ,$$
$$r_{y_3}(2) = ab = r_{y_1}(2) ,$$
$$r_{y_3}(\tau) = 0 , \qquad \tau > 2 .$$

As $y_1(t)$, $y_2(t)$, and $y_3(t)$ have identical covariance functions, their spectra will coincide as well.

Next calculate the third-moment sequences for the processes and consider only the nonzero elements of these sequences. In order to make the calculations general, write the processes in the jointly valid form

$$y(t) = h_0 e(t) + h_1 e(t - 1) + h_2 e(t - 2) ,$$

which covers all the three processes $y_1(t)$, $y_2(t)$, $y_3(t)$. Note first that it is sufficient to consider the elements $R(m, n)$ with $0 \leq n \leq m \leq 2$. This gives

$$R(0, 0) = \mathbf{E}\, y^3(t) = \beta[h_0^3 + h_1^3 + h_2^3] ,$$
$$R(1, 0) = \mathbf{E}\, y(t + 1)y^2(t) = \mathbf{E}\, [h_0 e(t + 1) + h_1 e(t) + h_2 e(t - 1)]$$
$$\times [h_0 e(t) + h_1 e(t - 1) + h_2 e(t - 2)]^2$$
$$= \beta[h_1 h_0^2 + h_2 h_1^2] ,$$
$$R(1, 1) = \mathbf{E}\, y^2(t + 1)y(t)$$
$$= \beta[h_1^2 h_0 + h_2^2 h_1] ,$$

Table 4.1. Third-moment sequence elements for some MA(2) models

$R(m,n)/\beta$		y_1	y_2	y_3
m	n	(minimum-phase)	(mixed-phase)	(maximum-phase)
0	0	$1+(a+b)^3+a^3b^3$	$b^3+(ab+1)^3+a^3$	$a^3b^3+(a+b)^3+1$
1	0	$(a+b)+ab(a+b)^2$	$b^2(ab+1)+a(ab+1)^2$	$(a+b)a^2b^2+(a+b)^2$
1	1	$(a+b)^2+a^2b^2(a+b)$	$(ab+1)^2b+a^2(ab+1)$	$(a+b)^2ab+(a+b)$
2	0	ab	ab^2	a^2b^2
2	1	$ab(a+b)$	$ab(1+ab)$	$ab(a+b)$
2	2	a^2b^2	a^2b	ab

$$
\begin{aligned}
R(2,0) &= \mathbf{E}\, y(t+2)y^2(t) = \mathbf{E}\left[h_0 e(t+2)+h_1 e(t+1)+h_2 e(t)\right] \\
&\quad \times\left[h_0 e(t)+h_1 e(t-1)+h_2 e(t-2)\right]^2 \\
&= \beta h_2 h_0^2 \,, \\
R(2,1) &= \mathbf{E}\, y(t+2)y(t+1)y(t) \\
&= \beta h_2 h_1 h_0 \,, \\
R(2,2) &= \mathbf{E}\, y^2(t+2)y(t) \\
&= \beta h_2^2 h_0 \,.
\end{aligned}
$$

When applying these calculations to the particular processes in (4.143)–(4.145), the results shown in Table 4.1 are obtained.

Apparently, the three processes have *different third-moment* sequences, and hence also different bispectra. □

One would expect that for a given process, the same type of filter representation will appear for the power spectrum and for the bispectrum. This is *not* so in general, as illustrated by the following example.

Example 4.11 Consider a process consisting of summing two independent and real-valued AR processes

$$
y(t) = \frac{1}{A(q)}e(t) + \frac{1}{C(q)}v(t) \,, \tag{4.146}
$$

$e(t)$ being Gaussian white noise and $v(t)$ non-Gaussian white noise. Both sequences are assumed to have unit variance, and $\mathbf{E}\, v^3(t) = 1$.

The Gaussian process will not contribute to the bispectrum. Further, $B_v(z_1, z_2) \equiv 1$, and, according to (4.141), the bispectrum will be

$$
\begin{aligned}
B_y(z_1, z_2) &= H(z_1^{-1}z_2^{-1})H(z_1)H(z_2)B_v(z_1, z_2) \\
&= \frac{1}{C(z_1^{-1}z_2^{-1})}\frac{1}{C(z_1)}\frac{1}{C(z_2)} \,,
\end{aligned} \tag{4.147}
$$

so

$$
H(z) = 1/C(z) \tag{4.148}
$$

is the relevant filter representation as far as the bispectrum is concerned. However, the power spectrum becomes

$$\phi(z) = \frac{1}{A(z)A(z^{-1})} + \frac{1}{C(z)C(z^{-1})} \,, \tag{4.149}$$

and in this case it will have a spectral factor of the form

$$H(z) = \frac{B(z)}{A(z)C(z)} \,, \tag{4.150}$$

where

$$B(z)B(z^{-1}) \equiv A(z)A(z^{-1}) + C(z)C(z^{-1}) \tag{4.151}$$

due to the spectral factorization. Clearly, the two filter representations of (4.148) and (4.150) differ. □

4.8 Algorithms for Covariance Calculations and Sampling

In this section some algorithms, most of which are based on the positive real part of the spectrum, are given. They will be useful for determining the covariance function. Another one can be used for sampling a continuous-time model. In the last section, algorithms for solving the Lyapunov equation are discussed. The case of real-valued processes is considered.

4.8.1 ARMA Covariance Function

Consider an ARMA process (scalar case)

$$\begin{aligned} A(q)y(t) &= C(q)e(t) \,, \\ \mathbf{E}\, e^2(t) &= \lambda^2 \,, \end{aligned} \tag{4.152}$$

and the problem of finding the covariance function $r(k) = \mathbf{E}\, y(t+k)y(t)$. It is solved here by means of the positive real part of the spectrum. First compute $D(z)$ from (4.110):

$$\lambda^2 C(z)C(z^{-1}) \equiv D(z)A(z^{-1}) + A(z)D(z^{-1}) \,. \tag{4.153}$$

Then, by (4.117), (3.27) and (3.28):

$$\tilde{\phi}(z) = \frac{D(z)}{A(z)} = \sum_{k=1}^{\infty} r(k)z^{-k} + \frac{1}{2}r(0) \,. \tag{4.154}$$

Setting $z^{-1} = 0$ easily gives

$$d_0 = r(0)/2 \,. \tag{4.155}$$

By multiplication in (4.154) by $A(z)$:

$$(d_0 z^n + d_1 z^{n-1} + \ldots + d_n) = d_0(z^n + a_1 z^{n-1} + \ldots + a_n)$$

$$+ (z^n + a_1 z^{n-1} + \ldots + a_n) \sum_{k=1}^{\infty} r(k) z^{-k} . \qquad (4.156)$$

By equating the different powers of z, the covariance elements are found as

$$\begin{cases} r(1) = d_1 - d_0 a_1 , \\ r(2) = d_2 - d_0 a_2 - a_1 r(1) , \\ \quad \vdots \\ r(n) = d_n - d_0 a_n - a_1 r(n-1) - a_2 r(n-2) \ldots - a_{n-1} r(1) , \end{cases} \qquad (4.157)$$

and (the Yule–Walker equations!)

$$r(k) = -\sum_{i=1}^{n} a_i r(k-i) , \qquad k > n . \qquad (4.158)$$

Alternatively, from (4.154) it can be seen that, except for $k = 0$, $r(k)$ is precisely the impulse response of the filter $D(z)/A(z)$. This may be a more convenient formulation of the relations in (4.157) for some forms of computer implementation.

Example 4.12 Consider a first-order ARMA process

$$y(t) + ay(t-1) = e(t) + ce(t-1) , \qquad \mathbf{E}\, e^2(t) = \lambda^2 .$$

According to Example 4.8:

$$d_0 = \frac{1}{2} \frac{\lambda^2}{1-a^2} (1 + c^2 - 2ac) ,$$

$$d_1 = \frac{1}{2} \frac{\lambda^2}{1-a^2} (2c - a - ac^2) .$$

Therefore, from (4.155):

$$r(0) = 2d_0 = \frac{\lambda^2}{1-a^2} (1 + c^2 - 2ac) .$$

Next, (4.157) implies that

$$r(1) = d_1 - ad_0 = \frac{\lambda^2}{1-a^2} (c-a)(1-ac) ,$$

and from (4.158) one finally obtains

$$r(k) = (-a)^{k-1} \frac{\lambda^2}{1-a^2} (c-a)(1-ac) .$$

Recall that the same results were established in Example 4.5.   □

4.8.2 ARMA Cross-Covariance Function

Consider the following situation. Assume that two (real-valued) ARMA processes are given:

$$A(q)y(t) = B(q)e(t) ,$$
$$A(q) = q^n + a_1 q^{n-1} + \ldots + a_n , \tag{4.159}$$
$$B(q) = b_0 q^n + b_1 q^{n-1} + \ldots + b_n ,$$

and

$$C(q)w(t) = D(q)e(t) ,$$
$$C(q) = q^m + c_1 q^{m-1} + \ldots + c_m , \tag{4.160}$$
$$D(q) = d_0 q^m + d_1 q^{m-1} + \ldots + d_m .$$

Assume that $e(t)$ is the same in (4.159) and (4.160) and that it is white noise of zero mean and unit variance. Find the covariance elements

$$r(k) = \mathbf{E}\, y(t+k)w(t) \tag{4.161}$$

for a number of arguments k. The cross-covariance function $r(k)$ is related to the cross-spectrum $\phi_{yw}(z)$ as (see (3.24) and (4.3))

$$
\begin{aligned}
\phi_{yw}(z) &= \sum_{k=-\infty}^{\infty} r(k)z^{-k} \\
&= \frac{B(z)}{A(z)} \frac{D^*(z^{-*})}{C^*(z^{-*})} \\
&= \frac{B(z)}{A(z)} \frac{D(z^{-1})}{C(z^{-1})} .
\end{aligned}
\tag{4.162}
$$

Introduce the two polynomials

$$F(z) = f_0 z^n + f_1 z^{n-1} + \ldots + f_n ,$$
$$G(z^{-1}) = g_0 z^{-m} + g_1 z^{-(m-1)} + \ldots + g_{m-1} z^{-1} , \tag{4.163}$$

through

$$\frac{B(z)}{A(z)} \frac{D(z^{-1})}{C(z^{-1})} \equiv \frac{F(z)}{A(z)} + z\frac{G(z^{-1})}{C(z^{-1})} , \tag{4.164}$$

or, equivalently:

$$B(z)D(z^{-1}) \equiv F(z)C(z^{-1}) + zA(z)G(z^{-1}) . \tag{4.165}$$

Since $A(z)$ and $C(z^{-1})$ are coprime, (4.165) has a unique solution. Note that, as a linear system of equations, (4.165) has $n+m+1$ equations and the same number of unknowns. The coprimeness condition ensures that the matrix appearing in the system of equations has full rank. Equations (4.162) and (4.164) now give

$$\sum_{k=-\infty}^{\infty} r(k)z^{-k} = \frac{F(z)}{A(z)} + z\frac{G(z^{-1})}{C(z^{-1})} . \tag{4.166}$$

The two terms in the right-hand side of (4.166) can be identified with each part of the sum. In fact:

$$\frac{F(z)}{A(z)} = \sum_{k=0}^{\infty} r(k)z^{-k} , \tag{4.167}$$

$$\frac{zG(z^{-1})}{C(z^{-1})} = \sum_{k=-\infty}^{-1} r(k)z^{-k} . \tag{4.168}$$

From (4.167), one obtains

$$F(z) \equiv A(z) \sum_{k=0}^{\infty} r(k)z^{-k} . \tag{4.169}$$

Equating the powers of z gives

$$\begin{cases} r_{yw}(0) = f_0 , \\ r_{yw}(1) = f_1 - a_1 r(0) , \\ \quad \vdots \\ r_{yw}(k) = f_k - \sum_{j=1}^{k} a_j r(k-j) , \ (2 \le k \le n) , \\ r_{yw}(k) = -\sum_{j=1}^{n} a_j r(k-j) , \quad (k > n) . \end{cases} \tag{4.170}$$

Note that the last part of (4.170) is a Yule–Walker type of equation.

Similarly, from (4.168):

$$zG(z^{-1}) \equiv C(z^{-1}) \sum_{k=-\infty}^{-1} r(k)z^{-k} , \tag{4.171}$$

and

$$\begin{cases} r_{yw}(-1) = g_0 , \\ r_{yw}(-k) = g_{k-1} - \sum_{j=1}^{k-1} c_j r(-k+j) , & (2 \le k \le m) , \\ r_{yw}(-k) = -\sum_{j=1}^{m} c_j r(-k+j) , & (k > m) . \end{cases} \tag{4.172}$$

□

Example 4.13 Consider again a first-order ARMA process

$$y(t) + ay(t-1) = e(t) + ce(t-1) , \qquad \mathbf{E}\,e^2(t) = 1 .$$

In this case the autocovariance function is sought. Hence choose $z(t) \equiv y(t)$ and thus $A(q) = q+a$, $B(q) = q+c$, $C(q) = q+a$, $D(q) = q+c$. The identity in (4.165) becomes

$$(z+c)(z^{-1}+c) \equiv (f_0 z + f_1)(z^{-1}+a) + z(z+a)(g_0 z^{-1}) .$$

Equating the powers of z leads to

$$\begin{pmatrix} a & 0 & 1 \\ 1 & a & a \\ 0 & 1 & 0 \end{pmatrix} \begin{pmatrix} f_0 \\ f_1 \\ g_0 \end{pmatrix} = \begin{pmatrix} c \\ 1 + c^2 \\ c \end{pmatrix} .$$

The solution of this system is easily found to be

$$f_0 = \frac{1 + c^2 - 2ac}{1 - a^2} , \qquad f_1 = c , \qquad g_0 = \frac{(c - a)(1 - ac)}{1 - a^2} .$$

Hence (4.170) implies that

$$r(0) = f_0 = \frac{1 + c^2 - 2ac}{1 - a^2} ,$$

$$r(1) = f_1 - ar(0) = \frac{(c - a)(1 - ac)}{1 - a^2} ,$$

$$r(k) = (-a)^{k-1} r(1) , \qquad k \geq 1 .$$

while (4.172) gives

$$r(-1) = g_0 = \frac{(c - a)(1 - ac)}{1 - a^2} ,$$

$$r(-k) = (-a)^{k-1} r(-1) , \qquad k \geq 1 .$$

Needless to say, these expressions for the covariance function are the same as those derived in previous examples. □

4.8.3 Continuous-Time Covariance Function

In order to compute the covariance function in continuous time, the procedure is a little different than in the discrete-time case. Assume that a model of the form

$$y(t) = \frac{B(p)}{F(p)} e(t) ,$$

$$B(p) = b_1 p^{n-1} + \ldots + b_n , \qquad (4.173)$$

$$F(p) = p^n + f_1 p^{p-1} + \ldots + f_n ,$$

where $e(t)$ is continuous-time white noise, $\mathbf{E}\, e(t)e(s) = \delta(t - s)$, is given. Find the covariance function $r(\tau)$ of $y(t)$. First compute $D(s)$ by solving (4.134):

$$B(s)B(-s) = D(s)F(-s) + D(-s)F(s) \qquad (4.174)$$

with respect to

$$D(s) = d_1 s^{n-1} + \ldots + d_n .$$

From (4.131):

$$\tilde{\phi}(s) = \mathcal{L}r(t) = \frac{D(s)}{F(s)} , \tag{4.175}$$

where $\mathcal{L}$ denotes the Laplace transform.

Next choose H, F_c and x_0 so that

$$H(sI - F_c)^{-1}x_0 = \frac{D(s)}{F(s)} . \tag{4.176}$$

This can be done, for example, by using any canonical form. Using

$$\mathcal{L}r(t) = H(sI - F_c)^{-1}x_0 , \tag{4.177}$$

a state space model governing the covariance function is readily obtained:

$$\dot{x} = F_c x , \quad x(0) = x_0 , \\ r = Hx . \tag{4.178}$$

Hence the covariance function can be written as

$$r(t) = H e^{F_c t}x_0 , \quad t \geq 0 . \tag{4.179}$$

It should be emphasized that the state space approach outlined above may be useful for computer implementation. For hand calculations, it seems more feasible to use $\mathcal{L}r(t) = D(s)/F(s)$ and a table of Laplace transforms in order to find the covariance function $r(t)$.

4.8.4 Sampling

Assume that a continuous-time process is given in the form

$$A(p)y(t) = B(p)e(t) , \tag{4.180}$$

where

$$A(p) = p^n + a_1 p^{n-1} + \ldots + a_n , \\ B(p) = b_1 p^{n-1} + \ldots + b_n , \\ \mathbf{E}\, e(t)e(s) = \delta(t - s) .$$

Find the sampled form of the model of (4.180). The following algorithm is an alternative to the state space approach described in Section 4.5.2.

Step 1 Compute $\tilde{\phi}_c(s) = D(s)/A(s)$ where

$$D(s) = d_1 s^{n-1} + \ldots + d_n$$

by solving (4.134):

$$B(s)B(-s) \equiv D(s)A(-s) + D(-s)A(s) . \tag{4.181}$$

Step 2 Next use the relations (see (4.131) and (3.27))

$$\tilde{\phi}_{\text{c}}(s) = \mathcal{L}\{r(t)\}\,, \tag{4.182}$$

$$\tilde{\phi}_{\text{d}}(z) = \left[\mathcal{Z}\{r(kh)\} - \frac{1}{2}r(0)\right] \tag{4.183}$$

to compute $\tilde{\phi}_{\text{d}}(z)$ as a fraction:

$$\tilde{\phi}_{\text{d}}(z) = h\frac{D_{\text{d}}(z)}{A_{\text{d}}(z)}\,. \tag{4.184}$$

The details of this step are dealt with after the description of Step 3.

Step 3 Finally, perform a spectral factorization

$$\lambda^2 C(z)C(z^{-1}) = D_{\text{d}}(z)A_{\text{d}}(z^{-1}) + A_{\text{d}}(z)D_{\text{d}}(z^{-1})\,, \tag{4.185}$$

to find λ^2 and $C(z)$. This gives the ARMA model

$$\begin{aligned} A_{\text{d}}(q^{-1})y(t) &= C(q^{-1})v(t)\,, \\ \mathbf{E}\,v(t)v(s) &= \lambda^2\delta_{t,s}\,. \end{aligned} \tag{4.186}$$

$\square$

In order to complete the algorithm, a procedure for Step 2 on how to go from a Laplace transform to the corresponding z-transform is needed. Fortunately, this is easy for a function that is a sum of exponentials. Such a function $f(t)$ can always be expressed as the solution to

$$\begin{aligned} \dot{x} &= Fx\,, \quad x(0) = x_0\,, \\ f &= Hx\,, \end{aligned} \tag{4.187}$$

giving

$$\begin{aligned} f(t) &= H\,e^{Ft}x_0\,, \\ \mathcal{L}\{f(t)\} &= H[sI - F]^{-1}x_0\,. \end{aligned} \tag{4.188}$$

The z-transform of the discretized function is easily obtained as

$$f(kh) = H\,e^{Fkh}\,x_0\,, \tag{4.189}$$

$$\mathcal{Z}\{f(kh)\} = H(I - z^{-1}\,e^{Fh})^{-1}x_0\,, \tag{4.190}$$

$$\begin{aligned} x(kh + h) &= e^{Fh}\,x(kh)\,, \quad x(0) = x_0\,, \\ f(kh) &= Hx(kh)\,. \end{aligned} \tag{4.191}$$

One can regard $f(t)$ as the impulse response of the system in (4.187), while $f(kh)$ is the impulse response of the discrete-time system in (4.191).

When applying this idea to Step 2, represent $\tilde{\phi}_{\text{c}}(s) = D(s)/A(s)$ in (some) state space form as (4.188). Then (4.191) can easily be calculated, and its impulse response evaluated as in (4.189). In this way the representation in (4.184) can be calculated.

4.8.5 Solving the Lyapunov Equation

When the solution to the Lyapunov equation

$$P(t+1) = FP(t)F^T + R_1 , \qquad (4.192)$$

is needed for all t, it has to be iterated. In many cases, it is enough to seek the stationary solution, which, of course, satisfies

$$P = FPF^T + R_1 . \qquad (4.193)$$

It is usually a bad idea to iterate (4.192) to convergence. An alternative is to convert (4.193) into an explicit set of linear equations. This will, however, often give rise to a high-order problem, as the number of unknowns will be $n(n+1)/2$, taking the symmetry of P into account. The specific structure of the problem can be taken into account by some such algorithm.

There are several numerically efficient algorithms for solving the Lyapunov equation of (4.192). An acceleration algorithm is presented here. The idea is to compute the solution of (4.192) with *increasing steps*, more precisely at times $t = 1, 2, 4, 8, 16 \ldots$. As the time interval between the points is doubled repeatedly, the convergence is fast.

Assume that $P(2^k)$ is known, and let $P(0) = 0$. Then (see (4.23)):

$$P(2^{k+1}) = \sum_{j=0}^{2^{k+1}-1} F^j R_1 F^{T^j}$$

$$= P(2^k) + \sum_{j=2^k}^{2^{k+1}-1} F^j R_1 F^{T^j}$$

$$= P(2^k) + F^{(2^k)}[\sum_{j=0}^{2^k(2-1)-1} F^j R_1 F^{T^j}]F^{T^{(2^k)}}$$

$$= P(2^k) + F^{(2^k)} P(2^k) F^{T^{(2^k)}} . \qquad (4.194)$$

This gives the following algorithm.

Step 1 Set $P_0 = R_1 , \qquad F_0 = F$.

Step 2 Iterate

$$\begin{aligned} P_{k+1} &= F_k P_k F_k^T + P_k , \\ F_{k+1} &= F_k^2 , \end{aligned} \qquad (4.195)$$

until convergence. □

4.A Appendix. Auxiliary Lemmas

Lemma 4.6 *Consider the scalar function*

$$f(z) = \sum_{|k| \le n} f_k z^k, \tag{4.196}$$

$\{f_k\}$ *being complex-valued coefficients. Assume that $f(e^{i\omega})$ is real-valued and non-negative (≥ 0) for all ω. Then*

(i) *There exists a real-valued, strictly positive constant C and a monic polynomial*

$$g(z) = z^n + g_1 z^{n-1} + \ldots + g_n,$$

with all zeros inside or on the unit circle, such that

$$f(z) = Cg(z)g^*(z^{-*}). \tag{4.197}$$

(ii) *In the case where $f(e^{i\omega}) > 0$, $\forall \omega$, $g(z)$ has no zeros on the unit circle.*

(iii) *In the case where $f(e^{i\omega}) \equiv f(e^{-i\omega})$, the coefficients $\{g_j\}$ are real-valued.*

Proof The constraint that $f(e^{i\omega})$ is real-valued for all ω implies

$$0 \equiv \text{Im} f(e^{i\omega}) = \text{Im} \left[f_0 + \sum_{k=1}^{n} (f_k e^{ik\omega} + f_{-k} e^{-ik\omega}) \right],$$

and hence $\text{Im}(f_0) = 0$, and

$$\text{Im}[f_k e^{ik\omega} + f_{-k} e^{-ik\omega}] \equiv 0,$$

giving $f_{-k} = \overline{f}_k$. Let z_1 be a zero of $f(z)$. Then $z_1 \ne 0$. Further:

$$f(\overline{z}_1^{-1}) = \sum_{|k| \le n} f_k \overline{z}_1^{-k} = \sum_{|k| \le n} f_{-k} \overline{z}_1^{k}$$

$$= \overline{\sum_{|k| \le n} \overline{f_{-k}} z_1^{k}} = \overline{\sum_{|k| \le n} f_k z_1^{k}} = \overline{f(z_1)} = 0.$$

Hence $\overline{z}_1^{-1}$ is also a zero. As $f(z)$ has $2n$ zeros, it follows that they can be characterized as $z_1 \ldots z_n$ fulfilling $0 < |z_i| \le 1$ and $z_{n+i} = \overline{z}_i^{-1}$ satisfying $1 \le |z_{n+i}| < \infty$, $i = 1, \ldots, n$. Set

$$g(z) = \prod_{i=1}^{n}(z - z_i), \qquad C = \frac{(-1)^n f_n}{\prod_{i=1}^{n} \overline{z}_i}.$$

This gives $g^*(z^{-*}) = \prod_{i=1}^{n}(z^{-*} - z_i) = \prod_{i=1}^{n}(z^{-1} - \overline{z}_i)$ and

$$f(z) - Cg(z)g^*(z^{-*}) = z^{-n} \sum_{k=-n}^{n} f_k z^{k+n} - C \prod_{i=1}^{n}(z - z_i) \prod_{j=1}^{n}(z^{-1} - \overline{z}_j)$$

$$= z^{-n} f_n \prod_{i=1}^{n}(z - z_i)(z - \overline{z}_i^{-1}) - \frac{(-1)^n f_n}{\prod_{i=1}^{n} \overline{z}_i} \prod_{i=1}^{n}(z - z_i) z^{-n} \prod_{j=1}^{n}(1 - z\overline{z}_j)$$

$$= z^{-n} f_n \prod_{i=1}^{n} (z - z_i) \left[\prod_{j=1}^{n} (z - \overline{z}_j^{-1}) - \frac{(-1)^n}{\prod_{k=1}^{n} \overline{z}_k} \prod_{j=1}^{n} (1 - z\overline{z}_j) \right] = 0 .$$

This proves the relation in (4.197). Next, consider the identity for $z = e^{i\omega}$. By assumption $f(e^{i\omega})$ is real-valued and nonnegative. Further:

$$g(e^{i\omega})\overline{g}(e^{i\omega}) = |g(e^{i\omega})|^2 \geq 0 .$$

It thus follows that C is real-valued and positive.

To prove part (ii), note simply that in this case

$$f(e^{i\omega}) = C|g(e^{i\omega})|^2 > 0 , \qquad \forall\, \omega .$$

Finally, consider part (iii). The symmetry condition $f(e^{i\omega}) \equiv f(e^{-i\omega})$ implies not only that $f_{-k} = \overline{f}_k$ but that the $\{f_k\}$ are real-valued. Hence the zeros z_j are real-valued or appear in complex conjugated pairs. Therefore, $g(z)$ will have real-valued coefficients. ∎

Lemma 4.7 *The following holds (in a distribution sense):*

$$\sum_{k=-\infty}^{\infty} e^{ikx} = 2\pi \sum_{j=-\infty}^{\infty} \delta(x - 2\pi j) . \qquad (4.198)$$

Proof Introduce a periodic function $f(x)$ through

$$f(x) = \frac{\pi^2}{6} - \frac{1}{2}(x - \pi)^2 , \qquad 0 \leq x \leq 2\pi ,$$
$$f(x + 2\pi) = f(x) .$$

The function $f(x)$ is displayed in Figure 4.5.

As $f(x)$ is periodic, it can be developed in a Fourier series

$$f(x) \sim \sum_{n=-\infty}^{\infty} c_n e^{inx} .$$

The Fourier coefficients $\{c_n\}$ are given by

$$c_n = \frac{1}{2\pi} \int_0^{2\pi} f(x) e^{-inx} \, dx .$$

By straightforward computing,

$$c_0 = \frac{1}{2\pi} \int_0^{2\pi} \left[\frac{\pi^2}{6} - \frac{1}{2}(x - \pi)^2 \right] dx$$
$$= \frac{\pi^2}{6} - \frac{1}{4\pi} \left[\frac{(x - \pi)^3}{3} \right]_0^{2\pi} = 0 ,$$

and for $n \neq 0$:

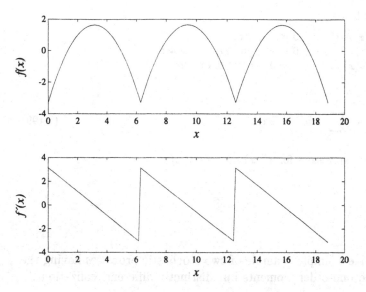

Fig. 4.5. The function $f(x)$ and its derivative $f'(x)$

$$c_n = \frac{1}{2\pi} \int_0^{2\pi} \left[\frac{\pi^2}{6} - \frac{1}{2}(x-\pi)^2 \right] e^{inx} \, dx = -\frac{1}{4\pi} \int_0^{2\pi} (x-\pi)^2 e^{-inx} \, dx$$

$$= -\frac{1}{4\pi} \left(\left[(x-\pi)^2 \frac{e^{-inx}}{-in} \right]_0^{2\pi} - \int_0^{2\pi} 2(x-\pi) \frac{e^{-inx}}{-in} \, dx \right)$$

$$= -\frac{1}{2\pi in} \int_0^{2\pi} (x-\pi) e^{-inx} \, dx$$

$$= -\frac{1}{2\pi in} \left(\left[(x-\pi) \frac{e^{-inx}}{-in} \right]_0^{2\pi} - \int_0^{2\pi} \frac{e^{-inx}}{-in} \, dx \right)$$

$$= -\frac{1}{2\pi in} \left(\frac{2\pi}{-in} - 0 \right) = -\frac{1}{n^2} \ .$$

The Fourier series is uniformly convergent. Hence write

$$f(x) = \sum_{\substack{n=-\infty \\ n \neq 0}}^{\infty} \frac{-1}{n^2} e^{inx} \ .$$

By termwise differentiation,

$$f'(x) = \sum_{\substack{n=-\infty \\ n \neq 0}}^{\infty} -\frac{i}{n} e^{inx} \ ,$$

$$f''(x) = \sum_{\substack{n=-\infty \\ n \neq 0}}^{\infty} e^{inx} \ .$$

The following also holds:

$$f'(x) = -(x - \pi), \qquad 0 < x < 2\pi,$$
$$f''(x) = -1, \qquad 0 < x < 2\pi,\ 2\pi < x < 4\pi,$$
$$f''(x) = 2\pi\delta(x), \qquad x = 0,\ \pm 2\pi,\ \pm 4\pi,\ etc.,$$

or, more precisely:

$$f''(x) = -1 + \sum_{j=-\infty}^{\infty} 2\pi\delta(x - 2\pi j). \tag{4.199}$$

Combining the two expressions for $f''(x)$ gives (4.198).   ∎

Exercises

Exercise 4.1 Give a simple example of two stochastic processes having the same first- and second-order moments but distinctly different realizations.

Exercise 4.2 Consider the transfer function operator

$$H(q) = I + H(qI - F)^{-1}K,$$

and represent it in state space form as

$$x(t+1) = Fx(t) + Ku(t),$$
$$y(t) = Hx(t) + u(t).$$

Show that the inverse system from $y(t)$ to $u(t)$ can be represented as

$$x(t+1) = (F - KH)x(t) + Ky(t),$$
$$u(t) = -Hx(t) + y(t),$$

and obtain (4.51), that is

$$H^{-1}(q) = I - H(qI - F + KH)^{-1}K.$$

Exercise 4.3 Consider an ARMA process observed in uncorrelated noise:

$$y(t) = \frac{q + 0.5}{q - 0.8}v(t) + e(t),$$

where $v(t)$ and $e(t)$ are uncorrelated white noise sequences with variances

$$\mathbf{E}\,v^2(t) = 18, \qquad \mathbf{E}\,e^2(t) = 125.$$

Find the innovations form of $y(t)$.

Exercise 4.4 Consider a first-order model

$$(p + a)y = e, \qquad \mathbf{E}\,e(t)e(s) = r\delta(t - s).$$

Determine the variance of y in two ways:

(a) by integrating the spectral density,
(b) by solving the Lyapunov equation.

Exercise 4.5 Consider the sampling of a state space model. Let the sampled system be characterized by the state transition matrix $F(h)$ and the noise covariance matrix $R_d(h)$ where h is the sampling interval. Prove the *doubling algorithm*

$$F(2h) = F^2(h) \,,$$
$$R_d(2h) = R_d(h) + F(h)R_d(h)F^T(h) \,.$$

Exercise 4.6 Show that the following is a simple (but not necessarily computationally efficient) way to compute a sampled system on state space form. Set

$$\overline{A} = \begin{pmatrix} A & R_c \\ 0 & -A^T \end{pmatrix} \,, \qquad \overline{F} = e^{\overline{A}h} = \begin{pmatrix} F_{11} & F_{12} \\ F_{21} & F_{22} \end{pmatrix} \,.$$

Prove that

$$F = F_{11} \,, \qquad R_d = F_{12}F_{22}^{-1} \,.$$

Exercise 4.7 Consider the continuous-time Lyapunov equation (4.91):

$$0 = AP + PA^T + R_c \,. \tag{4.200}$$

(a) Show that the bilinear transformation (α is an arbitrary positive scalar)

$$F = (\alpha I - A)^{-1}(A + \alpha I) \,,$$
$$R_d = 2\alpha(\alpha I - A)^{-1}R_c(\alpha I - A^T)^{-1} \tag{4.201}$$

converts it into the discrete-time Lyapunov equation

$$P = FPF^T + R_d \,, \tag{4.202}$$

that is, (4.200) and (4.202) have the same solution P.
(b) Verify that the inverse transformation (4.201) is given by

$$A = \alpha(F - I)(I + F)^{-1} \,,$$
$$R_c = 2\alpha(F + I)^{-1}R_d(F^T + I)^{-1} \,.$$

Exercise 4.8 Consider a function

$$\tilde{\phi}(z) = \frac{d_0 z + d_1}{z + a} \,,$$

where $|a| < 1$. Determine for what values of d_0 and d_1 the function $\tilde{\phi}(z)$ is positive real.

(a) Use the methodology outlined in Section 4.6.1.
(b) Use the methodology outlined in Section 4.6.2.

Exercise 4.9 Let $y(t)$ be an AR(p) process

$$y(t) + a_1 y(t-1) + \ldots + a_p y(t-p) = e(t) .$$

Assume that $e(t)$ is zero mean, non-Gaussian white noise with $\mathbf{E}\, e^3(t) = \beta$. Show that the third-moment sequence of $y(t)$, $R_y(m,n)$, satisfies

$$R_y(-k,-\ell) + \sum_{i=1}^{p} a_i R_y(i-k, i-\ell) = \beta \delta(k,\ell)$$

for $k \geq 0$, $\ell \geq 0$.

Exercise 4.10 Let $y(t)$ be a stationary process with zero mean and bispectrum $B(z_1, z_2)$. Prove that

$$\frac{1}{2\pi i} \oint B(z_1, z_2) \frac{dz_2}{z_2} = \phi_{y,y^2}(z_1) ,$$

where the integration is counterclockwise around the unit circle and $\phi_{y,y^2}(z)$ denotes the cross-spectrum between $y(t)$ and $y^2(t)$.

Exercise 4.11 Consider an AR(2) process

$$y(t) + a_1 y(t-1) + a_2 y(t-2) = e(t) , \quad \mathbf{E}\, e^2(t) = \lambda^2 .$$

(a) Determine its covariance function $r(\tau)$ for $\tau = 0, 1, 2$ using the algorithm of Section 4.8.1.
(b) Determine the same covariance elements by using the Yule–Walker equations in (4.30).
(c) Determine the same covariance elements by representing the process in state space form and solving the Lyapunov equation.

Exercise 4.12 Use the algorithm of Section 4.8.2 to compute the cross-covariance function between the two AR processes

$$y(t) = \frac{1}{1 + aq^{-1}} e(t) , \quad w(t) = \frac{1}{1 + cq^{-1}} e(t) .$$

Exercise 4.13 Consider the second-order model

$$(p + \alpha_1)(p + \alpha_2) y(t) = e(t) ,$$
$$\mathbf{E}\, e(t)e(s) = \delta(t-s) ,$$

where $\alpha_1 \neq \alpha_2$, $\alpha_1 > 0$, $\alpha_2 > 0$.

(a) Represent the model in state space form.
(b) Determine its covariance function using the algorithm of Section 4.8.3.
(c) Use the state space form to determine its covariance function.

Exercise 4.14 Consider the sampling of a first-order model

$$\dot{y} + ay = e \quad \longrightarrow \quad y(t) + fy(t-h) = v(t) ,$$
$$\mathbf{E}\, e(t)e(s) = r_c\delta(t-s) \qquad \mathbf{E}\, v^2(t) = \lambda^2 .$$

(a) Determine f and λ^2 as functions of a and r_c by using the state space approach. Also find explicit forms for the spectral densities $\phi_c(\omega)$ and $\phi_d(e^{i\omega})$.

(b) Determine f and λ^2 by using the algorithm of Section 4.8.4. Also determine $\tilde{\phi}_c(s)$ and $\tilde{\phi}_d(z)$ and verify the relations

$$2\,\mathrm{Re}\,\tilde{\phi}_c(i\omega) = \phi_c(\omega) ,$$
$$2\,\mathrm{Re}\,\tilde{\phi}_d(e^{i\omega h}) = \phi_d(e^{i\omega h}) .$$

Exercise 4.15 Consider the acceleration algorithm (4.195) for solving the Lyapunov equation (4.192). Introduce the error

$$E_k = P - P_k .$$

(a) Show that

$$E_k = F^{2^k} P (F^T)^{2^k} .$$

(b) Show that P_k essentially converges quadratically in the sense that

$$\| E_k \| \le e_k ,$$

where

$$e_{k+1} = ce_k^2 ,$$

c being a constant that depends on the original Lyapunov equation.

Exercise 4.16 Consider a state space model

$$x(t+1) = Ax(t) + v(t) , \qquad\qquad (4.203)$$

where $A \in \mathbf{C}$, $\mathbf{E}\, v(t)v^*(s) = R\delta_{t,s}$, $R \in \mathbf{R}$, $\mathbf{E}\, v(t)v^T(s) = 0$. Introduce the vector

$$\tilde{x}(t) = \begin{pmatrix} \mathrm{Re}\ x(t) \\ \mathrm{Im}\ x(t) \end{pmatrix} .$$

Derive a stochastic real-valued state space model for $\tilde{x}(t)$ of the form

$$\tilde{x}(t+1) = \tilde{A}\tilde{x}(t) + \tilde{v}(t) . \qquad\qquad (4.204)$$

Establish the links between the models in (4.203) and (4.204). Show how the Lyapunov equations giving the state covariance matrices $P = \mathbf{E}\, x(t)x^*(t)$ and $\tilde{P} = \mathbf{E}\, \tilde{x}(t)\tilde{x}^*(t)$ are related. Derive the Lyapunov equation for P from the Lyapunov equation for $\tilde{P}$.

Exercise 4.17 Consider the process

$$y(t) = \frac{1}{q^2 - 0.5q + 0.5} \begin{pmatrix} q^2 - 2q & q \\ -0.5q + 1 & q^2 - 0.5q \end{pmatrix} v(t) ,$$

where $v(t)$ is white noise of zero mean and covariance matrix

$$\mathbf{E}\, v(t)v^T(t) = \begin{pmatrix} 1 & 1 \\ 1 & 2 \end{pmatrix} .$$

(a) Show that the filter above is stable but that its inverse is unstable.

(b) Show that the system can be represented in state space form as

$$x(t+1) = \begin{pmatrix} 0.5 & 1 \\ -0.5 & 0 \end{pmatrix} x(t) + \begin{pmatrix} -1.5 & 1 \\ -0.5 & 0 \end{pmatrix} v(t) ,$$

$$y(t) = \begin{pmatrix} 1 & 0 \\ 0 & 1 \end{pmatrix} x(t) + \begin{pmatrix} 1 & 0 \\ 0 & 1 \end{pmatrix} v(t) .$$

(c) Determine the innovations form for $y(t)$.

Hint. The solution to the Riccati equation for the representation in (b) has the form

$$P = \begin{pmatrix} p & 0 \\ 0 & 0 \end{pmatrix} .$$

Remark Reasons for the solution to have this form are asked for in Exercise 6.9.   □

Exercise 4.18 Consider the sampling of a state space model

$$\dot{x} = Ax + e \quad \longrightarrow \quad x(t+h) = Fx(t) + v(t) ,$$
$$\mathbf{E}\, e(t)e^T(s) = R_c \delta(t-s) \qquad \mathbf{E}\, v(t)v^T(s) = R_d \delta_{t,s} .$$

Consider the following approach for determining R_d. The state covariance matrix, $P = \mathbf{E}\, x(t)x^T(t)$, must remain unchanged after the sampling. Then first determine P by solving

$$AP + PA^T + R_c = 0 .$$

with respect to P. Next determine R_d from

$$P = FPF^T + R_d .$$

Is this a sound approach? In particular, examine what numerical accuracy can be achieved.

Exercise 4.19 It is sometimes convenient to work with models where the cross-covariance matrix $R_{12} = 0$. It may under certain circumstances be possible to convert a model

$$x^{(1)}(t+1) = F^{(1)}x^{(1)}(t) + v^{(1)}(t) ,$$
$$y(t) = H^{(1)}x^{(1)}(t) + e^{(1)}(t) ,$$

$$\text{cov}\begin{pmatrix} v^{(1)}(t) \\ e^{(1)}(t) \end{pmatrix} = \begin{pmatrix} R_1^{(1)} & R_{12}^{(1)} \\ R_{21}^{(1)} & R_2^{(1)} \end{pmatrix} , \qquad R_{12}^{(1)} \neq 0 ,$$

(4.205)

into another model

$$x^{(2)}(t+1) = F^{(2)}x^{(2)}(t) + v^{(2)}(t) ,$$
$$y(t) = H^{(2)}x^{(2)}(t) + e^{(2)}(t) ,$$

$$\text{cov}\begin{pmatrix} v^{(2)}(t) \\ e^{(2)}(t) \end{pmatrix} = \begin{pmatrix} R_1^{(2)} & 0 \\ 0 & R_2^{(2)} \end{pmatrix} .$$

(4.206)

(a) Take $x^{(2)}(t)$ as an extension of $x^{(1)}(t)$; more precisely

$$x^{(2)}(t) = \begin{pmatrix} x^{(1)}(t) \\ e^{(1)}(t) \end{pmatrix} .$$

Show that this leads to a model of the form in (4.206) and give $F^{(2)}$, $H^{(2)}$, $R_1^{(2)}$, $R_2^{(2)}$, $v^{(2)}(t)$, $e^{(2)}(t)$.

(b) Consider a scalar case ($n = 1$) with

$$F^{(1)} = 0.8, \ H^{(1)} = 1, \ R_1^{(1)} = 0.25, \ R_{12}^{(1)} = 0.5, \ R_2^{(1)} = 1 .$$

Show by construction that there exists a *first-order* model of the form in (4.206) that preserves the spectrum [in the sense $\phi_{y^{(1)}}(z) \equiv \phi_{y^{(2)}}(z)$].

(c) Consider an MA(1) process

$$y(t) = e(t) + ce(t-1) ,$$

which may be represented in the form in (4.205) with

$$F^{(1)} = 0, \ H^{(1)} = 1, \ R_1^{(1)} = c^2, \ R_{12}^{(1)} = c, \ R_2^{(1)} = 1 .$$

Show that there *does not exist* any first-order model of the form in (4.206) that can represent the process.

Exercise 4.20 Consider a sequence $\{r(k)\}_{k=0}^{\infty}$ satisfying $r(k) = (-a)^{k-1}r(1)$, $k \geq 1$, $|a| < 1$. Under what conditions on $r(0)$ and $r(1)$ will $\{r(k)\}_{k=0}^{\infty}$ be a covariance sequence?

Exercise 4.21 Reconsider Example 4.10. Assume that the third-order sequence elements $R(m, n)$ are given for $0 \leq n \leq m \leq 2$. Is it possible to derive *uniquely* from these elements the model parameters h_0, h_1, h_2, β of the model

$$y(t) = h_0 e(t) + h_1 e(t-1) + h_2 e(t-2) ,$$
$$\mathbf{E}\, e^3(t) = \beta ?$$

If not, explain *what* ambiguity remains. It can be assumed that $y(t)$ is generated by such an MA(2) process, with unknown parameters.

Exercise 4.22 Consider the filtering

$$y(t) = \sum_{k=0}^{\infty} g_k u(t-k) , \qquad (4.207)$$

where $u(t)$ is a stationary process and $G(q) = \sum_0^{\infty} g_k q^{-k}$ is an asymptotically stable filter. It is *a priori* not obvious that the infinite sum in (4.207) exists. Use convergence in mean square and set

$$y_n(t) = \sum_{k=0}^{n} g_k u(t-k) .$$

It is said that $y_n(t) \to y(t)$ in mean square if $\mathbf{E} \parallel y_n(t) - y(t) \parallel^2 \to 0$ as $n \to \infty$. It can be proved that this is the case if $y_n(t)$ is a Cauchy sequence, that is, if it satisfies

$$\mathbf{E} \parallel y_n(t) - y_m(t) \parallel^2 \to 0 , \quad \text{as } \min(m,n) \to \infty . \qquad (4.208)$$

Prove that under the given assumptions the condition in (4.208) is satisfied.

Exercise 4.23 Consider the AR process

$$y(t) - ay(t-1) = e(t) , \qquad |a| < 1 ,$$

where $\{e(t)\}$ forms a sequence of independent random variables with zero mean and moments $\mathbf{E}\,[e(t)]^k = \beta_k$, $\beta_3 \neq 0$:

(a) Find the third-order moment

$$R(-\mu,-\nu) = \mathbf{E}\,y(t)y(t-\mu)y(t-\nu)$$

for $0 \leq \mu \leq \nu, \mu + \nu > 0$.

(b) Prove explicitly that (*cf.* Exercise 4.9)

$$R(-\mu,-\nu) - aR(1-\mu,1-\nu) = 0 .$$

(c) Consider the estimation of a from "measured" data $y(1),\dots,y(N)$ by the Yule–Walker equation

$$\hat{r}(\mu) - \hat{a}\hat{r}(\mu-1) = 0 , \qquad \mu > 0 ,$$

where

$$\hat{r}(\mu) = \frac{1}{N} \sum_{t=1}^{N} y(t)y(t+\mu) .$$

Determine the asymptotic variance of $\hat{a}$ as $N \to \infty$. How is the result influenced by μ? In particular, how should μ be chosen to make the variance small?

Hint. It holds that $\hat{r}(\mu) \to r(\mu)$, as $N \to \infty$. Hence:

$$\hat{a} - a = \frac{\hat{r}(\mu)}{\hat{r}(\mu - 1)} - a \approx \frac{\hat{r}(\mu) - a\hat{r}(\mu - 1)}{r(\mu - 1)}$$

$$= \frac{1}{r(\mu - 1)} \frac{1}{N} \sum_{t=1}^{N} e(t)y(t - \mu) .$$

(d) Consider the estimation of a using the equation, *cf.* (b):

$$\hat{R}(-\mu, -\nu) - \hat{a}\hat{R}(1 - \mu, 1 - \nu) = 0 , \qquad (0 < \mu \leq \nu) .$$

Here

$$\hat{R}(-\mu, -\nu) = \frac{1}{N} \sum_{t=1}^{N} y(t)y(t + \mu)y(t - \nu) .$$

Determine the asymptotic ($N \to \infty$) variance of $\hat{a}$. How is it influenced by μ and ν? In particular, how should μ and ν be chosen to make the variance small?

Hint. It holds that $\hat{R}(-\mu, -\nu) \to R(-\mu, -\nu)$ as $N \to \infty$.

(e) Compare the variances achieved in parts (c) and (d). Is one of them uniformly better than the other?

Exercise 4.24 Examine the possibilities to extend the Yule–Walker equation approach to compute the covariance function of a *complex-valued* AR process.

Exercise 4.25 Consider the ARE with

$$F = \begin{pmatrix} -a & 1 \\ 0 & 0 \end{pmatrix}, \qquad R_1 = \begin{pmatrix} 1 & c \\ c & c^2 \end{pmatrix} ,$$

$$H = (1 \ 0), \qquad R_2 = 0 ,$$

where $c^2 \neq 1$.

(a) Determine all symmetric solutions.
(b) Show that one solution (which one?) is largest in the sense expressed in (4.58).

Bibliography

There is a rich literature on continuous-time stochastic systems. An excellent book with profound mathematics combined with engineering application is

Åström, K.J., 1970. *Introduction to Stochastic Control Theory*. Academic Press, New York.

See also

Jazwinski. A.H., 1970. *Stochastic Processes and Filtering Theory.* Academic Press, New York.

Another book dealing with stochastic dynamic models is the classical text

Box. G.E.P., Jenkins. G.W., 1976. *Time Series Analysis: Forecasting and Control,* second ed. Holden-Day, San Francisco.

A comprehensive treatment is also given in

Brown, R.G., 1983. *Introduction to Random Signal Analysis and Kalman Filtering.* John Wiley & Sons, New York.

The Kalman–Yakubovich lemma is important in several system theoretic contexts. For examples, see

Anderson. B.D.O., Vongpanitlerd, S., 1973.: *Network Analysis and Synthesis.* Prentice-Hall, Englewood Cliffs, NJ.

Narendra. K.S., Taylor, J.H., 1973. *Frequency Domain Criteria for Absolute Stability.* Academic Press, New York.

Algorithms similar to the one presented in Section 4.8.2 are given in

Demeure, C.J., Mullis, C.T., 1989. The Euclid algorithm and fast computation of cross-covariance and autocovariance sequences. *IEEE Transactions on Acoustics, Speech and Signal Processing.* ASSP-37, 545–552.

The algorithm presented in Section 4.8.4 is based on

Ježek. J., 1990. Conversion of the polynomial continuous-time model to the delta discrete-time and *vice versa. Proc. 11th IFAC World Congress,* Tallinn, 1990.

For an efficient algorithm for spectral factorization in the polynomial approach, see

Kučera, V., 1979. *Discrete Linear Control.* John Wiley & Sons, Chichester.

5. Optimal Estimation

5.1 Introduction

In this chapter, the basis for optimal estimation methods is presented. By an optimal estimate we mean one that gives as small an error as possible. In many cases it is appropriate to measure the goodness of an estimate by the estimation error variance.

The initial sections cope with the problem of how to extract information about one stochastic variable, say x, when only another variable, say y, that is correlated with x, is available. In Secton 5.4 we discuss how the approaches and results can be used for estimating the states of a dynamic system when only, possibly noisy, measurements of the output are available. The case of Gaussian distributions is dealt with in much more detail in Chapter 6. The reason is twofold. First, it is often relevant to assume that errors are Gaussian distributed. Secondly, the assumption of Gaussian distributed disturbance eases the computational burden considerably in finding the optimal estimate.

More detailed results on the optimal estimation of dynamic systems will be given in the following two chapters. Chapter 6 deals with state space models, and transfer function models are treated in Chapter 7 using polynomial methods. Some comparisons are offered in Chapter 8, and suboptimal solutions and approaches are presented in Chapter 9.

5.2 The Conditional Mean

In this section, we address the problem of how to estimate the value of a vector x when another vector y, correlated to x, is observed. Both x and y are assumed to be real-valued. In Section 2.3 it was shown that the conditional mean is a natural estimate. In fact, under weak conditions, the *best* (i.e. the optimal) estimate is the conditional (or *a posteriori*) mean

$$\hat{x} \stackrel{\triangle}{=} \mathbf{E}\left[x|y\right] . \tag{5.1}$$

Recall that this is given by

$$\mathbf{E}\left[x|y\right] = \int_{-\infty}^{\infty} xp(x|y)\,\mathrm{d}x$$

$$= \int x\frac{p(x,y)}{p(y)}\,\mathrm{d}x$$

$$= \frac{\int xp(x,y)\,\mathrm{d}x}{\int p(x,y)\,\mathrm{d}x}\,,\tag{5.2}$$

where $p(x|y)$ is the conditional pdf of x, given y; $p(x,y)$ is the joint pdf of x and y; and $p(y)$ is the pdf of y.

In order to formulate the optimization problem, consider the following, matrix-valued, criterion:

$$Q(f(y)) = \mathbf{E}\left[\left(x - f(y)\right)\left(x - f(y)\right)^{T}|y\right]\,,\tag{5.3}$$

where $f(y)$ is an arbitrary estimate of x. It is, of course, a function of the observation y. It apparently holds that

$$\mathbf{E}\left[f(y)|y\right] = f(y)\,.\tag{5.4}$$

By an optimal estimate of x is meant a function $f(y)$ that makes the criterion Q "as small as possible". What is meant by this for the matrix-valued criterion Q is stated in the following lemma.

Lemma 5.1 *The optimal estimate that minimizes any scalar-valued, monotonically increasing function of Q (5.3) is the conditional mean $\hat{x} = \mathbf{E}\left[x|y\right]$.*

Proof We find, using (5.1)

$$Q(f(y)) = \mathbf{E}\left[xx^{T} - f(y)x^{T} - xf^{T}(y) + f(y)f^{T}(y)|y\right]$$

$$= \mathbf{E}\left[xx^{T}|y\right] - f(y)\hat{x}^{T} - \hat{x}f^{T}(y) + f(y)f^{T}(y)$$

$$= \mathbf{E}\left[(x - \hat{x})(x - \hat{x})^{T}|y\right] + (\hat{x} - f(y))(\hat{x} - f(y))^{T}$$

$$= Q(\hat{x}) + (\hat{x} - f(y))(\hat{x} - f(y))^{T} \geq Q(\hat{x})\,.\tag{5.5}$$

Hence, it can be concluded that the *conditional mean* $\hat{x}$ is the *optimal estimate*.   ∎

Remark Examples of "scalar-valued, monotonically increasing functions of Q" are $\det(Q)$, $\operatorname{tr}(Q)$, and $\operatorname{tr}(WQ)$ with W a positive definite weighting matrix.   □

The conditional mean turns out to be optimal even under partly weaker conditions than those in Lemma 5.1.

Lemma 5.2 *Consider the criterion of the form*

$$Q(f(y)) = \mathbf{E}\left[\ell(x - f(y))|y\right]\,.\tag{5.6}$$

and make the following assumptions:

(a) The function $\ell(x)$ is symmetric $(\ell(-x) = \ell(x))$ and increasing for $x > 0$.
(b) The conditional pdf $p(x|y)$ is symmetric around the conditional mean

$$\hat{x} = \int xp(x|y)\,\mathrm{d}x\,.$$

(c) The conditional pdf $p(x|y)$ is decreasing for $x > \hat{x}$.

Then the conditional mean $\hat{x}$ is optimal in the sense that it minimizes the criterion in (5.6).

Proof In order to prove this statement, Lemma 5.4 of the appendix (Section 5.A) will be used. Take

$$h(x) = p(x + \hat{x}|y)\,, \quad g(x) = \ell(x)\,,$$

which gives (set $t = x - \hat{x}$)

$$\begin{aligned}
Q(f(y)) = \mathbf{E}\left[\ell(x - f(y))|y\right] &= \int_{-\infty}^{\infty} \ell(x - f(y))p(x|y)\,\mathrm{d}x \\
&= \int_{-\infty}^{\infty} \ell(t + \hat{x} - f(y))p(t + \hat{x}|y)\,\mathrm{d}t \\
&= \int_{-\infty}^{\infty} g(t + \hat{x} - f(y))h(t)\,\mathrm{d}t \\
&\geq \int_{-\infty}^{\infty} g(t)h(t)\,\mathrm{d}t = \int_{-\infty}^{\infty} \ell(x - \hat{x})p(x|y)\,\mathrm{d}x = Q(\hat{x})\,.
\end{aligned}$$

The minimum of the criterion given in (5.6) is hence achieved by the conditional mean of (5.1). ∎

Recall that, in Section 2.4 it was shown that for jointly Gaussian distributed random variables the conditional mean is also Gaussian distributed.

An example is now presented where the assumptions of the two lemmas are violated, and $\hat{x}$ happens not to be the optimal estimate.

Example 5.1 Let x be a scalar and consider the criterion

$$Q(f(y)) = \mathbf{E}\left[\|x - f(y)\| | y\right]\,. \tag{5.7}$$

Find the estimate $f(y)$ that minimizes this function. Apparently:

$$\begin{aligned}
Q(f(y)) &= \int_{-\infty}^{\infty} |x - f(y)|p(x|y)\,\mathrm{d}x \\
&= \int_{-\infty}^{f(y)} [f(y) - x]p(x|y)\,\mathrm{d}x + \int_{f(y)}^{\infty} [x - f(y)]p(x|y)\,\mathrm{d}x\,.
\end{aligned}$$

Now, recall the formula (Leibnitz' rule)

$$\frac{\mathrm{d}}{\mathrm{d}y} \int_{v_1(y)}^{v_2(y)} g(x,y)\,\mathrm{d}x = \int_{v_1(y)}^{v_2(y)} \frac{\partial}{\partial y} g(x,y)\,\mathrm{d}x + v_2'(y)g(v_2(y),y)$$
$$-v_1'(y)g(v_1(y),y)\,.$$

Applying this formula gives

$$\frac{\mathrm{d}Q}{\mathrm{d}f} = \int_{-\infty}^{f(y)} p(x|y)\,\mathrm{d}x - \int_{f(y)}^{\infty} p(x|y)\,\mathrm{d}x\,,$$
$$\frac{\mathrm{d}^2Q}{\mathrm{d}f^2} = 2p(f(y)|y) > 0\,.$$

Hence the optimal estimate is characterized by

$$\int_{-\infty}^{f(y)} p(x|y)\,\mathrm{d}x = \int_{f(y)}^{\infty} p(x|y)\,\mathrm{d}x\,,$$

which means that $f(y)$, the optimal estimate of x, is the *median* of the conditional pdf $p(x|y)$. If this pdf is *nonsymmetric*, the conditional median differs from the conditional mean $\hat{x}$.   □

Note that the result is not in conflict with Lemmas 5.1 and 5.2. The criterion in (5.7) differs from that of Lemma 5.1. Condition (b) of Lemma 5.2 is violated in this example.

5.3 The Linear Least Mean Square Estimate

As shown in Section 5.2, the conditional mean is, under weak assumptions, the mean square optimal estimate. It is fairly complicated to use for dynamic systems, as will be shown later in Section 5.4. There is an important exception, however: linear systems with Gaussian distributed disturbances.

There is thus a motivation for some simpler estimates. Restricting the analysis to consider *linear* estimates, the problem becomes much easier to handle. The corresponding estimate is sometimes called the linear least mean square (LLMS) estimate. It turns out to be identical to the conditional mean for Gaussian random variables.

The LLMS estimate is easy to derive. Let x and y be complex-valued. Postulate an estimate of x to be affine, that is, of the form

$$\hat{x} = Ay + b\,, \tag{5.8}$$

and find the matrix A and the vector b so as to minimize any appropriate scalar measure of the error covariance matrix. Note that $\hat{x}$ will then no longer (except for special cases) denote the conditional mean, but, rather, just some estimate.

Lemma 5.3 *Assume that x and y are correlated and*

$$\mathbf{E}\,x = m_x\,, \quad \mathbf{E}\,y = m_y\,,$$

$$\mathbf{E}\begin{pmatrix} x - m_x \\ y - m_y \end{pmatrix}(x^* - m_x^*\ \ y^* - m_y^*) = \begin{pmatrix} R_x & R_{xy} \\ R_{yx} & R_y \end{pmatrix}\,, \tag{5.9}$$

with R_y being nonsingular. The linear estimate of the form

$$\hat{x} = Ay + b\,, \tag{5.10}$$

that minimizes

$$J = \mathbf{E}\,(x - \hat{x})(x - \hat{x})^*\,,$$

in the sense $J - J_{\min}$ nonnegative definite, is given by

$$\boxed{\hat{x}_{\mathrm{LLMS}} = m_x + R_{xy}R_y^{-1}(y - m_y)\,.} \tag{5.11}$$

The minimal value is

$$\min J = R_x - R_{xy}R_y^{-1}R_{yx}\,. \tag{5.12}$$

The estimation error $x - \hat{x}_{\mathrm{LLMS}}$ is orthogonal to y in the sense that

$$\mathbf{E}\,(x - \hat{x}_{\mathrm{LLMS}})y^* = 0\,. \tag{5.13}$$

Proof Using standard rules for the expected value of a quadratic form, one obtains

$$\begin{aligned}
J &= \mathbf{E}\,(x - Ay - b)(x - Ay - b)^* \\
&= (m_x - Am_y - b)(m_x - Am_y - b)^* \\
&\quad + (I\ \ -A)\begin{pmatrix} R_x & R_{xy} \\ R_{yx} & R_y \end{pmatrix}\begin{pmatrix} I \\ -A^* \end{pmatrix} \\
&= (m_x - Am_y - b)(m_x - Am_y - b)^* \\
&\quad + (R_x - AR_{yx} - R_{xy}A^* + AR_yA^*) \\
&= (m_x - Am_y - b)(m_x - Am_y - b)^* \\
&\quad + (A - R_{xy}R_y^{-1})R_y(A - R_{xy}R_y^{-1})^* \\
&\quad + (R_x - R_{xy}R_y^{-1}R_{yx})\,.
\end{aligned}$$

We find that

$$J \geq R_x - R_{xy}R_y^{-1}R_{yx}\,,$$

and that J is minimized by the choice $A = R_{xy}R_y^{-1}$, $b = m_x - Am_y$. Hence the LLMS estimate is

$$\hat{x}_{\mathrm{LLMS}} = m_x + R_{xy}R_y^{-1}(y - m_y)\,,$$

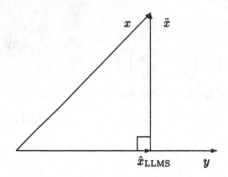

Fig. 5.1. Geometrical illustration of the LLMS estimate

which is (5.11). The estimation error $\tilde{x} = x - \hat{x}_{\text{LLMS}}$ is uncorrelated to "the measurements" y, as the following calculations show:

$$\mathbf{E}\left[x - \hat{x}_{\text{LLMS}}\right]y^* = \mathbf{E}\left[((x - m_x) - R_{xy}R_y^{-1}(y - m_y))\right]y^*$$
$$= R_{xy} - R_{xy}R_y^{-1}R_y = 0 \,,$$

which proves (5.13). ∎

The geometric interpretation is that the best estimate of x given y is the orthogonal projection of x on y. The random variables are then regarded as elements of a Hilbert space. The scalar product of x and y is then chosen as

$$<x|y> = \mathbf{E}\left(x - m_x\right)^*\left(y - m_y\right) \,.$$

Hence, two variables are orthogonal precisely when they are uncorrelated.

The optimal linear estimate of x is illustrated in Figure 5.1.

Intuitively, it is clear that the estimation error $\tilde{x}$ has minimum amplitude $\sqrt{<\tilde{x}|\tilde{x}>} = \sqrt{\mathbf{E}\parallel \tilde{x} \parallel^2}$ precisely when it is perpendicular (*i.e.* orthogonal) to the vector y.

5.4 Propagation of the Conditional Probability Density Function

Consider a nonlinear stochastic system of the form

$$x(t + 1) = f(x(t), v(t)) \,, \tag{5.14}$$
$$y(t) = h(x(t)) + e(t) \,, \tag{5.15}$$

where $v(t)$ and $e(t)$ are mutually independent white noise sequences of zero mean. From Section 5.2 it is known that, under mild conditions, the mean square optimal state estimate is

$$\hat{x}(t) = \mathbf{E}\left[x(t)|Y^t\right], \tag{5.16}$$

where Y^t denotes all the available measurements at time t, that is:

$$Y^t \triangleq \{y^T(t), \ y^T(t-1), \ \dots \}^T. \tag{5.17}$$

It is desirable to find some recursive formula for the state estimate of (5.16). For that purpose, it is first necessary to find a recursion for the conditional pdf $p(x(t)|Y^t)$. Using Bayes' rule, one finds that

$$p(x(t+1), y(t+1)|Y^t) = p(x(t+1)|y(t+1), Y^t)p(y(t+1)|Y^t) \tag{5.18}$$
$$= p(y(t+1)|x(t+1), Y^t)p(x(t+1)|Y^t). \tag{5.19}$$

As (5.15) is free from any dynamics and $e(t+1)$ is white, and in particular independent of Y^t, it is found that

$$p(y(t+1)|x(t+1), Y^t) = p(y(t+1)|x(t+1)), \tag{5.20}$$

and hence, by (5.18) and (5.19)

$$p(x(t+1)|Y^{t+1}) = \frac{p(y(t+1)|x(t+1))}{p(y(t+1)|Y^t)}p(x(t+1)|Y^t). \tag{5.21}$$

Now recall (2.14), which is rewritten in the form

$$p(x(t+1)) = \int p(x(t+1)|x(t))p(x(t))\,dx(t). \tag{5.22}$$

All the pdfs in (5.22) can be conditioned on Y^t, see Exercise 2.6. Hence, owing to (5.14) and the fact that $v(t)$ is white:

$$p(x(t+1)|Y^t) = \int p(x(t+1)|x(t), Y^t)p(x(t)|Y^t)\,dx(t)$$
$$= \int p(x(t+1)|x(t))p(x(t)|Y^t)\,dx(t). \tag{5.23}$$

Combining (5.21) and (5.23) gives

$$p(x(t+1)|Y^{t+1}) = \frac{p(y(t+1)|x(t+1))}{p(y(t+1)|Y^t)}$$
$$\times \int p(x(t+1)|x(t))p(x(t)|Y^t)\,dx(t). \tag{5.24}$$

Note that (5.24) gives a relation for how the conditional pdf propagates from $p(x(t)|Y^t)$ to $p(x(t+1)|Y^{t+1})$. The properties of the state equation $f(\cdot, \cdot)$ enters through the factor $p(x(t+1)|x(t))$, and $p(y(t+1)|x(t+1))$ reflects the measurement function $h(\cdot)$. Finally, the denominator $p(y(t+1)|Y^t)$ can be viewed as a normalizing factor that ensures the left-hand side of (5.24) is a pdf (and integrates to unity).

The denominator of (5.24) can be written as (use Bayes' rule and recall the assumption that $v(t)$ is white)

$$p(y(t+1)|Y^t) = \int \int p(y(t+1)|x(t+1), x(t), Y^t) p(x(t+1)|x(t), Y^t)$$
$$\times p(x(t)|Y^t) \, dx(t) \, dx(t+1)$$
$$= \int \int p(y(t+1)|x(t+1)) p(x(t+1)|x(t))$$
$$\times p(x(t)|Y^t) \, dx(t) \, dx(t+1) . \tag{5.25}$$

To summarize so far, it has been found that the conditional pdf propagates as

$$p(x(t+1)|Y^{t+1})$$

$$= \frac{\int p(y(t+1)|x(t+1)) p(x(t+1)|x(t)) p(x(t)|Y^t) \, dx(t)}{\int \int p(y(t+1)|x(t+1)) p(x(t+1)|x(t)) p(x(t)|Y^t) \, dx(t) \, dx(t+1)} . \tag{5.26}$$

Equation (5.26), or, equivalently, (5.24), describes how $p(x(t)|Y^t)$ propagates to $p(x(t+1)|Y^{t+1})$ as time increases from t to $t+1$. As it stands, it is not very practical. To evaluate all the pdfs involved in (5.26) requires a huge amount of computation. One therefore has to resort to special cases or approximations.

For linear systems with Gaussian disturbances

$$x(t+1) = Fx(t) + v(t) ,$$
$$y(t) = Hx(t) + e(t) , \tag{5.27}$$

the propagation equations have a much simpler form. All the conditional pdfs will then be Gaussian. As will be seen in Section 6.3, the conditional mean and covariance matrix then propagates according to the Kalman filter equations.

5.5 Relation to Maximum Likelihood Estimation

The maximum likelihood (ML) principle is a powerful statistical tool. ML estimates have strong statistical properties under mild conditions, such as minimal variance and consistency (convergence to the "true" values when the number of observations grows to infinity).

Let y be a stochastic variable whose distribution depends on an unknown vector θ. Assume that y is observed. The likelihood function $L(\theta)$ is the pdf of y given θ, evaluated using the observations

$$L(\theta) = p(y|\theta) . \tag{5.28}$$

The ML estimate of θ is the argument that maximizes the likelihood function

$$\hat{\theta}_{ML} = \arg \max_{\theta} L(\theta) . \tag{5.29}$$

Hence, $\hat{\theta}_{ML}$ is the θ value that makes the *a posteriori* probability (likelihood) of the observations as large as possible.

ML estimates are often difficult to calculate analytically in non-Gaussian cases. In the Gaussian case, though, the problem often becomes more tractable for linear systems.

Consider the special case of a linear dynamic system

$$\begin{aligned} x(t+1) &= Fx(t) + v(t)\,, \\ y(t) &= Hx(t) + e(t)\,, \end{aligned} \tag{5.30}$$

where $x(0)$, $v(t)$ and $e(t)$ are independent Gaussian random vectors with means m_0, 0 and 0, respectively, and covariance matrices R_0, R_1 and R_2 respectively. Assume, for simplicity, that all these covariance matrices are positive definite.

Consider the problem of estimating the states

$$X^N = \begin{pmatrix} x(0) \\ \vdots \\ x(N) \end{pmatrix}$$

from the measurements

$$Y^N = \begin{pmatrix} y(0) \\ \vdots \\ y(N) \end{pmatrix}.$$

The ML estimate of X^N given Y^N is maximizing the pdf

$$p(Y^N|X^N) = \frac{p(Y^N, X^N)}{p(X^N)}\,.$$

In case the denominator $p(X^N)$ can be neglected here, we should maximize the joint pdf $p(Y^N, X^N)$. This is otherwise an approximation of the (true) ML estimate.

Another interpretation of maximizing $p(X^N, Y^N)$ is to consider the maximum *a posteriori* (MAP) estimate, which is defined as the maximizing argument of the conditional pdf $p(X^N|Y^N)$. Note that

$$\begin{aligned} X^N_{\text{MAP}} &= \arg \max_{X^N} p(X^N|Y^N) \\ &= \arg \max_{X^N} \frac{p(X^N, Y^N)}{p(Y^N)} \\ &= \arg \max_{X^N} p(X^N, Y^N)\,, \end{aligned} \tag{5.31}$$

as the marginal pdf $p(Y^N)$ does not depend on X^N. In what follows we will examine the estimate in (5.31) more closely.

First, note that there are transformations

$$\begin{aligned} Y^N &\longleftrightarrow X^N, E^N\,, \\ X^N &\longleftrightarrow x(0), V^N\,, \end{aligned}$$

where

$$E^N = \begin{pmatrix} e(1) \\ \vdots \\ e(N) \end{pmatrix}, \qquad V^N = \begin{pmatrix} v(1) \\ \vdots \\ v(N) \end{pmatrix}.$$

Noting that $x(0)$, V^N and E^N are independent:

$$p(Y^N, X^N) = p(x(0))p(V^N)p(E^N). \tag{5.32}$$

It is therefore found that the MAP state estimates should maximize

$$J = \gamma(x(0); m_0, R_0) \left[\prod_{t=0}^{N-1} \gamma(v(t); x(t+1) - Fx(t), R_1) \right]$$

$$\times \left[\prod_{t=1}^{N} \gamma(e(t); y(t) - Hx(t), R_2) \right], \tag{5.33}$$

where $\gamma(x; m, P)$ is the pdf of a vector-valued Gaussian random variable

$$\gamma(x; m, P) = \frac{1}{(2\pi)^{n/2}(\det P)^{1/2}} \exp\left(-\frac{1}{2}[x - m]^T P^{-1}[x - m] \right). \tag{5.34}$$

Hence,

$$\log J = -\frac{1}{2}[x(0) - m_0]^T R_0^{-1}[x(0) - m_0]$$

$$-\frac{1}{2}\sum_{t=0}^{N-1}[x(t+1) - Fx(t)]^T R_1^{-1}[x(t+1) - Fx(t)]$$

$$-\frac{1}{2}\sum_{t=1}^{N}[y(t) - Hx(t)]^T R_2^{-1}[y(t) - Hx(t)]$$

$$-\frac{1}{2}\log[(2\pi)^n \det R_0]$$

$$-\frac{N}{2}\log[(2\pi)^n \det R_1] - \frac{N}{2}\log[(2\pi)^{ny} \det R_2]. \tag{5.35}$$

To maximize the function J with respect to $\{x(0), \ldots, x(N)\}$ is thus the same as minimizing the criterion

$$V_N = \frac{1}{2}[x(0) - m_0]^T R_0^{-1}[x(0) - m_0]$$

$$+\frac{1}{2}\sum_{t=0}^{N-1}[x(t+1) - Fx(t)]^T R_1^{-1}[x(t+1) - Fx(t)]$$

$$+\frac{1}{2}\sum_{t=1}^{N}[y(t) - Hx(t)]^T R_2^{-1}[y(t) - Hx(t)]. \tag{5.36}$$

This minimization problem will be examined in some detail in Section 6.6.

5.A Appendix. A Lemma for Optimality of the Conditional Mean

Lemma 5.4 *Let $g(x)$ and $h(x)$ be two integrable, real-valued, positive symmetric functions, such that $g(x)$ increases for $x \geq 0$, $h(x)$ decreases for $x \geq 0$. Then*

$$\int_{-\infty}^{\infty} g(x+a)h(x)\,dx \geq \int_{-\infty}^{\infty} g(x)h(x)\,dx , \tag{5.37}$$

provided the integrals exist.

Proof Assume that the integrals exist, and that $a > 0$. Then

$$\int_{-\infty}^{\infty} [g(x+a)h(x) - g(x)h(x)]\,dx$$

$$= \int_{-\infty}^{-a/2} [g(x+a) - g(x)]h(x)\,dx + \int_{-a/2}^{\infty} [g(x+a) - g(x)]h(x)\,dx$$

$$= \int_{a/2}^{\infty} [g(x-a) - g(x)]h(x)\,dx + \int_{a/2}^{\infty} [g(x) - g(x-a)]h(x-a)\,dx$$

$$= \int_{a/2}^{\infty} [g(x) - g(x-a)][h(x-a) - h(x)]\,dx .$$

For $\quad a/2 \leq x \leq a , \qquad g(x) - g(x-a) = g(x) - g(a-x) \geq 0 .$
For $\quad x \geq a , \qquad\qquad g(x) - g(x-a) \geq 0 .$
For $\quad x \geq a/2 , \qquad\quad h(x-a) - h(x) \geq 0 .$

Hence, the integrand above is positive, which proves the lemma for the case where $a > 0$.

To complete the proof that (5.37) also holds for negative a, let $a > 0$ and make the substitutions

$$g_1(y) = -h(y) , \qquad h_1(y) = -g(y) .$$

These functions satisfy the assumptions. Further:

$$\int_{-\infty}^{\infty} g(x-a)h(x)\,dx = \int_{-\infty}^{\infty} g(x)h(x+a)\,dx = \int_{-\infty}^{\infty} h_1(y)g_1(y+a)\,dy$$

$$\geq \int_{-\infty}^{\infty} h_1(y)g_1(y)\,dy$$

$$= \int_{-\infty}^{\infty} g(x)h(x)\,dx . \qquad\blacksquare$$

Exercises

Exercise 5.1 Let $x \sim N(m_x, R_x)$ and $e \sim N(0, R_e)$ be independent Gaussian random vectors. Suppose one observes

$$y = Cx + e .$$

Determine the mean square optimal estimate of x based on the observation y. What is the variance of the estimate?

Exercise 5.2 Let x and v be random vectors and jointly Gaussian distributed:

$$\begin{pmatrix} x \\ v \end{pmatrix} \sim N \left(\begin{pmatrix} m \\ 0 \end{pmatrix}, \begin{pmatrix} R_x & 0 \\ 0 & R_v \end{pmatrix} \right) ,$$

and set

$$y = Ax + v .$$

Assume that $\dim y \geq \dim x$ and that A has full (column) rank.

(a) What is the conditional distribution of $(x|y)$? Find the conditional mean $\hat{x}$ and the conditional covariance matrix P.
(b) Suppose $R_v = \lambda^2 I$ and that the *a priori* information of x is inaccurate, so that $R_x/\lambda^2 \to \infty$ (or $\lambda^2 R_x^{-1} \to 0$). What expressions are then obtained for $\hat{x}$ and P?
(c) Generalize the situation in (b) to correlated v (R_v nondiagonal) but $R_x^{-1} \to 0$.

   Hint. For rectangular matrices of compatible dimensions

   $$A(I + BA)^{-1} = (I + AB)^{-1}A$$

   holds.
(d) Apply the result of (b) to the following situation. Consider the first-order system

   $$x(t + 1) = ax(t) + bu(t) + v(t) ,$$

   where a is known, $x(t)$ is measurable, and $v(t)$ is Gaussian white noise of zero mean and variance λ^2. Let b be Gaussian but let its variance tend to infinity. Find

   $$\mathbf{E}\left[b|X^t\right] \quad \text{and} \quad \text{var}[b|X^t] .$$

   Compare with the theory of linear regression.

Exercise 5.3 Let x be a uniformly distributed random variable, with pdf

$$f_x(x) = \begin{cases} 1, & 0 \leq x \leq 1, \\ 0, & \text{elsewhere.} \end{cases}$$

(a) Assume that neither measurements nor any further information are available. What is the best estimate of x in a mean square sense? (That is, determine a number $\hat{x}_{MS}$ so that $\mathbf{E}\,(\hat{x}_{MS} - x)^2$ is minimal.)

In what follows assume that a noisy measurement is available as

$$y = x + v\,,$$

where v is noise with a uniform distribution:

$$f_v(v) = \begin{cases} 1/(2r)\,, & |v| \leq r\,, \\ 0\,, & \text{elsewhere.} \end{cases}$$

The variables x and v are independent, and $0 < r \leq 0.5$.

(b) Find the conditional mean $\mathbf{E}\,[x|y]$.
(c) Find the LLMS estimate of x.
(d) Illustrate graphically how the estimates in (a)–(c) depend on y and r.

Bibliography

The results reported in this chapter can often be found in comprehensive textbooks on stochastic systems and estimation. For some examples, see

Åström, K.J., 1970. *Introduction to Stochastic Control*. Academic Press, New York.

Anderson, B.D.O., Moore, J.B., 1979. *Optimal Filtering*. Prentice Hall, Englewood Cliffs, NJ.

Kailath, T., A. H. Sayed. A.H., Hassibi. B., 2000. *Linear Estimation*. Prentice Hall, Upper Saddle River, NJ.

Lewis, F.L., 1986. *Optimal Estimation*. John Wiley & Sons, New York.

6. Optimal State Estimation for Linear Systems

6.1 Introduction

The general state estimation problem for linear systems is formulated and discussed in this section.

Consider the state space model

$$x(t+1) = Fx(t) + Gu(t) + v(t) ,$$
$$y(t) = Hx(t) + e(t) ,$$

(6.1)

where $v(t)$ and $e(t)$ are white noise sequences with zero mean and covariance matrix

$$\mathbf{E} \begin{pmatrix} v(t) \\ e(t) \end{pmatrix} (v^*(s) \ \ e^*(s)) = \begin{pmatrix} R_1 & R_{12} \\ R_{21} & R_2 \end{pmatrix} \delta_{t,s} .$$

(6.2)

At time t, the available information is

$$Y^t = \{y^T(t), u^T(t), y^T(t-1), u^T(t-1), \dots , y^T(t_0), u^T(t_0)\}^T .$$

Now find the optimal state estimate of $x(t)$ given the measurements Y^s. Based on the relations between t and s, one can distinguish three cases:

- If $t > s$, it is a *prediction* problem.
- If $t = s$, it is a *filtering* problem.
- If $t < s$, it is a *smoothing* (or interpolation) problem.

The optimal estimate of $x(t)$ will be denoted by $\hat{x}(t|s)$. The covariance matrix of the estimation error will be denoted by

$$P(t|s) = \mathbf{E} \left[x(t) - \hat{x}(t|s) \right] [x(t) - \hat{x}(t|s)]^* .$$

(6.3)

Assume that the input signal to the system is either a known deterministic signal, or that it is determined by a feedback in such a way that $u(t)$ is completely determined from Y^s.

The optimal state estimation can be derived under some additional assumptions. Two different approaches will be considered:

- Section 6.2 is devoted to a study of the linear least mean square (LLMS) filter. Then the system is allowed to be complex-valued. Lemma 5.3 will be an important tool in the derivation.

- In Section 6.3, it will be assumed that the state space model is real-valued and that the noise sequences are Gaussian. It is known from Lemmas 5.1 and 5.2 that the optimal estimate is the conditional mean

$$\hat{x}(t|s) = \mathbf{E}\left[x(t)|Y^s\right].$$

The results of Sections 5.3 and 5.4 will be used to derive the optimal state estimates.

The two approaches will lead to the same filter, known as the Kalman filter. This is not surprising, as it is already known from Chapter 5 that the LLMS estimate coincides with the conditional mean for Gaussian distributed data.

Some further results related to the prediction and filtering cases will be given in Section 6.4, and smoothing is discussed in Section 6.5.

6.2 The Linear Least Mean Square One-Step Prediction and Filter Estimates

Consider now the system of (6.1) for the one-step prediction case, that is, $t = s + 1$. Find the LLMS estimate. Assume that an initial state estimate $\hat{x}(t_0|t_0 - 1)$ and its covariance matrix

$$R(t_0) = \mathbf{E}\left[x(t_0) - \hat{x}(t_0|t_0 - 1)\right]\left[x(t_0) - \hat{x}(t_0|t_0 - 1)\right]^*$$

are available, and that $\hat{x}(t_0|t_0-1)$ is independent of the noise sequences $\{e(t)\}$, $\{v(t)\}$. The initial estimate and its covariance may be chosen as the mean, if known, of the initial value $x(t_0)$ and its covariance matrix respectively.

Introduce the *output innovations* $\{\tilde{y}(t)\}$ as the one-step prediction errors

$$\tilde{y}(t) = y(t) - \hat{y}(t|t - 1) .\tag{6.4}$$

The term "innovation" reflects that $\tilde{y}(t)$ is the new piece of information in $y(t)$ that was not known at time $t - 1$. It follows from Lemma 5.3 that the $\{\tilde{y}(t)\}$ is a sequence of uncorrelated random variables. It can also be noted that there is a 1–1 transformation between $\{\tilde{y}(t)\}$ and $\{y(t)\}$. Utilizing this transformation, the conditional expectations $\mathbf{E}\left[x(t)|Y^t\right]$ and $\mathbf{E}\left[x(t)|\tilde{Y}^t\right]$ will, therefore, be the same.

As $e(t)$ is uncorrelated with Y^{t-1},

$$\hat{y}(t|t - 1) = H\hat{x}(t|t - 1) .\tag{6.5}$$

Now set

$$\Lambda(t) \stackrel{\triangle}{=} \mathbf{E}\,\tilde{y}(t)\tilde{y}^*(t) .\tag{6.6}$$

In what follows, notation such as $R_{x(t)Y^t}$ will be used to denote the co-variance matrix between the state $x(t)$ and the collected observations Y^t. This will follow the notational conventions of Chapter 2 and Section 5.3.

The input signal $u(t)$ will be regarded as known and deterministic. It will then play a role corresponding to the mean values of Section 5.3. Further, the observation that $\{\tilde{y}(t)\}$ and $\{y(t)\}$ have the same span can be reformulated as that there exists a nonsingular matrix Q_0 and a vector q_1 such that

$$Y^t = Q_0 \tilde{Y}^t + q_1 , \tag{6.7}$$

where

$$\tilde{Y}^t = \{\tilde{y}^T(t), \tilde{y}^T(t-1), \ldots \tilde{y}^T(t_0)\}^T . \tag{6.8}$$

The vector q_1 will be the mean value of Y^t. See also Exercise 6.18 for more details.

It may be said that $\tilde{Y}^t$ carries the same information as Y^t and hence one can equivalently base the estimate of $x(t+1)$ on $\tilde{Y}^t$ instead of Y^t. The main advantage of using $\tilde{Y}^t$ is that the components (the individual $\tilde{y}(t)$ values) are uncorrelated. Hence, $R_{\tilde{Y}^t}$ will be block-diagonal, which will simplify the final results considerably. In order to see the details, apply Lemma 5.3. The deterministic input will give the mean value contribution. Proceeding in this way:

$$
\begin{aligned}
\hat{x}(t+1|t) &= \mathbf{E}\, x(t+1) + R_{x(t+1)Y^t} R_{Y^t}^{-1}[Y^t - \mathbf{E}\, Y^t] \\
&= \mathbf{E}\, x(t+1) \\
&\quad + \operatorname{cov}[x(t+1), Q_0 \tilde{Y}^t + q_1][\mathbf{E}\, \{Q_0 \tilde{Y}^t \tilde{Y}^{t*} Q_0^*\}]^{-1}[Q_0 \tilde{Y}^t] \\
&= \mathbf{E}\, x(t+1) + \operatorname{cov}[x(t+1), \tilde{Y}^t][\mathbf{E}\, \tilde{Y}^t \tilde{Y}^{t*}]^{-1} \tilde{Y}^t \\
&= \mathbf{E}\, x(t+1) + \sum_{s=t_0}^{t} [\mathbf{E}\, x(t+1)\tilde{y}^*(s)]\Lambda^{-1}(s)\tilde{y}(s) .
\end{aligned}
\tag{6.9}
$$

Next, set

$$K(t) \triangleq R_{x(t+1)\tilde{y}(t)}\Lambda^{-1}(t) , \tag{6.10}$$

and observe that

1. $\mathbf{E}\, x(t+1) = F\mathbf{E}\, x(t) + Gu(t)$.
2. $v(t)$ is uncorrelated with $\tilde{y}(s)$ for $s < t$.

This gives

$$
\begin{aligned}
\hat{x}(t+1|t) &= F\mathbf{E}\, x(t) + Gu(t) + K(t)\tilde{y}(t) \\
&\quad + \sum_{s=t_0}^{t-1} [\mathbf{E}\, x(t+1)\tilde{y}^*(s)]\Lambda^{-1}(s)\tilde{y}(s) \\
&= F\mathbf{E}\, x(t) + Gu(t) + K(t)\tilde{y}(t)
\end{aligned}
$$

$$+ \sum_{s=t_0}^{t-1} \mathbf{E}\left\{[Fx(t) + Gu(t) + v(t)]\tilde{y}^*(s)\right\}\Lambda^{-1}(s)\tilde{y}(s)$$

$$= Gu(t) + K(t)\tilde{y}(t)$$

$$+F[\mathbf{E}\, x(t) + \sum_{s=t_0}^{t-1}[\mathbf{E}\, x(t)\tilde{y}^*(s)]\Lambda^{-1}(s)\tilde{y}(s)]$$

$$= F\hat{x}(t|t-1) + Gu(t) + K(t)[y(t) - H\hat{x}(t|t-1)] \,. \qquad (6.11)$$

Set

$$\tilde{x}(t) = x(t) - \hat{x}(t|t-1) \,. \qquad (6.12)$$

Hence (see (6.3)),

$$P(t|t-1) = \mathbf{E}\, \tilde{x}(t)\tilde{x}^*(t) \,.$$

As

$$\tilde{y}(t) = y(t) - \hat{y}(t|t-1)$$
$$= Hx(t) + e(t) - H\hat{x}(t|t-1)$$
$$= H\tilde{x}(t) + e(t) \,, \qquad (6.13)$$

and $\tilde{x}(t)$ and $e(t)$ are uncorrelated, it follows that

$$\Lambda(t) = HP(t|t-1)H^* + R_2 \,. \qquad (6.14)$$

Next the remaining part of the gain $K(t)$ must be evaluated. Recall from Lemma 5.3 that $\tilde{x}(t)$ and $\hat{x}(t|t-1)$ are uncorrelated. Hence

$$\mathbf{E}\, x(t+1)\tilde{y}^*(t) = \mathbf{E}\,[Fx(t) + Gu(t) + v(t)][\tilde{x}^*(t)H^* + e^*(t)]$$
$$= \mathbf{E}\,[Fx(t)\tilde{x}^*(t)H^*] + \mathbf{E}\,[v(t)e^*(t)]$$
$$= \mathbf{E}\, F[\tilde{x}(t) + \hat{x}(t|t-1)]\tilde{x}^*(t)H^* + R_{12}$$
$$= FP(t|t-1)H^* + R_{12} \,. \qquad (6.15)$$

It remains to derive an equation for the state prediction error covariance matrix $P(t|t-1)$. To do so, first establish a difference equation for the error $\tilde{x}(t)$:

$$\tilde{x}(t+1) = Fx(t) + Gu(t) + v(t) - [F\hat{x}(t|t-1) + Gu(t) + K(t)\tilde{y}(t)]$$
$$= F\tilde{x}(t) + v(t) - K(t)\tilde{y}(t) \,. \qquad (6.16)$$

Next, observe

$$\mathbf{E}\, \tilde{x}(t)v^*(t) = 0 \,,$$
$$\mathbf{E}\, \tilde{x}(t)\tilde{y}^*(t) = \mathbf{E}\, \tilde{x}(t)[\tilde{x}^*(t)H^* + e^*(t)] = P(t|t-1)H^* \,, \qquad (6.17)$$
$$\mathbf{E}\, v(t)\tilde{y}^*(t) = \mathbf{E}\, v(t)[\tilde{x}^*(t)H^* + e^*(t)] = R_{12} \,,$$

from which one obtains

$$P(t+1|t) = \mathbf{E}\,\tilde{x}(t+1)\tilde{x}^*(t+1)$$
$$= \mathbf{E}\,[F\tilde{x}(t) + v(t) - K(t)\tilde{y}(t)][F\tilde{x}(t) + v(t) - K(t)\tilde{y}(t)]^*$$
$$= FP(t|t-1)F^* - FP(t|t-1)H^*K^*(t) + R_1$$
$$\quad - R_{12}K^*(t) - K(t)HP(t|t-1)F^* - K(t)R_{21}$$
$$\quad + K(t)\Lambda(t)K^*(t)$$
$$= FP(t|t-1)F^* + R_1 - [FP(t|t-1)H^* + R_{12}]K^*(t)$$
$$\quad - K(t)[HP(t|t-1)F^* + R_{21}] + K(t)\Lambda(t)K^*(t)\ .$$

Inserting (6.10), (6.14) and (6.15) finally gives the *Riccati equation*

$$P(t+1|t) = FP(t|t-1)F^* + R_1 - [FP(t|t-1)H^* + R_{12}]$$
$$\times[HP(t|t-1)H^* + R_2]^{-1}[HP(t|t-1)F^* + R_{21}]\ . \quad (6.18)$$

The predictor is to be initialized at time t_0. An initial estimate $\hat{x}(t_0|t_0-1)$ was assumed available. Further, $P(t_0|t_0-1) = R(t_0)$.

Next, the LLMS *filter* estimate of $x(t)$ given Y^t is derived. One can proceed along the same lines as before and obtain

$$\hat{x}(t|t) = \mathbf{E}\,x(t) + R_{x(t)Y^t}R_{Y^t}^{-1}[Y^t - \mathbf{E}\,Y^t]$$
$$= \mathbf{E}\,x(t) + \mathrm{cov}[x(t), Q_0\tilde{Y}^t + q_1][Q_0\mathbf{E}\,\tilde{Y}^t\tilde{Y}^{t*}Q_0^*]^{-1}Q_0\tilde{Y}^t$$
$$= \mathbf{E}\,x(t) + \sum_{s=t_0}^{t}[\mathbf{E}\,x(t)\tilde{y}^*(s)]\Lambda^{-1}(s)\tilde{y}(s)$$
$$= \mathbf{E}\,x(t) + \sum_{s=t_0}^{t-1}[\mathbf{E}\,x(t)\tilde{y}^*(s)]\Lambda^{-1}(s)\tilde{y}(s)$$
$$\quad + [\mathbf{E}\,x(t)\tilde{y}^*(t)]\Lambda^{-1}(t)\tilde{y}(t)$$
$$\triangleq \hat{x}(t|t-1) + K_f(t)\tilde{y}(t)$$
$$= \hat{x}(t|t-1) + K_f(t)[y(t) - H\hat{x}(t|t-1)]\ , \quad (6.19)$$

where the *filter gain* $K_f(t)$ can be evaluated as follows:

$$K_f(t) = \mathbf{E}\,x(t)\tilde{y}^*(t)\Lambda^{-1}(t)$$
$$= [\mathbf{E}\,\{\tilde{x}(t) + \hat{x}(t|t-1)\}\{H\tilde{x}(t) + e(t)\}^*]\Lambda^{-1}(t)$$
$$= P(t|t-1)H^*\Lambda^{-1}(t)\ . \quad (6.20)$$

It remains to find the filter covariance $P(t|t)$. Note that, from (6.19):

$$x(t) - \hat{x}(t|t) = \tilde{x}(t) - K_f(t)\tilde{y}(t)\ ,$$

and (6.17) gives the correlation between $\tilde{x}(t)$ and $\tilde{y}(t)$. Hence:

$$P(t|t) = \mathbf{E}\,[x(t) - \hat{x}(t|t)][x(t) - \hat{x}(t|t)]^*$$
$$= P(t|t-1) + K_f(t)\Lambda(t)K_f^*(t)$$
$$\quad - P(t|t-1)H^*K_f^*(t) - K_f(t)HP(t|t-1)$$
$$= P(t|t-1) - P(t|t-1)H^*\Lambda^{-1}HP(t|t-1)$$

$$= P(t|t-1)$$
$$-P(t|t-1)H^*[HP(t|t-1)H^* + R_2]^{-1}HP(t|t-1) . \qquad (6.21)$$

Summing up, the following result has been derived.

Lemma 6.1 *Consider the state space model*

$$x(t+1) = Fx(t) + Gu(t) + v(t) ,$$
$$y(t) = Hx(t) + e(t) ,$$
$$\mathbf{E}\begin{pmatrix} v(t) \\ e(t) \end{pmatrix}(v^*(s) \ \ e^*(s)) = \begin{pmatrix} R_1 & R_{12} \\ R_{21} & R_2 \end{pmatrix}\delta_{t,s} \qquad (6.22)$$
$$\mathbf{E}\,x(t_0) = m_0 , \qquad \mathrm{cov}(x(t_0)) = R_0 .$$

The LLMS prediction estimate is

$$\boxed{\begin{aligned} \hat{x}(t+1|t) &= F\hat{x}(t|t-1) + Gu(t) \\ &\quad + K(t)[y(t) - H\hat{x}(t|t-1)] , \\ \hat{x}(t_0|t_0-1) &= m_0 , \end{aligned}} \qquad (6.23)$$

and the LLMS filter estimate is

$$\boxed{\hat{x}(t|t) = \hat{x}(t|t-1) + K_f(t)[y(t) - H\hat{x}(t|t-1)] .} \qquad (6.24)$$

The predictor and the filter gains are

$$\boxed{\begin{aligned} K(t) &= [FP(t|t-1)H^* + R_{12}]\Lambda^{-1}(t) , \\ K_f(t) &= P(t|t-1)H^*\Lambda^{-1}(t) , \end{aligned}} \qquad (6.25)$$

where

$$\boxed{\begin{aligned} \Lambda(t) &= HP(t|t-1)H^* + R_2 , \\ P(t+1|t) &= FP(t|t-1)F^* + R_1 \\ &\quad - K(t)[R_{21} + HP(t|t-1)F^*] , \\ P(t_0|t_0-1) &= R_0 . \end{aligned}} \qquad (6.26)$$

These matrices relate to the accuracy of the estimates in the following way:

$$\mathrm{cov}[\tilde{x}(t+1|t)] = P(t+1|t) ,$$
$$\mathrm{cov}[\tilde{x}(t|t)] = P(t|t-1) - K_f(t)HP(t|t-1) , \qquad (6.27)$$
$$\mathrm{cov}[y(t) - H\hat{x}(t|t-1)] = \Lambda(t) .$$

∎

The optimal state estimators are now illustrated by two simple examples.

Example 6.1 Consider the estimation of a constant from noisy and independent measurements. The following model is used to describe this situation:

$$x(t+1) = x(t) , \qquad [= x(0)]$$
$$y(t) = x(t) + e(t) , \qquad \mathbf{E}\, e^2(t) = r_2 .$$

Suppose the *a priori* information about $x(0)$ is given as an $\mathbf{N}(x_0, R_0)$ distribution. Find the optimal estimate of $x(0)$ given the measurements $y(0), \cdots ,$ $y(t)$. As the state is constant, identical results are obtained whether one looks for the optimal predictor $\hat{x}(t+1|t)$ or the optimal filter estimate $\hat{x}(t|t)$.

The Riccati equation (6.18) simplifies in this case to

$$P(t+1|t) = P(t|t-1) - \frac{P^2(t|t-1)}{P(t|t-1) + r_2}$$

$$= \frac{P(t|t-1)r_2}{P(t|t-1) + r_2} , \qquad P(0|-1) = R_0 .$$

Inverting this gives a *linear* relationship

$$P^{-1}(t+1|t) = \frac{1}{r_2} + P^{-1}(t|t-1) ,$$

which is easy to solve. It can readily be found that

$$P^{-1}(t|t-1) = \frac{t}{r_2} + P^{-1}(0|-1) = \frac{tR_0 + r_2}{r_2 R_0} ,$$

so (6.25) implies that

$$K(t) = \frac{P(t|t-1)}{P(t|t-1) + r_2} = \frac{R_0}{(t+1)R_0 + r_2} .$$

The optimal predictor estimate can now be written as (see (6.23))

$$\hat{x}(t+1|t) = \hat{x}(t|t-1) + \frac{R_0}{(t+1)R_0 + r_2}[y(t) - \hat{x}(t|t-1)] .$$

To simplify this recursion, introduce

$$z(t) \triangleq (tR_0 + r_2)\hat{x}(t|t-1) .$$

Then

$$z(t+1) = z(t) + R_0 y(t) ,$$

and hence

$$z(t+1) = z(0) + R_0[y(0) + \ldots + y(t)] .$$

Backsubstitution finally gives

$$\hat{x}(t+1|t) = \frac{1}{(t+1)R_0 + r_2} z(t+1)$$

$$= \frac{1}{(t+1)R_0 + r_2}[R_0 \sum_{s=0}^{t} y(s) + r_2 \hat{x}(0|-1)] .$$

Note that this result shows explicitly how the optimal estimate is a weighted average of the arithmetic mean of the observations $\frac{1}{t+1} \sum_{s=0}^{t} y(s)$ and the a priori estimate $\hat{x}(0|-1)$. In particular, if the a priori estimate becomes very uncertain ($R_0 \to \infty$), the optimal estimate will converge to the arithmetic mean of the observations.   □

Example 6.2 Consider the estimation of an exponential decay from noisy measurements. This is modelled as

$$x(t+1) = fx(t) ,$$
$$y(t) = x(t) + e(t) .$$

Applying (6.23) it is found that the optimal prediction estimate will be

$$\hat{x}(t+1|t) = f\hat{x}(t|t-1) + K(t)[y(t) - f\hat{x}(t|t-1)] ,$$

$$K(t) = \frac{fP(t|t-1)}{P(t|t-1) + r_2} ,$$

$$P(t+1|t) = f^2 P(t|t-1) - \frac{f^2 P^2(t|t-1)}{P(t|t-1) + r_2} ,$$

$$\hat{x}(0|-1) = x_0 ,$$

$$P(0|-1) = R_0 .$$

The performance of the above filter may be compared with one with a constant gain:

$$\hat{x}(t+1) = f\hat{x}(t) + K[y(t) - \hat{x}(t|t-1)] .$$

Such a filter is of course simpler but may have a degraded performance. It is instructive to compare the behaviour of the estimation error $\tilde{x}(t) = x(t) - \hat{x}(t)$ of such a filter with the Kalman filter. It is easy to derive

$$\tilde{x}(t+1) = f\tilde{x}(t) - K[\tilde{x}(t) + e(t)]$$
$$= (f - K)\tilde{x}(t) - Ke(t) .$$

The choice of the gain K should meet two conflicting objectives.

- One objective is to have the error $\tilde{x}(t)$ converge rapidly, which requires $f - K$ to be "small". In some sense this means that K should be "large".
- Another objective is that $\tilde{x}(t)$ should have a small steady state variance. This variance turns out to be

$$\text{var}[\tilde{x}(t)] = \frac{K^2 r_2}{1 - (f - K)^2} .$$

It will hence be small when K is small.

With a time-invariant filter, both objectives cannot be met simultaneously. The Kalman predictor makes the optimal trade-off between the objectives by minimizing the time-varying mean square error. When t is small, the gain $K(t)$ is "large" in order to obtain a fast decay of the transient, whereas $K(t)$ decreases with time so that the variance contribution is small as well.

The optimal and the time-invariant estimators are illustrated in Figure 6.1. In this case, the parameters are

$$
\begin{aligned}
x(0) &= 1 , & R_0 &= 1 , \\
\hat{x}(0|-1) &= 0 , & & \\
f &= 0.9 , & r_2 &= 0.04 .
\end{aligned}
$$

$\square$

6.3 The Conditional Mean

In this section the optimal state estimate will be derived based on the results given in Section 5.4 on the propagation of the conditional mean.

The result is formulated as a theorem.

Theorem 6.1 *Consider the system*

$$
\begin{aligned}
x(t+1) &= Fx(t) + Gu(t) + v(t) , \\
y(t) &= Hx(t) + e(t) ,
\end{aligned}
\tag{6.28}
$$

where $v(t)$ and $e(t)$ are mutually independent Gaussian white noise sequences with zero means and

$$
\mathbf{E} \begin{pmatrix} v(t) \\ e(t) \end{pmatrix} (v^T(s) \ \ e^T(s)) = \begin{pmatrix} R_1 & 0 \\ 0 & R_2 \end{pmatrix} \delta_{t,s} .
\tag{6.29}
$$

Assume that the initial state $x(t_0)$ is Gaussian distributed, $x(t_0) \sim \mathbf{N}(m_0, R_0)$, and independent of the noise sequences. Recall that the input signal is assumed to be either deterministic or determined by a feedback such that $u(t)$ is completely known from Y^{t-1}.

Then the conditional distributions of the state are given as

$$
\boxed{
\begin{aligned}
p(x(t)|Y^t) &\sim \mathbf{N}(\hat{x}(t|t), P(t|t)) , \\
p(x(t+1)|Y^t) &\sim \mathbf{N}(\hat{x}(t+1|t), P(t+1|t)) .
\end{aligned}
}
\tag{6.30}
$$

The conditional means and covariances are propagating as

$$
\begin{aligned}
\hat{x}(t+1|t) &= F\hat{x}(t|t-1) + Gu(t) + K_p(t)[y(t) - H\hat{x}(t|t-1)] , \\
\hat{x}(t+1|t+1) &= \hat{x}(t+1|t) + K_f(t+1)[y(t+1) - H\hat{x}(t+1|t)] ,
\end{aligned}
\tag{6.31}
$$

$$
\begin{aligned}
K_p(t) &= FP(t|t-1)H^T[HP(t|t-1)H^T + R_2]^{-1} , \\
K_f(t) &= P(t|t-1)H^T[HP(t|t-1)H^T + R_2]^{-1} ,
\end{aligned}
\tag{6.32}
$$

$$
\begin{aligned}
P(t+1|t) &= FP(t|t)F^T + R_1 , \\
P(t|t) &= P(t|t-1) - P(t|t-1)H^T \\
&\quad \times [HP(t|t-1)H^T + R_2]^{-1}HP(t|t-1) .
\end{aligned}
\tag{6.33}
$$

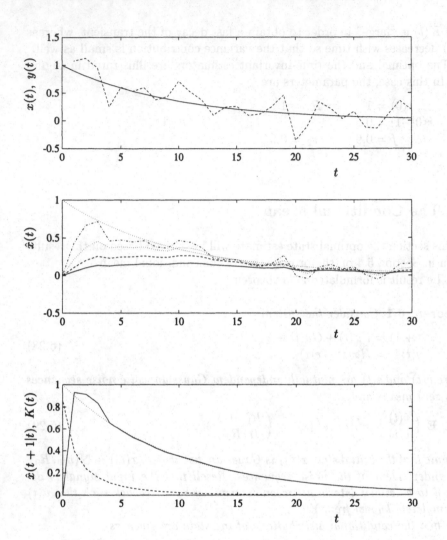

Fig. 6.1. Illustration of optimal and time-invariant estimators; Example 6.2. Top: exponential decay $x(t)$ (*solid line*); measurements $y(t)$ (*dotted line*). Middle: estimates based on time-invariant gains $K = 0.05$ (*solid line*); 0.1 (*dashed line*); 0.2 (*dotted line*); 0.4 (*dash–dotted line*). Bottom: Optimal estimate $\hat{x}(t+1|t)$ (*solid line*); optimal time-varying gain $K(t)$ (*dashed line*)

The initial values are

$$\hat{x}(t_0|t_0 - 1) = m_0 , \qquad P(t_0|t_0 - 1) = R_0 .$$

Proof It follows from Lemma 2.3 that the LLMS estimate and its error covariance are in fact also the mean value and the covariance matrix, respectively, of the conditioned pdf. The stated results then follow from Section 6.2.

In Section 6.A.2, a more direct proof based on the developments of Chapter 5 is presented. It consists of lengthy calculations.   ■

Remark 1 Note that $K_f(t)$ in (6.32) is the *filter gain* and $K_p(t)$ is the *predictor gain*.   □

Remark 2 Note that the equation for $P(t+1|t)$ and $P(t|t)$ can be combined to give the Riccati equation

$$P(t + 1|t) = FP(t|t - 1)F^T + R_1$$
$$-FP(t|t - 1)H^T[HP(t|t - 1)H^T + R_2]^{-1}HP(t|t - 1)F^T .   (6.34)$$

□

6.4 Optimal Filtering and Prediction

Reconsider the system of (6.28) with the additional assumptions stated in Theorem 6.1. In particular let $R_{12} = 0$. The present aim is first to establish some further links between the estimates $\hat{x}(t + 1|t)$ and $\hat{x}(t|t)$. In the second half of this section, the optimal general predictor $\hat{x}(t + k|t)$, $k > 0$, will be derived.

We first note from (6.32) that

$$K_p(t) = FK_f(t) .   (6.35)$$

As a consequence:

$$\hat{x}(t + 1|t) = F\hat{x}(t|t) + Gu(t)   (6.36)$$

(see (6.31)).

One can establish a direct difference equation for $\hat{x}(t|t)$, as is shown next.

Lemma 6.2 *The optimal filter estimate $\hat{x}(t|t)$ satisfies*

$$\hat{x}(t + 1|t + 1) = [I - K_f(t + 1)H]F\hat{x}(t|t)$$
$$+[I - K_f(t + 1)H]Gu(t) + K_f(t + 1)y(t + 1) .   (6.37)$$

Proof By direct calculation from (6.19) and (6.36):

$$\hat{x}(t + 1|t + 1) = [I - K_f(t + 1)H]\hat{x}(t + 1|t) + K_f(t + 1)y(t + 1)$$
$$= [I - K_f(t + 1)H][F\hat{x}(t|t) + Gu(t)] + K_f(t + 1)y(t + 1) .$$

■

Then consider the general prediction problem of estimating $x(t+k)$ from Y^t, $k > 0$. Recall that the future input values $u(s)$, $t + 1 \leq s$, that appear in the following expressions are assumed to be known at time t. For $k \geq 1$:

$$x(t+k) = Fx(t+k-1) + Gu(t+k-1) + v(t+k-1) . \tag{6.38}$$

Therefore, since $v(t+k-1)$ is uncorrelated with Y^t,

$$\begin{aligned}
\hat{x}(t+k|t) &= \mathbf{E}\left[x(t+k)|Y^t\right] \\
&= F\mathbf{E}\left[x(t+k-1)|Y^t\right] + G\mathbf{E}\left[u(t+k-1)|Y^t\right] \\
&\quad + \mathbf{E}\left[v(t+k-1)|Y^t\right] \\
&= F\hat{x}(t+k-1|t) + Gu(t+k-1) .
\end{aligned} \tag{6.39}$$

Equation (6.39) can be iterated to give

$$\hat{x}(t+k|t) = F^{k-1}\hat{x}(t+1|t) + \sum_{s=t+1}^{t+k-1} F^{t+k-1-s}Gu(s)$$

or

$$\hat{x}(t+k|t) = F^k\hat{x}(t|t) + \sum_{s=t}^{t+k-1} F^{t+k-1-s}Gu(s) . \tag{6.40}$$

It is also of interest to evaluate the conditional covariance matrix. Straightforward calculation gives

$$\begin{aligned}
P(t+k|t) &= \mathbf{E}\left[x(t+k) - \hat{x}(t+k|t)\right]\left[x(t+k) - \hat{x}(t+k|t)\right]^* \\
&= \mathbf{E}\left[F\{x(t+k-1) - \hat{x}(t+k-1|t)\} + v(t+k-1)\right] \\
&\quad \times \left[F\{x(t+k-1) - \hat{x}(t+k-1|t)\} + v(t+k-1)\right]^* \\
&= FP(t+k-1|t)F^* + R_1 .
\end{aligned} \tag{6.41}$$

This equation can be iterated to yield

$$P(t+k|t) = F^k P(t|t)F^{*^k} + \sum_{s=0}^{k-1} F^s R_1 F^{*^s} . \tag{6.42}$$

6.5 Smoothing

Consider again a discrete-time stochastic system (6.28). For simplicity, drop the input term.

$$\begin{aligned}
x(t+1) &= Fx(t) + v(t) , \\
y(t) &= Hx(t) + e(t) ,
\end{aligned} \tag{6.43}$$

where $x(t)$ is the state vector and $v(t)$ and $e(t)$ are white noise sequences with covariance matrices

$$\mathbf{E}\, v(t)v^*(t) = R_1 , \quad \mathbf{E}\, v(t)e^*(t) = 0 , \quad \mathbf{E}\, e(t)e^*(t) = R_2 . \tag{6.44}$$

The smoothing estimate of $x(t)$ is $\hat{x}(t|s)$ where $s > t$. In this section we will treat two different situations:

- Fixed point smoothing. Then t is fixed, while s is changing.
- Fixed lag smoothing. In this case the time difference $s - t$ is kept fixed.

6.5.1 Fixed Point Smoothing

We modify the notation slightly as follows. The purpose is to compute the optimal fixed point smoothing estimate $\hat{x}(t_0|t)$, $t > t_0$. Note that here t_0 denotes an arbitrary fixed time, *not* the initial time. Consider the following augmented system, where $x_{t_0}(t) = x(t_0)$:

$$z(t) \triangleq \begin{pmatrix} x(t) \\ x_{t_0}(t) \end{pmatrix}, \qquad t \geq t_0 ,$$

$$z(t+1) = \begin{pmatrix} F & 0 \\ 0 & I \end{pmatrix} z(t) + \begin{pmatrix} v(t) \\ 0 \end{pmatrix}, \qquad z(t_0) = \begin{pmatrix} x(t_0) \\ x(t_0) \end{pmatrix}, \qquad (6.45)$$

$$y(t) = (H \ \ 0)z(t) + e(t) .$$

By applying a standard Kalman filter to this system, one can obtain the optimal estimate $\hat{z}(t+1|t)$. The lower half of this vector is precisely $\hat{x}(t_0|t)$.

In order to obtain a more detailed algorithm, it is necessary to partition the Kalman filter for $\hat{z}(t+1|t)$. The Riccati equation associated with (6.45) can be written as

$$\begin{pmatrix} P_{11}(t+1|t) & P_{12}(t+1|t) \\ P_{12}^*(t+1|t) & P_{22}(t+1|t) \end{pmatrix}$$

$$= \begin{pmatrix} F & 0 \\ 0 & I \end{pmatrix} \begin{pmatrix} P_{11}(t|t-1) & P_{12}(t|t-1) \\ P_{12}^*(t|t-1) & P_{22}(t|t-1) \end{pmatrix} \begin{pmatrix} F^* & 0 \\ 0 & I \end{pmatrix} + \begin{pmatrix} R_1 & 0 \\ 0 & 0 \end{pmatrix}$$

$$- \begin{pmatrix} K_1(t) \\ K_2(t) \end{pmatrix} (H \ \ 0) \begin{pmatrix} P_{11}(t|t-1) & P_{12}(t|t-1) \\ P_{12}^*(t|t-1) & P_{22}(t|t-1) \end{pmatrix} \begin{pmatrix} F^* & 0 \\ 0 & I \end{pmatrix}, \qquad (6.46)$$

where the Kalman (predictor) gain is

$$\begin{pmatrix} K_1(t) \\ K_2(t) \end{pmatrix} = \begin{pmatrix} F & 0 \\ 0 & I \end{pmatrix} \begin{pmatrix} P_{11}(t|t-1) & P_{12}(t|t-1) \\ P_{12}^*(t|t-1) & P_{22}(t|t-1) \end{pmatrix}$$

$$\times \begin{pmatrix} H^* \\ 0 \end{pmatrix} (H P_{11}(t|t-1)H^* + R_2)^{-1} . \qquad (6.47)$$

Examining the equations for $P_{11}(t|t-1)$ and $K_1(t)$, it can easily be seen that they are precisely (as they should be!) the standard equations for the one-step optimal predictor. Hence $P_{11}(t+1|t) = P(t+1|t)$. Equating the other blocks of the Riccati equation gives

$$P_{12}(t+1|t) = [F - K_1(t)H]P_{12}(t|t-1) , \qquad (6.48)$$

$$P_{22}(t+1|t) = P_{22}(t|t-1) - K_2(t)H P_{12}(t|t-1)$$

$$= P_{22}(t|t-1) - P_{12}^*(t|t-1)H^*$$

$$\times [H P_{11}(t|t-1)H^* + R_2]^{-1} H P_{12}(t|t-1) . \qquad (6.49)$$

As

$$z(t_0) = \begin{pmatrix} x(t_0) \\ x(t_0) \end{pmatrix}, \qquad (6.50)$$

it is concluded that the initial values for (6.48) and (6.49) are given by

$$P_{12}(t_0|t_0 - 1) = P_{22}(t_0|t_0 - 1) = P_{11}(t_0|t_0 - 1) = P(t_0|t_0 - 1) . \quad (6.51)$$

The gain $K_2(t)$ is given by

$$K_2(t) = P_{12}^*(t|t - 1)H^*[HP_{11}(t|t - 1)H^* + R_2]^{-1} . \quad (6.52)$$

The fixed point smoothing estimate will thus be

$$\hat{x}(t_0|t) = \hat{x}(t_0|t - 1) + K_2(t)[y(t) - H\hat{x}(t|t - 1)] . \quad (6.53)$$

This equation can be iterated to give

$$\hat{x}(t_0|t) = \hat{x}(t_0|t_0) + \sum_{s=t_0+1}^{t} K_2(s)[y(s) - H\hat{x}(s|s - 1)] . \qquad (6.54)$$

Equation (6.54) demonstrates how the smoothing estimate of $x(t_0)$ based on Y^t $(t > t_0)$ is equal to the filter estimate $\hat{x}(t_0|t_0)$ plus a weighted sum of the future output innovations $\tilde{y}(s)$, $s = t_0 + 1, \ldots, t$.

6.5.2 Fixed Lag Smoothing

We set for convenience

$$m = s - t \qquad (6.55)$$

and seek the smoothing estimate

$$\hat{x}(t|t + m) . \qquad (6.56)$$

The solution can be derived in different ways. One possibility is to start with the fixed point smoothing estimate. Here, we will follow another route, and use an extended state space, partly as in (6.45). As a consequence, we will get not only the estimate in (6.56), but also

$$\hat{x}(t|t + i) , \quad i = 0, \ldots, m . \qquad (6.57)$$

Introduce the extended state vector

$$z(t) = \begin{pmatrix} x(t) \\ \vdots \\ x(t - m) \end{pmatrix} . \qquad (6.58)$$

Then it is easily seen that

$$\begin{aligned} z(t + 1) &= \overline{F}z(t) + \overline{v}(t) \\ y(t) &= \overline{H}z(t) + e(t) \end{aligned} \qquad (6.59)$$

with

$$\overline{F} = \begin{pmatrix} F & & & 0 \\ I & & & \\ & \ddots & & \\ 0 & & I & 0 \end{pmatrix}, \quad \overline{R}_1 = \mathbf{E}\,\overline{v}(t)\overline{v}^T(t) = \begin{pmatrix} R_1 & 0 & \dots & 0 \\ 0 & & & \vdots \\ \vdots & & & \\ 0 & \dots & & 0 \end{pmatrix},$$

(6.60)

$$\overline{H} = \begin{pmatrix} H & 0 & \dots & 0 \end{pmatrix}.$$

Further, the filter estimate of $z(t)$ must be

$$\hat{z}(t|t) = \begin{pmatrix} \hat{x}(t|t) \\ \hat{x}(t-1|t) \\ \vdots \\ \hat{x}(t-m|t) \end{pmatrix},$$

(6.61)

and updating $\hat{z}(t|t)$ thus provides all the smoothing estimates in (6.57). The estimate in (6.57) can be updated according to Lemma 6.2 as

$$\hat{z}(t|t) = \overline{F}\hat{z}(t-1|t-1) + \overline{K}_f(t)[y(t) - \overline{H}\,\overline{F}\hat{z}(t-1|t-1)],$$

(6.62)

and the gain matrix $\overline{K}_f(t)$ will be written in partioned form as

$$\overline{K}_f(t) = \begin{pmatrix} K^{(0)}(t) \\ \vdots \\ K^{(m)}(t) \end{pmatrix}.$$

(6.63)

Spelling out the details, (6.62) and (6.63) mean

$$\hat{x}(t|t) = F\hat{x}(t-1|t-1) + K^{(0)}(t)[y(t) - HF\hat{x}(t-1|t-1)],$$ (6.64)

$$\hat{x}(t-i|t) = \hat{x}(t-i|t-1) + K^{(i)}(t)[y(t) - HF\hat{x}(t-1|t-1)],$$

$$i = 1,\dots,m.$$ (6.65)

Note that (6.64) is nothing but the usual filter update, so $K^{(0)}(t) = K_f(t)$ must hold. Further note that

$$y(t) - HF\hat{x}(t-1|t-1) = y(t) - H\hat{x}(t|t-1)$$

(6.66)

is the usual innovation or prediction error $\tilde{y}(t)$ at time t.

Note the similarity between (6.54) and (6.65). Both formulas show how the smoothing estimate $\hat{x}(t-i|t)$ is a linear combination of the filter estimate $\hat{x}(t-i|t-i)$ and the future output innovations $\tilde{y}(t-i+1),\dots,\tilde{y}(t)$.

It remains to find expressions for the gains $K^{(i)}(t)$, $i = 1,\dots,m$. Naturally, $\overline{K}_f(t)$ is linked to the solution of a Riccati equation as

$$\overline{K}_f(t) = \overline{P}(t)\overline{H}^T[\overline{H}\,\overline{P}(t)\overline{H}^T + R_2]^{-1},$$

(6.67)

$$\overline{P}(t+1) = \overline{R}_1 + \overline{F}\left[\overline{P}(t) - \overline{P}(t)\overline{H}^T[\overline{H}\,\overline{P}(t)\overline{H}^T + R_2]^{-1}\overline{H}\,\overline{P}(t)\right]\overline{F}^T.$$

(6.68)

To proceed, we partition also $\overline{P}(t)$ as

$$\overline{P}(t) = \begin{pmatrix} P^{(00)}(t) & P^{(01)}(t) & \cdots & P^{(0m)}(t) \\ P^{(10)}(t) & P^{(11)}(t) & \cdots & \\ \vdots & & & \\ P^{(m0)}(t) & P^{(m1)}(t) & \cdots & P^{(mm)}(t) \end{pmatrix}. \tag{6.69}$$

Spelling out (6.67) we have

$$\begin{aligned} K^{(0)} &= P^{(00)}(t)H^T[HP^{(00)}(t)H^T + R_2]^{-1}, \\ K^{(i)} &= P^{(i0)}(t)H^T[HP^{(00)}(t)H^T + R_2]^{-1}, \quad i = 1, \ldots, m. \end{aligned} \tag{6.70}$$

The first block of the Riccati equation in (6.68) becomes

$$\begin{aligned} P^{(00)}(t+1) = R_1 + F\,\big[P^{(00)}(t) - P^{(00)}(t)H^T \\ \times \{HP^{(00)}(t)H^T + R_2\}^{-1}HP^{(00)}(t)\big]\,F^T \end{aligned} \tag{6.71}$$

which indeed shows that

$$P^{(00)}(t) = P(t|t-1), \quad K^{(0)}(t) = K_f(t) \tag{6.72}$$

as already known, as the first block of $\hat{z}(t|t)$ is $\hat{x}(t|t)$. In what follows we will use the abbreviated notation

$$P(t) = P(t|t-1). \tag{6.73}$$

Equating (6.68) for blocks $(0, i)$ with $i = 1, \ldots, m$ gives

$$\begin{aligned} P^{(0i)}(t+1) = FP^{(0,i-1)}(t) - FP(t)H^T \\ \times [HP(t)H^T + R_2]^{-1}HP^{(0,i-1)}(t) \end{aligned} \tag{6.74}$$

and for blocks (i, j), $i, j = 1, \ldots, m$

$$\begin{aligned} P^{(ij)}(t+1) = P^{(i-1,j-1)}(t) - P^{(i-1,0)}(t)H^T \\ \times [HP(t)H^T + R_2]^{-1}HP^{(0,j-1)}(t). \end{aligned} \tag{6.75}$$

Equations (6.74) and (6.75) indicate how the block components of $\overline{P}(t)$ are updated from blocks of similar or lower indices i, j.

The initial values for the Riccati equation in (6.69) and the state estimator are

$$\overline{P}(0) = \begin{pmatrix} R_0 & 0 & \cdots & 0 \\ 0 & 0 & & \\ \vdots & & \ddots & \\ 0 & & & 0 \end{pmatrix}, \tag{6.76}$$

$$\hat{z}(0|-1) = \begin{pmatrix} m_0 \\ 0 \\ \vdots \\ 0 \end{pmatrix}, \tag{6.77}$$

where $\mathbf{N}(m_0, R_0)$ is the *a priori* distribution of the initial state vector $x(0)$. The associated initial value of the filter estimate in (6.61) is

$$\hat{z}(0|0) = \begin{pmatrix} m_1 \\ 0 \\ \vdots \\ 0 \end{pmatrix} \tag{6.78}$$

with

$$m_1 = \hat{x}(0|0) = m_0 + R_0 H^T (H R_0 H^T + R_2)^{-1} (y(0) - H m_0) . \tag{6.79}$$

6.6 Maximum *a posteriori* Estimates

In Section 5.5 it was stated that the MAP state estimates of $x(0), \ldots x(N)$, given the measurements $y(1), \ldots, y(N)$, are the minimizing elements of the criterion

$$V_N = \frac{1}{2}[x(0) - m_0]^T R_0^{-1}[x(0) - m_0]$$

$$+ \frac{1}{2} \sum_{t=0}^{N-1} [x(t+1) - Fx(t)]^T R_1^{-1}[x(t+1) - Fx(t)]$$

$$+ \frac{1}{2} \sum_{t=1}^{N} [y(t) - Hx(t)]^T R_2^{-1}[y(t) - Hx(t)] . \tag{6.80}$$

According to the previous findings in this chapter, the optimal estimates must be the smoothing estimates $\hat{x}(t|N)$, $t = 0, \ldots, N$, as given in Section 6.5. The aim of the present analysis is to show that the smoothing estimates do in fact minimize the criterion in (6.80).

As a starting point, the derivatives of V_N with respect to the state variables are evaluated. Straightforward calculation gives

$$z(0) \triangleq \left(\frac{\partial V_N}{\partial x(0)} \right)^T = R_0^{-1}[x(0) - m_0] - F^T R_1^{-1}[x(1) - Fx(0)] , \tag{6.81}$$

$$z(N) \triangleq \left(\frac{\partial V_N}{\partial x(N)} \right)^T = R_1^{-1}[x(N) - Fx(N-1)]$$

$$- H^T R_2^{-1}[y(N) - Hx(N)] . \tag{6.82}$$

and for $t = 1, \ldots N - 1$

$$z(t) \triangleq \left(\frac{\partial V_N}{\partial x(t)} \right)^T = R_1^{-1}[x(t) - Fx(t-1)] - F^T R_1^{-1}[x(t+1) - Fx(t)]$$

$$- H^T R_2^{-1}[y(t) - Hx(t)] . \tag{6.83}$$

The minimizing elements $\{x(t)\}_{t=0}^{N}$ of V_N correspond to $z(t) = 0$, for $t = 0, \ldots, N$ given all the measurements, that is Y^N. By taking conditional expectations of both sides of (6.81) - 6.83), we see that replacing $x(t)$ with the conditional expection $\mathbf{E}\left[x(t)|Y^N\right], t = 0, \ldots, N$ will make both sides equal to zero, if

$$\mathbf{E}\left[z(t)|Y^N\right] = 0, \quad t = 0, \ldots, N. \tag{6.84}$$

This, in turn, is equivalent to

$$\mathbf{E}\left[z(t)|y(s)\right] = 0, \quad t = 0, \ldots, N; \quad s = 1, \ldots, N. \tag{6.85}$$

or yet

$$\mathbf{E}\, z(t)y^T(s) = 0, \quad t = 0, \ldots, N; \quad s = 1, \ldots, N. \tag{6.86}$$

In order to verify (6.86), a first step is to rewrite $z(t)$ using the nominal model. This gives

$$z(0) = R_0^{-1}[x(0) - m_0] - F^T R_1^{-1} v(0), \tag{6.87}$$

$$z(N) = R_1^{-1} v(N-1) - H^T R_2^{-1} e(N), \tag{6.88}$$

$$z(t) = R_1^{-1} v(t-1) - F^T R_1^{-1} v(t) - H^T R_2^{-1} e(t), \tag{6.89}$$

$$(t = 1, \ldots, N-1).$$

As, by construction, $y(s)$ is a (linear) function of $e(s), v(s-1), e(s-1), \ldots,$ $v(0)$, $e(0)$, $x(0)$, it is directly seen that (6.86) is satisfied at least as soon as $s \leq t - 1$.

Consider next the case where $t = N, s = N$. Then

$$\begin{aligned}
\mathbf{E}\, z(N)y^T(N) &= \mathbf{E}\left[R_1^{-1} v(N-1) - H^T R_2^{-1} e(N)\right][Hx(N) + e(N)]^T \\
&= R_1^{-1}[\mathbf{E}\, v(N-1)x^T(N)]H^T - H^T R_2^{-1} R_2 \\
&= R_1^{-1}\mathbf{E}\, v(N-1)[Fx(N-1) + v(N-1)]^T H^T - H^T \\
&= R_1^{-1} R_1 H^T - H^T \\
&= 0.
\end{aligned}$$

Secondly, consider the case where $t = 0, s > t$. In this case

$$\begin{aligned}
\mathbf{E}\, z(0)y^T(s) &= \mathbf{E}\left[R_0^{-1}(x(0) - m_0) - F^T R_1^{-1} v(0)\right][Hx(s) + e(s)]^T \\
&= \mathbf{E}\left[R_0^{-1}(x(0) - m_0) - F^T R_1^{-1} v(0)\right][F^s x(0) + F^{s-1} v(0) \\
&\quad + \sum_{j=1}^{s-1} F^{s-1-j} v(j)]^T H^T \\
&= \mathbf{E}\left[R_0^{-1}(x(0) - m_0) - F^T R_1^{-1} v(0)\right] \\
&\quad \times [x^T(0)F^T + v^T(0)]F^{T^{s-1}} H^T \\
&= [R_0^{-1} R_0 F^T - F^T R_1^{-1} R_1]F^{T^{s-1}} H^T \\
&= 0.
\end{aligned}$$

As the third and last case, let $t = 1, \ldots, N$. If $s = t$:

$$\mathbf{E}\, z(t)y^T(t) = \mathbf{E}\,[R_1^{-1}v(t-1) - F^T R_1^{-1}v(t) - H^T R_2^{-1}e(t)]$$
$$\times [Hx(t) + e(t)]^T$$
$$= \mathbf{E}\,[R_1^{-1}v(t-1)x^T(t)]H^T - H^T R_2^{-1}R_2$$
$$= R_1^{-1}\mathbf{E}\,v(t-1)[Fx(t-1) + v(t-1)]^T H^T - H^T$$
$$= R_1^{-1}R_1 H^T - H^T$$
$$= 0\,,$$

and $s > t$ implies that

$$\mathbf{E}\, z(t)y^T(s) = \mathbf{E}\,[R_1^{-1}v(t-1) - F^T R_1^{-1}v(t) - H^T R_2^{-1}e(t)]$$
$$\times [Hx(s) + e(s)]^T$$
$$= \mathbf{E}\,[R_1^{-1}v(t-1) - F^T R_1^{-1}v(t) - H^T R_2^{-1}e(t)]$$
$$\times [F^{s-t}x(t) + F^{s-t-1}v(t) + \sum_{j=1}^{s-t-1} F^{s-t-1-j}v(t+j)]^T H^T$$
$$= \mathbf{E}\,[R_1^{-1}v(t-1) - F^T R_1^{-1}v(t) - H^T R_2^{-1}e(t)]$$
$$\times [Fx(t) + v(t)]F^{T^{s-t-1}} H^T$$
$$= [R_1^{-1}\mathbf{E}\,v(t-1)x^T(t)F^T - F^T R_1^{-1}\mathbf{E}\,v(t)v^T(t)]F^{T^{s-t-1}}H^T$$
$$= [R_1^{-1}\mathbf{E}\,v(t-1)[Fx(t-1) + v(t-1)]^T F^T - F^T R_1^{-1}R_1]$$
$$\times F^{T^{s-t-1}} H^T$$
$$= [R_1^{-1}R_1 F^T - F^T]F^{T^{s-t-1}} H^T$$
$$= 0\,.$$

Thus (6.86) has been established, which means that the conditional means $\{\hat{x}(t|N)\}_{t=0}^N$ minimize the criterion V_N (6.80). In other words, the optimal state estimates can also be interpreted as maximum *a posteriori* estimates.

6.7 The Stationary Case

The stationary case is studied in this section. Let $t_0 \to -\infty$, so that an infinite set of data is available. Under weak conditions, the optimal state estimators will then be time-invariant. The following result holds, but will not be proved here (for a comprehensive treatment see the book edited by Bittanti *et al.* (1991) listed in the Bibliography section at the end of this chapter.

Theorem 6.2 *Consider the Riccati equation*

$$P(t+1|t) = FP(t|t-1)F^T + R_1$$
$$- FP(t|t-1)H^T[HP(t|t-1)H^T + R_2]^{-1}HP(t|t-1)F^T\,.$$
$$(6.90)$$

Assume that R_2 is positive definite and that the nonnegative definite matrix R_1 is factorized as

$$R_1 = BB^T .$$

Assume, further, that the pair (F, B) is stabilizable, and that the pair (F, H) is detectable. Then the solution to (6.90) converges, as $t - t_0 \to \infty$, to a positive definite matrix P that satisfies the ARE

$$P = FPF^T + R_1 - FPH^T(HPH^T + R_2)^{-1}HPF^T . \tag{6.91}$$

■

As a further consideration, the following lemma holds.

Lemma 6.3 *Consider the ARE*

$$P = FPF^T + R_1 - FPH^T(HPH^T + R_2)^{-1}HPF^T , \tag{6.92}$$

and let it have a positive definite solution P. Let

$$K = FPH^T(HPH^T + R_2)^{-1} . \tag{6.93}$$

Assume that (F, B) is controllable, where $R_1 = BB^T$ and R_2 is nonsingular. (We may, as an alternative to these two assumptions, assume that R_1 is positive definite.) Then the matrix

$$\tilde{F} = F - KH \tag{6.94}$$

has all its eigenvalues strictly inside the unit circle.

Proof (The proof will follow the path given in Theorem 4.2.) Examine the stability properties of the system

$$x(t + 1) = \tilde{F}^T x(t)$$

by means of the candidate Lyapunov function

$$V(x(t)) = x^T(t)Px(t) .$$

Clearly, $V(x(t)) \geq 0$. Further:

$$\begin{aligned}
\Delta V(x) &\triangleq V(x(t+1)) - V(x(t)) \\
&= x^T(t)[\tilde{F}P\tilde{F}^T - P]x(t) \\
&= x^T(t)[FPF^T - KHPF^T - FPH^TK^T \\
&\quad + KHPH^TK^T - P]x(t) \\
&= -x^T(t)[R_1 + KR_2K^T]x(t) \leq 0 .
\end{aligned} \tag{6.95}$$

It remains to show that $\Delta V(x) = 0 \Rightarrow x(t) = 0$.

However:

$$\Delta V(x(t)) = 0 \Longrightarrow R_1 x(t) = 0, \quad K^T x(t) = 0 . \tag{6.96}$$

Hence, $B^T x(t) = 0$ and the dynamics become

$$x(t + 1) = (F^T - H^T K^T)x(t) = \tilde{F}^T x(t) \,. \tag{6.97}$$

If (6.96) and (6.97) hold for any t, it is required that

$$\begin{pmatrix} B^T \\ B^T F^T \\ \vdots \\ B^T (F^T)^{n-1} \end{pmatrix} x(0) = 0 \,.$$

As (F, B) is controllable, the pair (F^T, B^T) is observable. This implies that $x(0) = 0$ and hence that $x(t) \equiv 0$. This completes the proof that $\tilde{F}$ is asymptotically stable. ∎

It is appropriate to recapitulate briefly the optimal state estimator for the stationary case, assuming that the solution to the Riccati equation converges to P. The Kalman prediction gain is given by (6.93). The optimal one-step predictor is

$$\hat{x}(t + 1|t) = F\hat{x}(t|t - 1) + Gu(t) + K[y(t) - H\hat{x}(t|t - 1)] \,, \tag{6.98}$$

which can also be written as

$$\hat{x}(t|t - 1) = [qI - (F - KH)]^{-1}[Gu(t) + Ky(t)] \,, \tag{6.99}$$

where it is emphasized how the predictor depends on all the available past data points. The optimal filter estimate can be written as (see (6.37))

$$\hat{x}(t + 1|t + 1) = [I - K_f H]F\hat{x}(t|t) + [I - K_f H]Gu(t) + K_f y(t + 1) \,,$$

or

$$\hat{x}(t|t) = [qI - (F - K_f HF)]^{-1}[(I - K_f H)Gu(t) + K_f y(t + 1)] \,. \tag{6.100}$$

In another form, using (6.36) and (6.99), the filter estimate can be written as

$$\begin{aligned} \hat{x}(t|t) &= (I - K_f H)\hat{x}(t|t - 1) + K_f y(t) \\ &= (I - K_f H)[qI - (F - KH)]^{-1} \\ &\quad \times [Gu(t) + Ky(t)] + K_f y(t) \,. \end{aligned} \tag{6.101}$$

Here $K_f = PH^T(HPH^T + R_2)^{-1}$ is the stationary filter gain.

Consider also the fixed point smoothing estimate. From (6.48),

$$P_{12}(t|t - 1) = (F - KH)^{t-t_0} P_{12}(t_0|t_0 - 1) \,.$$

Note, however, that

$$P_{12}(t_0|t_0 - 1) = \text{cov}[x(t_0)|Y^{t_0-1}] = P \,.$$

Hence

$$P_{12}(t|t - 1) = (F - KH)^{t-t_0} P \,. \tag{6.102}$$

and the gains become (see (6.52)),

$$K_2(t) = P(F - KH)^{T^{t-t_0}} H^T [HPH^T + R_2]^{-1} . \qquad (6.103)$$

The accuracy of the smoothing estimate can be expressed by $P_{22}(t)$. In the stationary case, (6.49) and (6.102) give

$$P_{22}(t+1|t) = P_{22}(t|t-1) - P(F - KH)^{T^{t-t_0}}$$
$$\times H^T (HPH^T + R_2)^{-1} H (F - KH)^{t-t_0} P . \qquad (6.104)$$

The initial value is the predictor accuracy

$$P_{22}(t_0|t_0 - 1) = P . \qquad (6.105)$$

Equation (6.104) shows the successive improvement in accuracy when a new piece of data is used for the smoothing estimate. By iterating (6.104),

$$P_{22}(t|t-1) = P - P \sum_{s=0}^{t-t_0-1} (F - KH)^{T^s} H^T (HPH^T + R_2)^{-1}$$
$$\times H (F - KH)^s P . \qquad (6.106)$$

The matrices $\{P_{22}(t|t-1)\}$, $t \geq t_0$, form a bounded and decreasing sequence. They therefore converge. Next, examine the limit as $t \to \infty$. The limit will be the covariance matrix obtainable when estimating the stable $x(t_0)$ from data ranging from $t = -\infty$ to $t = \infty$. Set

$$Q(t) = \sum_{s=0}^{t} (F - KH)^{T^s} H^T (HPH^T + R_2)^{-1} H (F - KH)^s . \qquad (6.107)$$

Apparently:

$$P_{22}(t|t-1) = P - PQ(t - t_0 - 1)P . \qquad (6.108)$$

The matrix $Q(t)$ obeys the Lyapunov equation

$$Q(t+1) = (F - KH)^T Q(t)(F - KH) + H^T (HPH^T + R_2)^{-1} H . \qquad (6.109)$$

As the matrix $F - KH$ has all its eigenvalues inside the unit circle, $Q(t)$ will converge as $t \to \infty$ (see Corollary 2 of Lemma 4.2). The limit, say Q, satisfies the algebraic Lyapunov equation

$$Q = (F - KH)^T Q (F - KH) + H^T (HPH^T + R_2)^{-1} H . \qquad (6.110)$$

The ultimate smoothing accuracy is

$$P_{22} = \lim_{t \to \infty} P_{22}(t|t-1) = P - PQP . \qquad (6.111)$$

This section ends with a discussion of the role of the noise model, notably the covariance matrices R_1 and R_2 on the filter performance. One possibility is to regard R_1 and R_2 just as the user's tuning knobs for achieving certain filters. By adjusting these matrices, the bandwidth and the shape of the filters, *etc.*, can be changed.

It can be seen from the ARE that it is the relation between R_1 and R_2 that plays a role, not the absolute values of these matrices. More precisely, if R_1 and R_2 are changed into αR_1, αR_2 (α being a scaling factor), then P will change into αP but the predictor gain K will remain unchanged. In the scalar output case, it is thus the normalized covariance matrix R_1/R_2 that will have an impact on the filter. In general terms, when R_1/R_2 is increased, the measurements are given larger weights and the filter bandwidth will increase. The idea is illustrated by the following example.

Example 6.3 Consider an AR(2) process with measurement noise, modelled as

$$x(t+1) = \begin{pmatrix} -a_1 & 1 \\ -a_2 & 0 \end{pmatrix} x(t) + \begin{pmatrix} 1 \\ 0 \end{pmatrix} v(t) ,$$

$$y(t) = (1 \ \ 0)x(t) + e(t) .$$

In this case, set

$$R_1 = \begin{pmatrix} r & 0 \\ 0 & 0 \end{pmatrix} , \qquad R_2 = 1 ,$$

and regard r as a tuning variable. The filter for estimating the undisturbed AR process from the current measurements turns out to be (*cf.* (6.100))

$$\hat{x}_1(t|t) = Hq[qI - (F - K_f HF)]^{-1} K_f y(t)$$

$$\triangleq G(q)y(t) .$$

The frequency properties of $G(q)$ are illustrated graphically in Figure 6.2.

□

6.8 Algorithms for Solving the Algebraic Riccati Equation

6.8.1 Introduction

An important aspect when finding the optimal filter in steady state is solving the ARE. This section is devoted to discussing that problem and to providing an algorithm in Section 6.8.2. See also Exercises 6.11, 6.12, 6.21–6.24.

First, it is shown that the general problem with $R_{12} \neq 0$ can be reduced to the somewhat simpler case where $R_{12} = 0$.

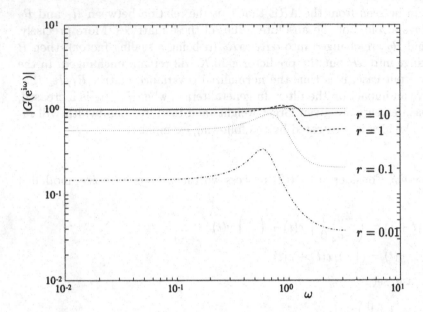

Fig. 6.2. Frequency function $G(e^{i\omega})$, Example 6.3, $a_1 = -1.5$, $a_2 = 0.7$, with r as a parameter

Lemma 6.4 *Consider the Riccati equation*

$$P(t+1) = FP(t)F^T + R_1 - [FP(t)H^T + R_{12}]$$
$$\times [HP(t)H^T + R_2]^{-1}[HP(t)F^T + R_{21}] . \qquad (6.112)$$

Set

$$\overline{F} = F - R_{12}R_2^{-1}H , \qquad (6.113)$$
$$\overline{R}_1 = R_1 - R_{12}R_2^{-1}R_{21} .$$

Then $P(t)$ satisfies

$$P(t+1) = \overline{F}P(t)\overline{F}^T + \overline{R}_1$$
$$-\overline{F}P(t)H^T[HP(t)H^T + R_2]^{-1}HP(t)\overline{F}^T . \qquad (6.114)$$

Proof It is apparently sufficient to show that the right-hand sides of (6.112) and (6.114) are identical for any $P(t)$. Set $P = P(t)$, $Q = HP(t)H^T + R_2$, $W = PH^TQ^{-1}HP$. The difference between the right-hand sides can then be evaluated as

$$FPF^T + R_1 - [FPH^T + R_{12}][HPH^T + R_2]^{-1}[HPF^T + R_{21}]$$
$$-\overline{F}P\overline{F}^T - \overline{R}_1 + \overline{F}PH^T[HPH^T + R_2]^{-1}HP\overline{F}^T$$

$$= FPF^T + R_1 - FWF^T - R_{12}Q^{-1}HPF^T - FPH^TQ^{-1}R_{21}$$
$$\quad -R_{12}Q^{-1}R_{21} - [FPF^T - R_{12}R_2^{-1}HPF^T - FPH^TR_2^{-1}R_{21}$$
$$\quad +R_{12}R_2^{-1}(Q - R_2)R_2^{-1}R_{21}]$$
$$\quad -[R_1 - R_{12}R_2^{-1}R_{21}]$$
$$\quad +[FWF^T - R_{12}R_2^{-1}HWF^T - FWH^TR_2^{-1}R_{21}$$
$$\quad +R_{12}R_2^{-1}HWH^TR_2^{-1}R_{21}]$$

$$= F[P - W - P + W]F^T$$
$$\quad +F[-PH^TQ^{-1} + PH^TR_2^{-1} - WH^TR_2^{-1}]R_{21}$$
$$\quad +R_{12}[-Q^{-1}HP + R_2^{-1}HP - R_2^{-1}HW]F^T$$
$$\quad +R_{12}[-Q^{-1} - R_2^{-1}QR_2^{-1} + R_2^{-1} + R_2^{-1} + R_2^{-1}HWH^TR_2^{-1}]R_{21}$$

$$= FPH^TQ^{-1}[-R_2 + Q - HPH^T]R_2^{-1}R_{21}$$
$$\quad +R_{12}R_2^{-1}[-R_2 + Q - HPH^T]Q^{-1}HPF^T$$
$$\quad +R_{12}R_2^{-1}[-R_2Q^{-1}R_2 - Q + 2R_2 + HPH^TQ^{-1}HPH^T]R_2^{-1}R_{21}$$

$$= R_{12}R_2^{-1}[-R_2Q^{-1}R_2 - Q + 2R_2 + (Q - R_2)Q^{-1}(Q - R_2)]R_2^{-1}R_{21}$$

$$= 0 \,. \qquad\qquad\qquad\qquad\qquad\qquad\qquad\qquad\qquad\qquad \blacksquare$$

When the time-varying problem is to be solved, a straightforward approach is to iterate the Riccati equation in (6.112) successively. However, such an attempt is not numerically sound, as rounding errors may accumulate in an uncontrollable fashion and destroy the result. In extreme cases the computed $P(t)$ matrix may even be indefinite. In such cases, the Kalman filter will deteriorate considerably. Instead, one should use square-root types of algorithm such as the U–D algorithm. The idea is that a positive definite covariance matrix $P(t)$ can be written as

$$P(t) = U(t)D(t)U^T(t) \,, \qquad\qquad\qquad\qquad (6.115)$$

where $D(t)$ is diagonal and $U(t)$ is lower triangular with unit diagonal elements. Instead of directly updating $P(t)$, the factors $U(t)$ and $D(t)$ are updated. In this way, the rounding errors are kept under control. By construction, $P(t)$ in (6.115) is constrained to be nonnegative definite.

6.8.2 An Algorithm Based on the Euler Matrix

It will first be shown how the Riccati equation is closely tied to a $2n$-dimensional linear system. In order not to complicate the analysis, the case of nonsingular F and R_2 is considered.

Lemma 6.5 *Consider the Riccati equation*

$$P(t+1) = FP(t)F^T + R_1 - FP(t)H^T[HP(t)H^T + R_2]^{-1}$$
$$\times HP(t)F^T , \qquad (6.116)$$
$$P(0) = P_0 .$$

Assume that F and R_2 are invertible. Consider also the system

$$\begin{pmatrix} Y(t+1) \\ Z(t+1) \end{pmatrix} = \begin{pmatrix} F^{-T} & F^{-T}H^T R_2^{-1}H \\ R_1 F^{-T} & F + R_1 F^{-T}H^T R_2^{-1}H \end{pmatrix} \begin{pmatrix} Y(t) \\ Z(t) \end{pmatrix} , \qquad (6.117)$$

with the initial condition

$$Y(0) = I , \qquad Z(0) = P_0 . \qquad (6.118)$$

Then

$$P(t) = Z(t)Y^{-1}(t) . \qquad (6.119)$$

Remark The matrix

$$\phi \overset{\triangle}{=} \begin{pmatrix} F^{-T} & F^{-T}H^T R_2^{-1}H \\ R_1 F^{-T} & F + R_1 F^{-T}H^T R_2^{-1}H \end{pmatrix} \qquad (6.120)$$

appearing in (6.117) is called the *Euler matrix*.   □

Proof Equation (6.119) is proved by induction. Apparently, it is true for $t = 0$. Assume that it is true for time t. Set $P = P(t)$. Using the matrix inversion lemma (Lemma 6.7), it is found that

$$Z(t+1)Y^{-1}(t+1)$$
$$= [R_1 F^{-T}Y(t) + \{F + R_1 F^{-T}H^T R_2^{-1}H\}PY(t)]$$
$$\times [F^{-T}Y(t) + F^{-T}H^T R_2^{-1}HPY(t)]^{-1}$$
$$= [R_1 F^{-T} + \{F + R_1 F^{-T}H^T R_2^{-1}H\}P][I + H^T R_2^{-1}HP]^{-1}F^T$$
$$= [R_1 F^{-T} + \{F + R_1 F^{-T}H^T R_2^{-1}H\}P]$$
$$\quad \times [I - H^T(R_2 + HPH^T)^{-1}HP]F^T$$
$$= R_1 + FPF^T + R_1 F^{-T}H^T R_2^{-1}HPF^T$$
$$\quad - [R_1 F^{-T} + \{F + R_1 F^{-T}H^T R_2^{-1}H\}P]H^T(R_2 + HPH^T)^{-1}HPF^T$$
$$= R_1 + FPF^T - FPH^T(R_2 + HPH^T)^{-1}HPF^T + R_1 F^{-T}H^T$$
$$\quad \times [R_2^{-1} - (R_2 + HPH^T)^{-1} - R_2^{-1}HPH^T(R_2 + HPH^T)^{-1}]HPF^T$$
$$= P(t+1) + R_1 F^{-T}H^T R_2^{-1}[(R_2 + HPH^T)$$
$$\quad - R_2 - HPH^T](R_2 + HPH^T)^{-1}HPF^T$$
$$= P(t+1) . \qquad ∎$$

Apparently, the solution to the ARE is closely tied to the properties of the system of (6.117) and the Euler matrix. In order to proceed, some properties of the Euler matrix are first investigated.

Lemma 6.6 *Consider the Euler matrix*

$$\phi \triangleq \begin{pmatrix} F^{-T} & F^{-T}H^T R_2^{-1}H \\ R_1 F^{-T} & F + R_1 F^{-T}H^T R_2^{-1}H \end{pmatrix}. \tag{6.121}$$

(a) Set $R_1 = GG^T$, $A(z) = \det(zI - F)$ and $H(zI - F)^{-1}G = B(z)/A(z)$. The characteristic polynomial of ϕ can be written as

$$\det[zI - \phi] = \frac{(-z)^n}{\det F} A(z)A(z^{-1}) \det\left[I + \frac{B^T(z^{-1})}{A(z^{-1})} R_2^{-1} \frac{B(z)}{A(z)} \right]. \tag{6.122}$$

(b) Let λ be a nonzero eigenvalue of ϕ. Then λ^{-1} is also an eigenvalue of ϕ.

Proof In order to prove part (a), apply Lemma 6.8, which gives

$$\det(zI - \phi) = \det\begin{pmatrix} zI - F^{-T} & -F^{-T}H^T R_2^{-1}H \\ -GG^T F^{-T} & zI - F - GG^T F^{-T}H^T R_2^{-1}H \end{pmatrix}$$

$$= \det(zI - F^{-T}) \det[zI - F - GG^T F^{-T}H^T R_2^{-1}H$$
$$\qquad - GG^T F^{-T}(zI - F^{-T})^{-1}F^{-T}H^T R_2^{-1}H]$$

$$= \det(F^{-1}) \det(zF^T - I)$$
$$\qquad \times \det[zI - F - GG^T F^{-T}(zI - F^{-T})^{-1}zH^T R_2^{-1}H]$$

$$= \frac{(-z)^n}{\det F} A(z^{-1})A(z)$$
$$\qquad \times \det[I - (zI - F)^{-1}GG^T F^{-T}(zI - F^{-T})^{-1}zH^T R_2^{-1}H]$$

$$= \frac{(-z)^n}{\det F} A(z^{-1})A(z)$$
$$\qquad \times \det[I - G^T(F^T - z^{-1}I)^{-1}H^T R_2^{-1}H(zI - F)^{-1}G]$$

$$= \frac{(-z)^n}{\det F} A(z^{-1})A(z) \det\left[I + \frac{B^T(z^{-1})}{A(z^{-1})} R_2^{-1} \frac{B(z)}{A(z)} \right],$$

which is (6.122). In the penultimate equality, Lemma 6.9 was applied. Part (b) is immediate from part (a). ∎

As

$$\begin{pmatrix} Y(t) \\ Z(t) \end{pmatrix} = \phi^t \begin{pmatrix} Y(0) \\ Z(0) \end{pmatrix} = \phi^t \begin{pmatrix} I \\ P_0 \end{pmatrix},$$

it can be seen that the asymptotic (as $t \to \infty$) properties of the solution to (6.117) and to the ARE are determined by the modes associated with eigenvalues with magnitude larger than one. Let

$$\begin{pmatrix} U_{11} & U_{12} \\ U_{21} & U_{22} \end{pmatrix}$$

denote a $2n|2n$ matrix, whose columns contain the eigenvectors of ϕ sorted so that the associated eigenvalues are decreasing.

As a consequence of Lemma 6.6, the modes associated with eigenvalues with magnitude larger than one constitute the matrix

$$\begin{pmatrix} U_{11} \\ U_{21} \end{pmatrix} .$$

Assume that U_{11} is invertible. It then follows from Lemma 6.5 that the solution to the Riccati equation converges to

$$P = U_{21} U_{11}^{-1} . \qquad (6.123)$$

The development so far has given a basis for the following algorithm.

Algorithm for solving the ARE

1. Compute the Euler matrix and perform an eigendecomposition. Collect the n eigenvectors associated with the eigenvalues of largest modulus in the matrix

$$\begin{pmatrix} U_{11} \\ U_{21} \end{pmatrix} .$$

2. Assuming that the inverse exists, set

$$P = U_{21} U_{11}^{-1} . \qquad (6.124)$$

□

See also Exercise 6.21 for some further details on the connection between the Euler matrix and the ARE.

The algorithm sketched above is limited in the sense that it is constrained to the case with both F and R_2 being nonsingular.

Standard software for solving the ARE is rather based on a *generalized* eigenvalue problem. Assume R_2 to be nonsingular but allow F to be arbitrary. Consider the matrix pencil

$$\lambda L - M \qquad (6.125)$$

where

$$L = \begin{pmatrix} I & H^T R_2^{-1} H \\ 0 & F \end{pmatrix} , \qquad M = \begin{pmatrix} F^T & 0 \\ -R_1 & I \end{pmatrix} . \qquad (6.126)$$

Then determine orthogonal matrices U and V such that

$$V(\lambda L - M)U = \lambda \hat{L} - \hat{M} \qquad (6.127)$$

with $\hat{L}$ and $\hat{M}$ being quasi-upper triangular (triangular if all eigenvalues are real-valued; complex-valued eigenvalues leads to 2×2 real-valued blocks on the diagonal). This is a standard step when solving generalized eigenvalue problems numerically.

Assume further that the generalized eigenvalues of (6.125) are sorted so that those inside the unit circle are in the upper block. Decompose the matrix U as

$$U = \begin{pmatrix} U_{11} & U_{12} \\ U_{21} & U_{22} \end{pmatrix} . \tag{6.128}$$

Then the sought solution of the ARE is

$$P = U_{21} U_{11}^{-1} . \tag{6.129}$$

Note the similarity with (6.124). Further, the corresponding generalized eigenvalues of the pencil (6.125) are precisely the eigenvalues of $F - KH$.

Remark There are techniques to handle the case of a singular R_2 matrix. One then starts with the pencil

$$\lambda \begin{pmatrix} I & 0 & 0 \\ 0 & F & 0 \\ 0 & -H & 0 \end{pmatrix} - \begin{pmatrix} F & 0 & H^T \\ -R_1 & I & 0 \\ 0 & 0 & R_2 \end{pmatrix}$$

and proceed by first compressing it to an "equivalent" matrix pencil of order $(2n)|(2n)$.   □

6.A Appendix. Proofs

6.A.1 The Matrix Inversion Lemma

The following lemma is a useful tool in linear system and estimation theory.

Lemma 6.7 *Assuming that the involved matrices have compatible dimensions and that the indicated inverses exist:*

$$(A + BCD)^{-1} = A^{-1} - A^{-1} B (C^{-1} + DA^{-1} B)^{-1} DA^{-1} \tag{6.130}$$

holds.

Proof By direct calculations:

$$(A + BCD)[A^{-1} - A^{-1} B (C^{-1} + DA^{-1} B)^{-1} DA^{-1}]$$
$$= I + BCDA^{-1} - B(C^{-1} + DA^{-1} B)^{-1} DA^{-1}$$
$$\quad - BCDA^{-1} B (C^{-1} + DA^{-1} B)^{-1} DA^{-1}$$
$$= I + B[C(C^{-1} + DA^{-1} B) - I - CDA^{-1} B]$$
$$\quad \times (C^{-1} + DA^{-1} B)^{-1} DA^{-1}$$
$$= I ,$$

which verifies (6.130).   ■

6.A.2 Proof of Theorem 6.1

That the conditional pdfs are Gaussian is proved by induction. According to the assumptions, this is at least true for $t = t_0$. In that case:

$$\hat{x}(t_0|t_0) = m , \qquad P(t_0|t_0) = R_0 . \tag{6.131}$$

Assuming that (6.30) holds for a particular value of t, it will be seen that it holds for $t + 1$ as well. Recursions for the conditional means $\hat{x}(t|t)$, $\hat{x}(t + 1|t)$ and the conditional covariance matrices $P(t|t)$, $P(t + 1|t)$ are derived. Recall that if x is Gaussian distributed, $x \sim N(m, P)$, then the notation

$$p(x) = \gamma(x; m, P)$$

is used.

Time update

Consider first the *time update*, that is, examine $p(x(t + 1)|Y^t)$. In order to do so, the results of Section 5.4 will be used.

First note that

$$\begin{aligned}
p(x(t + 1)|x(t)) &= p_{v(t)}(x(t + 1) - Fx(t) - Gu(t)) \\
&= \gamma(x(t + 1); Fx(t) + Gu(t) , R_1) .
\end{aligned}$$

Hence, according to (5.23):

$$p(x(t + 1)|Y^t) = \int p(x(t + 1)|x(t))p(x(t)|Y^t)\, dx(t)$$

$$= \int \frac{1}{(2\pi)^{n/2}(\det R_1)^{1/2}}$$

$$\exp\left(-\frac{1}{2}[x(t + 1) - Fx - Gu(t)]^T R_1^{-1}[x(t + 1) - Fx + Gu(t)]\right)$$

$$\times \frac{1}{(2\pi)^{n/2}(\det P(t|t))^{1/2}} \exp\left(-\frac{1}{2}[x - \hat{x}(t|t)]^T P^{-1}(t|t)[x - \hat{x}(t|t)]\right) dx . \tag{6.132}$$

This relation can be rewritten as

$$p(x(t + 1)|Y^t) = \frac{1}{(2\pi)^n (\det R_1)^{1/2}(\det P(t|t))^{1/2}}$$

$$\times \int \exp\left(-\frac{1}{2}[x^T Q^{-1}x - x^T b - b^T x + c]\right) dx , \tag{6.133}$$

where

$$\begin{aligned}
Q^{-1} &= P^{-1}(t|t) + F^T R_1^{-1}F , \\
b &= F^T R_1^{-1}(x(t + 1) - Gu(t)) + P^{-1}(t|t)\hat{x}(t|t) , \\
c &= (x^T(t + 1) - u^T(t)G^T)R_1^{-1}(x(t + 1) - Gu(t)) \\
&\quad + \hat{x}^T(t|t)P^{-1}(t|t)\hat{x}(t|t) .
\end{aligned} \tag{6.134}$$

Recalling the form of the Gaussian pdf, the integral in (6.133) can be evaluated as follows:

$$\int \exp\left(-\frac{1}{2}[x^T Q^{-1}x - x^T b - b^T x + c]\right) dx$$

$$= (2\pi)^{n/2}(\det Q)^{1/2} \int \frac{1}{(2\pi)^{n/2}(\det Q)^{1/2}}$$

$$\times \exp\left(-\frac{1}{2}[x - Qb]^T Q^{-1}[x - Qb]\right) dx \exp\left(\frac{1}{2}(b^T Qb - c)\right)$$

$$= (2\pi)^{n/2}(\det Q)^{1/2} \exp\left(\frac{1}{2}(b^T Qb - c)\right). \tag{6.135}$$

For the particular case given by (6.132), one obtains using (6.134) and (6.135):

$$p(x(t+1)|Y^t) = \frac{(\det Q)^{1/2}}{(2\pi)^{n/2}(\det R_1)^{1/2}(\det P(t|t))^{1/2}} e^{\frac{1}{2}(b^T Qb - c)}. \tag{6.136}$$

In this case

$$\frac{\det R_1 \times \det P(t|t)}{\det Q} = \det R_1 \times \det P(t|t) \times \det\left(P^{-1}(t|t) + F^T R_1^{-1} F\right)$$

$$= \det R_1 \det\left(I + P(t|t)F^T R_1^{-1} F\right)$$

$$= \det R_1 \det\left(I + F P(t|t)F^T R_1^{-1}\right)$$

$$= \det\left(F P(t|t)F^T + R_1\right)$$

$$= \det P(t+1|t) \tag{6.137}$$

also holds, where

$$P(t+1|t) \stackrel{\triangle}{=} F P(t|t)F^T + R_1 \tag{6.138}$$

was introduced. It is also true that

$$b^T Qb - c$$
$$= [(x^T(t+1) - u^T(t)G^T)R_1^{-1}F + \hat{x}^T(t|t)P^{-1}(t|t)][P^{-1}(t|t) + F^T R_1 F]^{-1}$$
$$\times [F^T R_1^{-1}(x(t+1) - Gu(t)) + P^{-1}(t|t)\hat{x}(t|t)]$$
$$- (x^T(t+1) - u^T(t)G^T)R_1^{-1}(x(t+1) - Gu(t)) - \hat{x}^T(t|t)P^{-1}(t|t)\hat{x}(t|t)$$
$$= (x(t+1) - Gu(t))^T \{R_1^{-1}F[P^{-1}(t|t) + F^T R_1^{-1} F]^{-1}$$
$$\times F^T R_1^{-1} - R_1^{-1}\}(x(t+1) - Gu(t))$$
$$+ (x(t+1) - Gu(t))^T R_1^{-1}F[P^{-1}(t|t) + F^T R_1^{-1} F]^{-1}P^{-1}(t|t)\hat{x}(t|t)$$
$$+ \hat{x}^T(t|t)P^{-1}(t|t)[P^{-1}(t|t) + F^T R_1^{-1} F]^{-1}F^T R_1^{-1}(x(t+1) - Gu(t))$$
$$+ \hat{x}^T(t|t)\{P^{-1}(t|t)[P^{-1}(t|t) + F^T R_1^{-1} F]^{-1}P^{-1}(t|t) - P^{-1}(t|t)\}\hat{x}(t|t).$$

Next evaluate the matrices of the above quadratic forms, using the matrix inversion lemma (Lemma 6.7) repeatedly:

$$R_1^{-1}F[P^{-1}(t|t) + F^T R_1^{-1} F]^{-1}F^T R_1^{-1} - R_1^{-1}$$
$$= -(R_1 + F P(t|t)F^T)^{-1}$$
$$= -P^{-1}(t+1|t)R_1^{-1}F[P^{-1}(t|t) + F^T R_1 F]^{-1}P^{-1}(t|t)$$
$$= R_1^{-1}F[I + P(t|t)F^T R_1^{-1} F]^{-1}$$
$$= [I + R_1^{-1}F P(t|t)F^T]^{-1}R_1^{-1}F$$
$$= P^{-1}(t+1|t)F P^{-1}(t|t)[P^{-1}(t|t) + F^T R_1^{-1} F]^{-1}P^{-1}(t|t) - P^{-1}(t|t)$$
$$= [P(t|t) + P(t|t)F^T R_1^{-1}F P(t|t)]^{-1} - P^{-1}(t|t)$$

$$= -F^T[R_1 + FP(t|t)F^T]^{-1}F$$
$$= -F^T P^{-1}(t+1|t)F \ .$$

Thus

$$b^T Qb - c = -[x(t+1) - Gu(t) - F\hat{x}(t|t)]^T P^{-1}(t+1|t) \\ \times [x(t+1) - Gu(t) - F\hat{x}(t|t)] \ . \tag{6.139}$$

Now, using (6.137), (6.138) and (6.139) in (6.136) gives

$$p(x(t+1)|Y^t) = \frac{1}{(2\pi)^{n/2}} \frac{1}{(\det P(t+1|t))^{1/2}}$$

$$\times \exp\left(-\frac{1}{2}[x(t+1) - \hat{x}(t+1|t)]^T P^{-1}(t+1|t)[x(t+1) - \hat{x}(t+1|t)]\right)$$

$$= \gamma(x(t+1); \hat{x}(t+1|t), P(t+1|t)) \ , \tag{6.140}$$

where

$$\hat{x}(t+1|t) = F\hat{x}(t|t) + Gu(t) \ . \tag{6.141}$$

Measurement update

The *measurement update* must be considered next, that is, how to find $p(x(t+1)|Y^{t+1})$. To do this, (5.21), rather than (5.26), will be used, as $p(x(t+1)|Y^t)$ is known. In fact, all the conditional pdfs appearing in (5.21) are known:

$$\begin{cases} p(y(t+1)|x(t+1)) = \gamma(y(t+1); Hx(t+1), R_2) \ , \\ p(y(t+1)|Y^t) = \gamma(y(t+1); H\hat{x}(t+1|t), HP(t+1|t)H^T + R_2) \ , \\ p(x(t+1)|Y^t) = \gamma(x(t+1); \hat{x}(t+1|t), P(t+1|t)) \ . \end{cases} \tag{6.142}$$

Inserting the explicit expressions for the pdfs, gives

$$p(x(t+1)|Y^{t+1}) = \frac{1}{(2\pi)^{ny/2}(\det R_2)}$$

$$\times \exp\left(-\frac{1}{2}[y(t+1) - Hx(t+1)]^T R_2^{-1}[y(t+1) - Hx(t+1)]\right)$$

$$\times \frac{(2\pi)^{ny/2}(\det[HP(t+1|t)H^T + R_2])^{1/2}}{D}$$

$$\times \frac{1}{(2\pi)^{n/2}(\det P(t+1|t))^{1/2}}$$

$$\times \exp\left(-\frac{1}{2}[x(t+1) - \hat{x}(t+1|t)]^T P^{-1}(t+1|t)[x(t+1) - \hat{x}(t+1|t)]\right) \ ,$$

where the denominator factor D is given by

$$D = \exp\left(-\frac{1}{2}[y(t+1) - H\hat{x}(t+1|t)]^T[HP(t+1|t)H^T + R_2]^{-1}\right.$$

$$\left. \times [y(t+1) - H\hat{x}(t+1|t)]\right) \ . \tag{6.143}$$

Now introduce, as in (6.30)–(6.33),

$$\begin{cases} \hat{x}(t+1|t+1) = \hat{x}(t+1|t) + K(t)[y(t+1) - H\hat{x}(t+1|t)] \,, \\ \quad K(t) = P(t+1|t)H^T[HP(t+1|t)H^T + R_2]^{-1} \,, \\ P(t+1|t+1) = P(t+1|t) - P(t+1|t) \\ \qquad \times H^T[HP(t+1|t)H^T + R_2]^{-1}HP(t+1|t) \,. \end{cases} \quad (6.144)$$

The "nonexponential factor" of (6.143) is examined first. Using (6.144), it is found that

$$\begin{aligned} \det[P(t+1|t+1)] &= \det(P(t+1|t)) \\ &\quad \times \det[I - H^T(HP(t+1|t)H^T + R_2)^{-1}HP(t+1|t)] \\ &= \det(P(t+1|t)) \\ &\quad \times \det[I - HP(t+1|t)H^T\{HP(t+1|t)H^T + R_2\}^{-1}] \\ &= \frac{\det(P(t+1|t))\det(R_2)}{\det(HP(t+1|t)H^T + R_2)} \,. \end{aligned}$$

Thus the "nonexponential factor" of (6.143) is precisely

$$\frac{1}{(2\pi)^{n/2}} \frac{1}{(\det P(t+1|t+1))^{1/2}} \,.$$

Next, the exponential factors are considered. For brevity, the notation

$$\begin{aligned} K &= K(t) \,, \\ Q &= HP(t+1|t)H^T + R_2 \,, \\ P &= P(t+1|t) \,, \\ \hat{x} &= \hat{x}(t+1|t) \,. \end{aligned}$$

is used. The total exponent appearing in (6.143), save for the factor $-1/2$, can be written as

$$\begin{aligned} &[y(t+1) - Hx(t+1)]^T R_2^{-1}[y(t+1) - Hx(t+1)] \\ &- [y(t+1) - H\hat{x}]^T Q^{-1}[y(t+1) - H\hat{x}] + [x(t+1) - \hat{x}]^T P^{-1}[x(t+1) - \hat{x}] \\ &= x^T(t+1)[H^T R_2^{-1}H + P^{-1}]x(t+1) - x^T(t+1)[H^T R_2^{-1}y(t+1) + P^{-1}\hat{x}] \\ &\quad + [H^T R_2^{-1}y(t+1) + P^{-1}\hat{x}]^T x(t+1) \\ &\quad + \{y^T(t+1)R_2^{-1}y(t+1) - [y(t+1) - H\hat{x}]^T Q^{-1}[y(t+1) - H\hat{x}] \\ &\quad + \hat{x}^T P^{-1}\hat{x}\} \,. \end{aligned} \quad (6.145)$$

Applying the matrix inversion lemma to $P(t+1|t+1)$ in (6.144) gives

$$P^{-1}(t+1|t+1) = P^{-1} + H^T R_2^{-1}H \,. \quad (6.146)$$

The right-hand side in this relation appears in the first term of the right-hand side of (6.145). To examine the second and third terms of (6.145), write

$$\begin{aligned} &H^T R_2^{-1}y(t+1) + P^{-1}\hat{x} \\ &= P^{-1}(t+1|t+1)[P - PH^TQ^{-1}HP][H^T R_2^{-1}y(t+1) + P^{-1}\hat{x}] \\ &= P^{-1}(t+1|t+1) \\ &\quad \times [PH^T(I - Q^{-1}HPH^T)R_2^{-1}y(t+1) + (I - PH^TQ^{-1}H)\hat{x}] \\ &= P^{-1}(t+1|t+1)[PH^TQ^{-1}(Q - HPH^T)R_2^{-1}y(t+1) + \hat{x} - KH\hat{x}] \\ &= P^{-1}(t+1|t+1)[Ky(t+1) + \hat{x} - KH\hat{x}] \\ &= P^{-1}(t+1|t+1)\hat{x}(t+1|t+1) \,. \end{aligned}$$

Finally, the last term of (6.145) is investigated. It can be written as

$$y^T(t+1)R_2^{-1}y(t+1) - [y(t+1) - H\hat{x}]^T Q^{-1}[y(t+1) - H\hat{x}] + \hat{x}^T P^{-1}\hat{x}$$

$$= y^T(t+1)[R_2^{-1} - Q^{-1}]y(t+1) + y^T(t+1)Q^{-1}H\hat{x} + \hat{x}^T H^T Q^{-1}y(t+1)$$

$$+\hat{x}^T[P^{-1} - H^T Q^{-1}H]\hat{x} . \tag{6.147}$$

Now compare this expression with what is expected, namely:

$$\hat{x}^T(t+1|t+1)P^{-1}(t+1|t+1)\hat{x}^T(t+1|t+1)$$

$$= [Ky(t+1) + (I - KH)\hat{x}]^T[P^{-1} + H^T R_2^{-1}H][Ky(t+1) + (I - KH)\hat{x}] . \tag{6.148}$$

It has to be shown that (6.147) and (6.148) are identical. To do so, examine the difference of the matrices in the quadratic forms. For the quadratic forms in $y(t+1)$:

$$K^T(P^{-1} + H^T R_2^{-1}H)K - R_2^{-1} + Q^{-1}$$

$$= Q^{-1}HP(P^{-1} + H^T R_2^{-1}H)PH^T Q^{-1} - R_2^{-1} + Q^{-1}$$

$$= Q^{-1}[HPH^T + HPH^T R_2^{-1}HPH^T - QR_2^{-1}Q + Q]Q^{-1}$$

$$= Q^{-1}[(Q - R_2) + (Q - R_2)R_2^{-1}(Q - R_2) - QR_2^{-1}Q + Q]Q^{-1}$$

$$= 0 .$$

For the quadratic form in $\hat{x}$:

$$(I - H^T K^T)(P^{-1} + H^T R_2^{-1}H)(I - KH) - P^{-1} + H^T Q^{-1}H$$

$$= P^{-1} + H^T R_2^{-1}H - H^T Q^{-1}H - H^T Q^{-1}HPH^T R_2^{-1}H$$

$$\quad -H^T Q^{-1}H - H^T R_2^{-1}HPH^T Q^{-1}H + H^T Q^{-1}HPH^T Q^{-1}H$$

$$\quad +H^T Q^{-1}HPH^T R_2^{-1}HPH^T Q^{-1}H - P^{-1} + H^T Q^{-1}H$$

$$= H^T Q^{-1}[QR_2^{-1}Q - HPH^T R_2^{-1}Q - Q - QR_2^{-1}HPH^T$$

$$\quad +HPH^T + HPH^T R_2^{-1}HPH^T]Q^{-1}H$$

$$= H^T Q^{-1}[QR_2^{-1}Q - (Q - R_2)R_2^{-1}Q - Q - QR_2^{-1}(Q - R_2)$$

$$\quad +(Q - R_2) + (Q - R_2)R_2^{-1}(Q - R_2)]Q^{-1}H$$

$$= 0 .$$

Similarly, for the mixed form:

$$K^T(P^{-1} + H^T R_2^{-1}H)(I - KH) - Q^{-1}H$$

$$= Q^{-1}[HP(P^{-1} + H^T R_2^{-1}H - H^T Q^{-1}H - H^T R_2^{-1}HPH^T Q^{-1}H) - H]$$

$$= Q^{-1}[I + HPH^T R_2^{-1} - HPH^T Q^{-1} - HPH^T R_2^{-1}HPH^T Q^{-1} - I]H$$

$$= Q^{-1}HPH^T[R_2^{-1} - Q^{-1} - R_2^{-1}(Q - R_2)Q^{-1}]H$$

$$= 0 .$$

Thus it has been established that the expressions in (6.147) and (6.148) are identical. Inserting the expressions found into (6.143), it can be seen that

$$p(x(t+1)|Y^{t+1}) = \gamma(x(t+1); \hat{x}(t+1|t+1), P(t+1|t+1)) , \tag{6.149}$$

where the conditional mean $\hat{x}(t+1|t+1)$ and the conditional covariance matrix $P(t+1|t+1)$ are given by (6.144).

This concludes the whole proof of the theorem.   ∎

6.A.3 Two Determinant Results

This section presents two results related to the determinant of a partitioned matrix.

Lemma 6.8 *Let A and D be square matrices and assume that A is invertible. Then*

$$\det \begin{pmatrix} A & B \\ C & D \end{pmatrix} = \det(A) \det(D - CA^{-1}B) . \tag{6.150}$$

Proof The result follows by straightforward calculation as follows:

$$\det \begin{pmatrix} A & B \\ C & D \end{pmatrix} = \det \begin{pmatrix} A & B \\ C & D \end{pmatrix} \begin{pmatrix} I & -A^{-1}B \\ 0 & I \end{pmatrix}$$

$$= \det \begin{pmatrix} A & 0 \\ C & D - CA^{-1}B \end{pmatrix}$$

$$= \det(A) \det(D - CA^{-1}B) . \tag{6.151}$$

∎

Lemma 6.9 *Let A be an $(n|p)$ matrix and B an $(p|n)$ matrix. Then*

$$\det(I_n + AB) = \det(I_p + BA) . \tag{6.152}$$

Proof Simple calculation gives

$$\det \begin{pmatrix} I_n & A \\ -B & I_p \end{pmatrix} = \det \left[\begin{pmatrix} I_n & A \\ -B & I_p \end{pmatrix} \begin{pmatrix} I_n & 0 \\ B & I_p \end{pmatrix} \right] = \det \begin{pmatrix} I_n + AB & A \\ 0 & I_p \end{pmatrix}$$

$$= \det(I_n + AB) \det I_p = \det(I_n + AB) .$$

However,

$$\det \begin{pmatrix} I_n & A \\ -B & I_p \end{pmatrix} = \det \left[\begin{pmatrix} I_n & 0 \\ B & I_p \end{pmatrix} \begin{pmatrix} I_n & A \\ -B & I_p \end{pmatrix} \right] = \det \begin{pmatrix} I_n & A \\ 0 & BA + I_p \end{pmatrix}$$

$$= \det I_n \det(BA + I_p) = \det(I_p + BA)$$

also holds.

∎

Exercises

Exercise 6.1 Consider a stochastic system with noise-free measurements

$$x(t + 1) = Fx(t) + v(t) ,$$
$$y(t) = Hx(t) ,$$

$v(t)$ being white noise and $\dim y < \dim x$. Show explicitly that

$$\hat{y}(t|t) = H\hat{x}(t|t) = y(t) .$$

Exercise 6.2 Prove the inequality

$$HP(t|t)H^T \le R_2$$

and give an interpretation. Under what conditions does equality hold?

Hint. One can show that for A, B symmetric and nonnegative definite

$$A - A(A + B)^{-1}A \le B .$$

Exercise 6.3 Consider the system

$$\dot{x} = Ax + e ,$$
$$y = Cx ,$$

where

$$\mathbf{E}\, e(t)e^T(s) = R\delta(t - s) .$$

Determine the LLMS optimal predictor $\hat{y}(t + \tau|t)$, $\tau > 0$, of $y(t + \tau)$ given the state vector up to time t.

Exercise 6.4 Consider the following very simplified problem of estimating the probability of the rendezvous of two vehicles. Both vehicles move in the x–y plane. One vehicle is known to move along the x-axis according to $x(t) = t$, $y(t) = 0$. The movement of the other vehicle is a bit uncertain and is modelled as

$$z(t + 1) = \begin{pmatrix} 1 & 1 & 0 & 0 \\ 0 & 1 & 0 & 0 \\ 0 & 0 & 1 & 1 \\ 0 & 0 & 0 & 1 \end{pmatrix} z(t) + v(t) ,$$

$$\begin{pmatrix} x(t) \\ y(t) \end{pmatrix} = \begin{pmatrix} 1 & 0 & 0 & 0 \\ 0 & 0 & 1 & 0 \end{pmatrix} z(t) + e(t) ,$$

where $v(t)$ and $e(t)$ are white noise sequences of zero mean and covariance matrices

$$R_1 = r_a \begin{pmatrix} 0 & 0 & 0 & 0 \\ 0 & 1 & 0 & 0 \\ 0 & 0 & 0 & 0 \\ 0 & 0 & 0 & 1 \end{pmatrix} , \qquad R_2 = r_e \begin{pmatrix} 1 & 0 \\ 0 & 1 \end{pmatrix} .$$

Simulate the system for $t = 0, 1, \ldots, N$. At each time t compute the estimates $\hat{z}(s|t)$, $s = t, t + 1, \ldots, N$ and assess the risk that the two vehicles will meet. Try, for example, the numerical values $N = 15$, $r_a = 0.001$, $r_e = 1$

$$\mathbf{E}\, z(0) = \begin{pmatrix} 0 \\ 1 \\ 10 \\ -1 \end{pmatrix} , \qquad \text{cov}(z(0)) = 10 \times I .$$

Exercise 6.5 Show algebraically that the following two representations of the stationary optimal filter are equivalent. Assume that $R_{12} = 0$:

(i) $\hat{x}(t|t) = [qI - (F - K_f H F)]^{-1}[(I - K_f H)Gu(t) + K_f y(t+1)]$.

(ii) $\hat{x}(t|t) = (I - K_f H)[qI - (F - KH)]^{-1}[Gu(t) + Ky(t)] + K_f y(t)$.

Exercise 6.6 Consider the system

$$x(t+1) = \begin{pmatrix} -a & 1 \\ 0 & 0 \end{pmatrix} x(t) + \begin{pmatrix} 1 \\ c \end{pmatrix} v(t) , \qquad E\, v(t)v(s) = \delta_{t,s} ,$$

$$y(t) = \begin{pmatrix} 1 & 0 \end{pmatrix} x(t) ,$$

where $c^2 \neq 1$.

(a) Determine *all* symmetric solutions to the ARE. Determine which solution is related to optimal prediction!

Hint. One can first show that

$$P = \begin{pmatrix} 1+q & c \\ c & c^2 \end{pmatrix} ,$$

and then derive a nonlinear equation in the *scalar* variable q.

(b) Assume that the Riccati equation

$$P(t+1) = FP(t)F^T + R_1 - FP(t)H^T[HP(t)H^T]^{-1}HP(t)F^T ,$$

$$P(0) = P_0 \geq 0$$

is iterated until convergence. What are the stability properties for the stationary solutions?

Hint. Show first that all solutions have the form

$$P(t) = \begin{pmatrix} 1+q(t) & c \\ c & c^2 \end{pmatrix} , \qquad t \geq 1 ,$$

where $q(t)$ satisfies a certain nonlinear difference equation, say $q(t+1) = f(q(t))$. Assume that $q(t) = \bar{q}$ is a stationary solution. Its stability properties are determined by the stability properties of the linearized model

$$q(t+1) - \bar{q} = \frac{\partial f(q)}{\partial q}\bigg|_{q=\bar{q}} (q(t) - \bar{q}) .$$

Exercise 6.7 In many cases measurement devices contain a bias. As an illustration of how to handle this, consider the system

$$x(t+1) = Fx(t) + v(t) ,$$

$$y(t) = Hx(t) + m + e(t) ,$$

where m denotes the measurement bias and $v(t)$ and $e(t)$ are uncorrelated white noise sequences of zero mean.

(a) Assume that m is unknown. Show that mean square optimal estimates of the states and m can be obtained by introducing an enlarged model using

$$\bar{x}(t) = \begin{pmatrix} x(t) \\ m \end{pmatrix}$$

as state vector.

(b) Consider explicitly the scalar case

$$F = -a ,$$
$$H = 1 ,$$
$$e(t) \equiv 0 ,$$
$$\mathbf{E}\, v^2(t) = r .$$

Describe the uncertainty of m as a random walk:

$$m(t+1) = m(t) + v_m(t) ,$$

where $v_m(t)$ is white noise of zero mean and variance r_b.

Show that optimal estimates of $x(t)$ and $m(t)$ can be obtained in the stationary phase as

$$\hat{x}(t+1|t) = \frac{-a(1-\alpha)(q-1)}{q + (a\alpha - 1 + \alpha)} y(t) ,$$

$$\hat{x}(t|t) = \frac{(1-\alpha)(q-1)}{q + (a\alpha - 1 + \alpha)} y(t) ,$$

$$\hat{m}(t+1|t) = \hat{m}(t|t) = \frac{\alpha(q+a)}{q + (a\alpha - 1 + \alpha)} y(t) ,$$

where α is a parameter. Show also how α depends on the model parameters a and r_b/r. What are the static gains of the filters?

(c) Compute numerically the frequency functions of the filters in part (b). Examine how the model parameters a and r_b/r influence the behaviour.

Exercise 6.8 Consider the usual state estimation problem for

$$x(t+1) = Fx(t) + v(t) ,$$
$$y(t) = Hx(t) + e(t) ,$$

where $v(t)$ and $e(t)$ are uncorrelated. In stationarity,

$$\hat{x}(t+1|t) = G_p(q)y(t) ,$$
$$\hat{x}(t|t) = G_f(q)y(t) .$$

Show how the transfer functions $G_p(q)$ and $G_f(q)$ depend on the model and the filter gain $K_f = PH^T(HPH^T + R_2)^{-1}$. Also prove the relation

$$G_p(q) = FG_f(q) .$$

Exercise 6.9 Reconsider Exercise 4.17, where the ARE was shown to have a positive semidefinite, and hence singular, solution. Find out why a positive definite solution as guaranteed in Theorem 6.2 does not exist in this case.

Hint. The transformation in (6.113) may be useful as a first step.

Exercise 6.10 Consider fixed lag smoothing in the the stationary case. Find explicit expressions for $\overline{K}_f$ (6.63) and $\overline{P}$ (6.69).

Exercise 6.11 Consider the ARE

$$P = FPF^T + R_1 - FPH^T(HPH^T + R_2)^{-1}HPF^T \ .$$

Assume that there are two positive definite solutions, P_1 and P_2, such that

$$\overline{F}_i = F - FP_iH^T(HP_iH^T + R_2)^{-1}H \qquad (i = 1, 2)$$

has all eigenvalues strictly inside the unit circle.

(a) Prove that

$$\overline{F}_1(P_1 - P_2)\overline{F}_2^T = (P_1 - P_2) \ .$$

(b) Use the result of (a) to prove that $P_1 = P_2$, that is, there exists at most one positive definite solution to the ARE.

Exercise 6.12 Consider the system

$$x(t + 1) = Fx(t) \ ,$$
$$y(t) = Hx(t) + e(t) \ ,$$

where $e(t)$ is white noise of zero mean and covariance R_2. Assume that all eigenvalues of F satisfy

$$0 < |\lambda_i(F)| < 1 \ .$$

Show that the solution to the associated Riccati equation can be written

$$P(t) = F^t[P^{-1}(0) + \sum_{j=0}^{t-1}(F^T)^jH^TR_2^{-1}HF^j]^{-1}(F^T)^t \ ,$$

and that $P(t) \to 0$ as $t \to \infty$.

Hint. One can first derive a linear difference equation for $P^{-1}(t)$.

Exercise 6.13 Consider the continuous-time system

$$\dot{x} = Ax + v \ ,$$
$$\mathbf{E}\,v(t)v^T(s) = R_c\delta(t - s) \ ,$$

with discrete-time measurements (e and v being independent)

$$y(kh) = Hx(kh) + e(kh) \quad (k \text{ integer}) ,$$
$$\mathbf{E}\, e(kh)e^T(jh) = R_2\delta_{k,j} .$$

Derive the "continuous–discrete" Kalman filter, that is, the optimal state estimate of $x(t)$ based on all available measurements of $y(kh)$ available at time t (*i.e.* all $k < t/h$).

Examine, for the stationary, time-invariant case, how the covariance matrix of the optimal state estimate varies over the sampling interval.

Exercise 6.14 It seems intuitively reasonable that a prediction k-steps ahead should have worse accuracy than a prediction $(k - 1)$-steps ahead. Consider the statements

- $\hat{x}(t + k|t)$ has worse accuracy than $\hat{x}(t + k|t + 1)$ (*i.e.* $P(t + k|t) \geq P(t+k\text{---}t+1)$).
- $\hat{x}(t + k|t)$ has worse accuracy than $\hat{x}(t + k - 1|t)$ (*i.e.* $P(t + k|t) \geq P(t+k-1\text{---}t)$).

For each statement, either prove it or construct a counterexample.

Exercise 6.15 Consider the dynamic system

$$x(t + 1) = \begin{pmatrix} 1 & h \\ 0 & 1 \end{pmatrix} x(t) + \begin{pmatrix} 0 \\ 1 \end{pmatrix} v(t) ,$$
$$y(t) = \begin{pmatrix} 1 & 0 \end{pmatrix} x(t) ,$$

where $v(t)$ is Gaussian white noise. Determine the optimal state estimates $\hat{x}(t|t - 1)$ and $\hat{x}(t|t)$ for arbitrary values of t.

Hint. When solving the Riccati equation, it is useful to introduce the scalar

$$\alpha(t) = p_{22}(t) - p_{12}^2(t)/p_{11}(t) .$$

Exercise 6.16 Consider the output prediction errors $\{\tilde{y}(t)\}$, (6.4). Prove that they are white, in the sense that they are uncorrelated

$$\mathbf{E}\, \tilde{y}(t + k)\tilde{y}(t) = 0 , \quad \text{for } k > 0 .$$

Exercise 6.17 Consider a scalar system

$$x(t + 1) = 0.8x(t) + v(t) ,$$
$$y(t) = x(t) + e(t) ,$$

with $r_1 = \mathbf{E}\, v^2(t) = 0.68$, $r_2 = \mathbf{E}\, e^2(t) = 1$.

(a) Determine the signal-to-noise ratio (SNR) $= \mathbf{E}\, x^2(t)/\mathbf{E}\, e^2(t)$.
(b) Determine the innovations form.

(c) Assume that a smoothing estimate $\hat{x}(t|t+m)$, $m > 0$ is sought. Show that it can be written as

$$\hat{x}(t|t+m) = \sum_{j=-m}^{\infty} h_j \tilde{y}(t-j)$$

where $\tilde{y}(t)$ is the innovation at time t. Determine the weighting coefficients $\{h_j\}$. At what rate does $\{h_j\}$ decay when $j \to \infty$, and what is the rate when $j \to -m$?

(d) Show that the smoothing estimate in part (c) can be written as

$$\hat{x}(t|t+m) = \sum_{j=-m}^{\infty} g_j y(t-j)$$

and determine the weighting coefficients $\{g_j\}$.

Exercise 6.18 Consider the system

$$x(t+1) = Fx(t) + v(t)\,,$$
$$y(t) = Hx(t) + e(t)\,,$$

where $v(t)$ and $e(t)$ are mutually independent and Gaussian distributed white noise sequences with zero mean and covariances R_1 and R_2 respectively. The initial value, $x(0)$, is assumed to be Gaussian distributed, $x(0) \sim \mathbf{N}(m_0, R_0)$, and independent of the noise sequences.

In the basic formulation of the optimal state estimation, the covariance matrix of total data vector

$$Y^t = \begin{pmatrix} y(0) \\ y(t) \\ \vdots \\ y(t) \end{pmatrix}$$

has an important role.

(a) Show that

$$Y^t = \begin{pmatrix} H \\ HF \\ \vdots \\ HF^t \end{pmatrix} x(0) + \begin{pmatrix} e(0) \\ \vdots \\ e(t) \end{pmatrix} + \begin{pmatrix} 0 & & & \\ H & 0 & & \\ HF & H & & \\ \vdots & & \ddots & \\ HF^{t-1} & & & H & 0 \end{pmatrix} \begin{pmatrix} v(0) \\ \vdots \\ v(t) \end{pmatrix}\,,$$

and hence

$$R \triangleq \mathbf{E}\, Y^t Y^{t^T} = \begin{pmatrix} H \\ HF \\ \vdots \\ HF^t \end{pmatrix} R_0 (H^T \dots F^{T^t} H^T) + \begin{pmatrix} R_2 & & 0 \\ & \ddots & \\ 0 & & R_2 \end{pmatrix}$$

$$+ \begin{pmatrix} 0 & & & \\ H & & & \\ HF & H & 0 & \\ \vdots & & \ddots & \\ HF^{t-1} & & & H\ 0 \end{pmatrix} \begin{pmatrix} R_1 & & 0 \\ & \ddots & \\ 0 & & R_1 \end{pmatrix}$$

$$\times \begin{pmatrix} 0 & H^T & F^T H^T & \cdots & F^{T^{t-1}} H^T \\ & 0 & H^T & & \\ & & & \ddots & \\ & & & & H^T \\ 0 & & & & 0 \end{pmatrix}.$$

(b) Show using the Kalman filter and the innovations $\{\tilde{y}(t)\}$ that

$$y(t) = \tilde{y}(t) + HK(t-1)\tilde{y}(t-1) + HFK(t-2)\tilde{y}(t-2)$$
$$+ \ldots + HF^{t-1}K(0)\tilde{y}(0) + HF^t \hat{x}(0|-1)$$

and hence that

$$Y^t = \begin{pmatrix} I & & & \\ HK(0) & I & & 0 \\ & & \ddots & \\ HF^{t-1}K(0) & & HK(t-1)\ I \end{pmatrix} \begin{pmatrix} \tilde{y}(0) \\ \vdots \\ \tilde{y}(t) \end{pmatrix}$$

$$+ \begin{pmatrix} H \\ HF \\ \vdots \\ HF^t \end{pmatrix} \hat{x}(0|-1).$$

(c) Prove, by explicit algebraic calculations, that the covariance matrix of Y^t as expressed in (b) coincides with R as given in (a).

Remark The result in (b) gives a *triangularization* of the covariance matrix R. As this admits rewriting R as $R = LDL^T$ with L lower triangular and D block diagonal, the inversion of R is highly facilitated. In fact, the result can also be viewed as a *time-varying generalization of spectral factorization*. Note that the representation in (b) is in principle the same as that given implicitly in (6.7).   □

Exercise 6.19 Consider the system

$$x(t+1) = Fx(t) + v(t) ,$$
$$y(t) = Hx(t) + e(t) ,$$

where $v(t) \sim \mathbf{N}(0, R_1)$ and $e(s) \sim \mathbf{N}(0, R_2)$ are mutually independent for all t and s. Show that

$$p(x(t+1)|Y^t) = p(x(t+1)|\hat{x}(t+1|t)) .$$

Remark This means that the conditional mean $\hat{x}(t+1|t)$ in this example is a *sufficient statistic*. □

Exercise 6.20 Consider the tracking of a moving object from discrete-time position measurements. Model this as

$$\dot{x} = \begin{pmatrix} 0 & 1 \\ 0 & 0 \end{pmatrix} x + \begin{pmatrix} 0 \\ 1 \end{pmatrix} w,$$

$$y = (1 \ \ 0)x,$$

$$\mathbf{E}\,w(t)w(s) = \delta(t-s).$$

(a) Sample the model. Let the sampling interval be h.
(b) Determine the stationary optimal predictor $\hat{x}(t+h|t)$ and filter $\hat{x}(t|t)$. Show that they can be written as

$$\hat{x}(t+h|t) = \frac{1}{q+(h\beta-1)} \left(\frac{q(1+h\beta)-1}{\beta(q-1)} \right) y(t),$$

$$\hat{x}(t|t) = \left(\frac{y(t)}{\frac{1}{q+(h\beta-1)}\beta(q-1)y(t)} \right),$$

for some parameter β. Also determine the covariance matrices $P(t+h|t)$ and $P(t|t)$ of these estimates.
(c) Assume that one wishes to estimate the state vector between the measurements. Let $0 \le \tau < h$. Determine the optimal estimate $\hat{x}(t+\tau|t)$ (t being a sampling instant) and its covariance matrix $P(t+\tau|t)$. Examine how $P(t+\tau|t)$ varies with τ. Is $\hat{x}(t+\tau|t)$ always inferior to $\hat{x}(t|t)$?
(d) Discuss the result and compare with simple heuristic ways of estimating the derivative $x_2(t) = \dot{y}(t)$.

Exercise 6.21 Consider the ARE

$$P = FPF^T + R_1 - FPH^T(HPH^T + R_2)^{-1}HPF^T.$$

Assume that F and R_2 are invertible. Introduce the Euler matrix

$$\phi = \begin{pmatrix} F^{-T} & F^{-T}H^T R_2^{-1} H \\ R_1 F^{-T} & F + R_1 F^{-T} H^T R_2^{-1} H \end{pmatrix}.$$

This problem establishes some links between the ARE and the Euler matrix.

(a) Assume that $e_1 \ldots e_n$ are eigenvectors to ϕ. Set

$$(e_1 \ldots e_n) = \begin{pmatrix} P \\ R \end{pmatrix},$$

and assume P to be invertible. Show that $X = RP^{-1}$ will be a solution to the ARE.

(b) Let X be an arbitrary symmetric solution to the ARE. Show that the eigenvalues of $F - FXH^T(HXH^T + R_2)^{-1}H$ are also eigenvalues of the Euler matrix ϕ.

Exercise 6.22 Consider the matrix pencil given by (6.125) and (6.126):

$$\lambda L - M$$

where

$$L = \begin{pmatrix} I & H^T R_2^{-1} H \\ 0 & F \end{pmatrix}, \qquad M = \begin{pmatrix} F^T & 0 \\ -R_1 & I \end{pmatrix}.$$

and the algorithm given by (6.127)–(6.129) for solving the ARE.

(a) Show that P as defined in (6.129) is a solution to the ARE.
(b) Consider the scalar case. Find U and V. Show that the upper left part of the pencil (6.127) vanishes, that is

$$\lambda \hat{L}_{11} - \hat{M}_{11} = 0$$

gives precisely

$$\lambda = F - KH, \quad K = FPH^T(HPH^T + R_2)^{-1}.$$

c) For the scalar case derive the equation in λ for the generalized eigenvalues. Show that, in general, if λ_1 is a solution, then so is λ_1^{-1}. Under what conditions will only one finite solution exist? Under what conditions will there be no solution strictly inside the unit circle?

Exercise 6.23 Consider the ARE

$$P = FPF^T + R_1 - FPH^T(HPH^T + R_2)^{-1}HPF^T.$$

Assume that P_k is an approximate solution. Seek an improved solution of the form $P = P_k + \Delta P$. We derive an update equation for ΔP by dropping all terms higher than linear in ΔP. Set

$$K_k = FP_kH^T(HP_kH^T + R_2)^{-1},$$
$$P_{k+1} = P_k + \Delta P.$$

(a) Show that

$$P_{k+1} = (F - K_kH)P_{k+1}(F - K_kH)^T + (R_1 + K_kR_2K_k^T).$$

(b) Set

$$K = FPH^T(HPH^T + R_2)^{-1}.$$

Show that

$$(P_k - P_{k+1}) - (F - K_kH)(P_k - P_{k+1})(F - K_kH)^T$$
$$= (K_k - K_{k-1})(HP_kH^T + R_2)(K_k - K_{k-1})^T$$

and

$$(P_{k+1} - P) - (F - K_k H)(P_{k+1} - P)(F - K_k H)^T$$

$$= (K_k - K)(HPH^T + R_2)(K_k - K)^T.$$

(c) Assume that $F - K_k H$ has all eigenvalues strictly inside the unit circle. Show that

$$P_k \geq P_{k+1} \geq P .$$

(d) Factorize R_1 as $R_1 = GG^T$. Assume that (F, G) is controllable and that R_2 and P_k are positive definite. Show that $F - K_k H$ has all eigenvalues strictly inside the unit circle.

Hint. Study the stability properties of the system

$$x(t + 1) = (F - K_k H)^T x(t)$$

by means of the Lyapunov function $V(x) = x^T P_k x$.

(e) Let $\{P_k\}$ be a sequence of symmetric matrices and $\overline{P}$ a symmetric matrix satisfying

$$P_0 \geq P_1 \geq P_2 \geq \ldots \geq P_n \geq P_{n+1} \geq \ldots \geq \overline{P} .$$

Show that the sequence $\{P_k\}$ is convergent.

(f) Assume that the ARE has a (unique) symmetric positive definite solution P. Factorize R_1 as $R_1 = GG^T$ and assume (F, G) to be controllable and that R_2 is positive definite. Let P_0 be symmetric and positive definite and assume that $F - K_0 H$ has all eigenvalues inside the unit circle. Use the previous parts of the problem to show that the sequence $\{P_k\}$ is decreasing and converges to P.

Exercise 6.24 Derive an acceleration algorithm for computing $\phi^{(2^k)}$, where the Euler matrix ϕ is defined in (6.120). Set

$$\phi^{(2^k)} \triangleq \begin{pmatrix} \alpha_k^{-1} & \alpha_k^{-1} \beta_k \\ \gamma_k \alpha_k^{-1} & \delta_k + \gamma_k \alpha_k^{-1} \beta_k \end{pmatrix},$$

where $\alpha_k, \beta_k, \gamma_k, \delta_k$ are all $(n|n)$ matrices. The initial values are readily found:

$$\alpha_0 = F^T , \qquad \beta_0 = H^T R_2^{-1} H , \qquad \gamma_0 = R_1 , \qquad \delta_0 = F .$$

Derive the following recursions:

$$\alpha_{k+1} = \alpha_k (I + \beta_k \gamma_k)^{-1} \alpha_k ,$$
$$\beta_{k+1} = \beta_k + \alpha_k (I + \beta_k \gamma_k)^{-1} \beta_k \delta_k ,$$
$$\gamma_{k+1} = \gamma_k + \delta_k \gamma_k (I + \beta_k \gamma_k)^{-1} \alpha_k ,$$
$$\delta_{k+1} = \delta_k (I + \gamma_k \beta_k)^{-1} \delta_k .$$

Remark In fact, it holds for the given initial values that $\delta_k = \alpha_k^T$, β_k and γ_k are symmetric. These relations can be used to simplify the algorithm further.

$\square$

Exercise 6.25 Consider fixed lag smoothing, as described in Section 6.5.2. Assume that $m = 2$, and examine the stationary case. Let P be a positive definite solution to the Riccati equation

$$P = FPF^T + R_1 - FPH^T(HPH^T + R_2)^{-1}HPF^T .$$

Determine the stationary values of the covariance matrix $\overline{P}(t)$ (6.69) and the filter gain $\overline{K}_f(t)$ (6.63).

Bibliography

The results given in this chapter can be found in various textbooks on stochastic systems and estimation.

For some historical benchmark papers, see

Kailath, T. (Ed.), 1977. *Linear Least-Squares Estimation*. Dowden, Hutchinson and Ross, Inc., Stroudsburg, PA.

Sorenson, H.W. (Ed.), 1985. *Kalman Filtering: Theory and Application*. IEEE Press, New York.

Alternative treatments and derivations of the optimal state estimate can be found in

Anderson, B.D.O., Moore, J.B., 1979. *Optimal Filtering*. Prentice Hall, Englewood Cliffs, NJ.

Åström, K.J., 1970. *Introduction to Stochastic Control*. Academic Press, New York.

Brown, R.G., 1983. *Introduction to Random Signal Analysis and Kalman Filtering*. John Wiley & Sons, New York.

Chui, C.K., Chen, G., 1987. *Kalman Filtering with Real-Time Application*. Springer–Verlag, Berlin.

Gelb, A. (Ed.), 1975. *Applied Optimal Estimation*. MIT Press, Cambridge, MA.

Grimble, M.J., Johnson, M.A., 1988. *Optimal Control and Stochastic Estimation*, vol. 2. John Wiley & Sons, Chichester, UK.

Kailath, T., Sayed, A.H., Hassibi, B., 2000. *Linear Estimation*. Prentice Hall, Upper Saddle River, NJ.

Kay, S.M., 1993. *Fundamentals of Statistical Signal Processing: Estimation Theory*. PTR Prentice Hall, Englewood Cliffs, NJ.

Lewis, F.L., 1986. *Optimal Estimation*. John Wiley & Sons, New York.

Maybeck, P.S., 1979-82. *Stochastic Models, Estimation and Control*, vols 1 and 2. Academic Press, New York.

A comprehensive treatment of the Riccati equation and its properties is given in

Bittanti, S., Laub, A.J., Willems, J.C., (Eds), 1991. *The Riccati Equation*. Springer-Verlag, Heidelberg.

Implementation aspects, notably the $U-D$ algorithm mentioned in Section 6.8.1, are treated in

Bierman, G.J., 1977. *Factorization Methods for Discrete Sequential Estimation*. Academic Press, New York.

Fundamental aspects on numerical algorithms for solving the ARE are given in

Arnold, III, W.F., Laub, A.J., 1984. Generalized eigenproblem algorithms and software for algebraic Riccati equations. *IEEE Proceedings*, vol 72, 1746–1754.

There seem to have been several persons who derived the optimal state estimate around 1960. The optimal filter is named after Kalman, who gave the fundamental description in

Kalman, R.E., 1960. A new approach to linear filtering and prediction problems. *Transactions of ASME, Journal of Basic Engineering*, Series D, vol 82, 342–345.

Kalman also cooperated with R. S. Bucy. The optimal filter of continuous-time systems is often called the Kalman–Bucy filter.

7. Optimal Estimation for Linear Systems by Polynomial Methods

7.1 Introduction

Many estimation and control problems can be phrased using either a state space approach or a transfer function framework.

The *state space methodology* is more complete, in the sense that it gives a more detailed view of what happens to individual state variables. It is also the preferable choice when transient phenomena are to be studied. As derived in Chapter 6, the optimal state estimate is governed by a time-varying Kalman filter. Only in stationary cases where there is a sufficiently long data record will a time-invariant filter give optimal performance.

On the other hand, in many cases the state variables are not interesting *per se*, but merely as a step for describing the input–output behaviour. If the system is operating in a stationary mode and transient effects can be dispensed with, it is appropriate to use a *transfer function formalism*, since it is simpler than a state space approach.

A rational transfer function can be written as a ratio of polynomials in the scalar case. In the multivariable case, one can use two polynomial matrices to form a matrix fraction decomposition. Hence, it is relevant to call the transfer function formalism a *polynomial approach*.

As a prelude, in this chapter the optimal (in a mean square sense) predictor of a time series is studied (Section 7.2). Then, the Wiener filtering techniques are developed in Section 7.3 and applied to a general estimation problem (Section 7.4). Section 7.5 shows how the techniques developed can be modified to take uncertainties in the model description into account, thus achieving a form of robust filter.

It is worth noting that many estimation problems can be phrased and solved with either a state space or a polynomial approach. An extensive example to illustrate both approaches is given in Chapter 8.

7.2 Optimal Prediction

7.2.1 Introduction

Consider a stationary stochastic process described by a state space model

$$x(t+1) = Fx(t) + v(t) ,$$
$$y(t) = Hx(t) + e(t) ,$$

(7.1)

where $v(t)$ and $e(t)$ are mutually uncorrelated white noise sequences with zero mean and covariances R_1 and R_2 respectively. Assume that data have been available since the infinite past (*i.e.* the initial time $t_0 \to -\infty$). The optimal (in the mean square sense) k-step predictor of the signal $y(t)$ is then given by (6.40) and (6.99):

$$\hat{y}(t+k|t) = H\hat{x}(t+k|t)$$
$$= HF^{k-1}\hat{x}(t+1|t)$$
$$= HF^{k-1}[I - Fq^{-1} + KHq^{-1}]^{-1}Ky(t) ,$$

(7.2)

where K denotes the stationary predictor gain. Obviously, (7.2) can be rewritten as

$$\hat{y}(t+k|t) = G(q)y(t) ,$$

(7.3)

where the filter $G(q)$ is given by

$$G(q) = HF^{k-1}[I - (F - KH)q^{-1}]^{-1}K .$$

(7.4)

The purpose of this section is to derive optimal predictors of the form of (7.3) using a polynomial formalism instead of the state space methodology.

As a starting point, consider a simple but illustrative example.

Example 7.1 Consider one-step prediction of a first-order ARMA process

$$y(t) + ay(t-1) = e(t) + ce(t-1) ,$$

(7.5)

$$|a| < 1, \ |c| < 1 ,$$

(7.6)

where $e(t)$ is white noise of zero mean and variance λ^2. The predictor $\hat{y}(t+1|t)$ should be a function of the data available at time t, that is, of Y^t. It is instructive to examine the quantity $y(t+1)$ to be predicted. First, write $y(t+1)$ as a weighted sum of old noise values:

$$y(t+1) = \frac{1 + cq^{-1}}{1 + aq^{-1}}e(t+1)$$
$$= (1 + cq^{-1})[1 - aq^{-1} + a^2q^{-2} + \dots(-a)^jq^{-j} + \dots]e(t+1)$$
$$= e(t+1) + [(c-a)e(t) - a(c-a)e(t-1) + \dots$$

(7.7)

$$+(-a)^j(c-a)e(t-j) + \dots] .$$

(7.8)

Here, the first term, $e(t+1)$, will be uncorrelated with all available data, Y^t. At best, one can possibly hope to reconstruct the second part (in the brackets). For this purpose, invert the underlying process description, which gives

$$e(t) = \frac{1 + aq^{-1}}{1 + cq^{-1}}y(t) .$$

Note that, to do so, it is crucial that the assumption $|c| < 1$ is satisfied. Proceeding with the details, it is found that

$$[(c - a)e(t) - a(c - a)e(t - 1) + \ldots + (-a)^j(c - a)e(t - j) + \ldots]$$

$$= (c - a)[1 - aq^{-1} + a^2q^{-2} + \ldots + (-a)^jq^{-j} + \ldots]e(t)$$

$$= (c - a)\frac{1}{1 + aq^{-1}}\frac{1 + aq^{-1}}{1 + cq^{-1}}y(t)$$

$$= \frac{c - a}{1 + cq^{-1}}y(t) .$$

Hence

$$y(t + 1) = e(t + 1) + \frac{c - a}{1 + cq^{-1}}y(t) . \tag{7.9}$$

Here, it can clearly be seen that the first part, that is, $e(t + 1)$, cannot be computed from data, whereas the second part certainly can. The optimal predictor is thus

$$\hat{y}(t + 1|t) = \frac{c - a}{1 + cq^{-1}}y(t) . \tag{7.10}$$

It will apparently be a weighted sum of all old output values:

$$\hat{y}(t + 1|t) = (c - a)y(t) - c(c - a)y(t - 1) + \ldots$$
$$+(-c)^j(c - a)y(t - j) + \ldots .$$

For implementation, however, it is more feasible to use the form in (7.10) of the optimal predictor or to rewrite it as a difference equation

$$\hat{y}(t + 1|t) = -c\hat{y}(t|t - 1) + (c - a)y(t) . \tag{7.11}$$

□

7.2.2 Optimal Prediction of ARMA Processes

Consider a complex-valued ARMA process:

$$A(q)y(t) = C(q)e(t) , \qquad \mathbf{E}\,|e(t)|^2 = \lambda^2 , \tag{7.12}$$

where

$$A(q) = q^n + a_1q^{n-1} + \ldots + a_n ,$$
$$C(q) = q^n + c_1q^{n-1} + \ldots + c_n ,$$

have all zeros inside the unit circle.

The clue to finding the optimal k-step predictor is similar to the method used in Example 7.1, namely to rewrite $y(t + k)$ in two terms. The first term is a weighted sum of future noise values, $\{e(t + j)\}_{j=1}^k$. As it is uncorrelated to all available data, it cannot be reconstructed in any way. The other term is a weighted sum of past noise values $\{e(t - s)\}_{s=0}^{\infty}$. By inverting the process model, this term can be written as a weighted sum of output values, $\{y(t - s)\}_{s=0}^{\infty}$. Hence, the second term can be computed exactly from data.

In order to proceed, introduce the *predictor identity*

$$z^{k-1}C(z) \equiv A(z)F(z) + L(z), \tag{7.13}$$

where

$$F(z) = z^{k-1} + f_1 z^{k-2} + \ldots + f_{k-1}, \tag{7.14}$$

$$L(z) = \ell_0 z^{n-1} + \ell_1 z^{n-2} + \ldots + \ell_{n-1}. \tag{7.15}$$

Then

$$
\begin{aligned}
y(t + k) &= \frac{C(q)}{A(q)} e(t + k) \\
&= \frac{q^{k-1}C(q)}{A(q)} e(t + 1) \\
&= \frac{A(q)F(q) + L(q)}{A(q)} e(t + 1) \\
&= F(q)e(t + 1) + \frac{qL(q)}{A(q)} e(t) \\
&= F(q)e(t + 1) + \frac{qL(q)}{A(q)} \frac{A(q)}{C(q)} y(t) \\
&= F(q)e(t + 1) + \frac{qL(q)}{C(q)} y(t). \tag{7.16}
\end{aligned}
$$

Note that $F(q)e(t + 1)$ is precisely the first term mentioned above. It is a weighted sum of *future* noise values, and $qL(q)/C(q)y(t)$ is a weighted sum of *available* measurements Y^t.

Now, let $\bar{\bar{y}}(t + k)$ be an *arbitrary* predictor of $y(t + k)$. As it is a function of Y^t, the prediction error variance will be

$$
\begin{aligned}
\mathbf{E}\,|y(t + k) - \bar{\bar{y}}(t + k)|^2 &= \mathbf{E}\left|F(q)e(t + 1) + \frac{qL(q)}{C(q)} y(t) - \bar{\bar{y}}(t + k)\right|^2 \\
&= \mathbf{E}\,|F(q)e(t + 1)|^2 + \mathbf{E}\left|\frac{qL(q)}{C(q)} y(t) - \bar{\bar{y}}(t + k)\right|^2 \\
&\geq \mathbf{E}\,|F(q)e(t + 1)|^2. \tag{7.17}
\end{aligned}
$$

Hence the *mean square optimal predictor* is given by

$$\hat{y}(t + k|t) = \frac{qL(q)}{C(q)} y(t), \tag{7.18}$$

and the associated prediction error is

$$\boxed{\begin{aligned}\tilde{y}(t+k) &= F(q)e(t+1)\\ &= e(t+k) + f_1 e(t+k-1) + \ldots + f_{k-1} e(t+1)\,,\end{aligned}} \tag{7.19}$$

and has variance

$$\mathbf{E}\,|\tilde{y}(t+k)|^2 = \lambda^2(1 + |f_1|^2 + \ldots + |f_{k-1}|^2)\,. \tag{7.20}$$

It is illustrative to compare the above derivation with that obtained by treating the problem in a state space setting. In order to simplify, consider only one-step prediction.

Example 7.2 Consider the ARMA process

$$A(q)y(t) = C(q)e(t)\,.$$

and represent it in state space form as

$$x(t+1) = \begin{pmatrix} -a_1 & 1 & & 0 \\ \vdots & & \ddots & \\ -a_n & 0 & & 1 \\ 0 & & & 0 \end{pmatrix} x(t) + \begin{pmatrix} 1 \\ c_1 \\ \vdots \\ c_n \end{pmatrix} v(t)\,,$$

$$y(t) = (1\ \ 0 \ldots 0)x(t)\,,$$

which is written in brief as

$$x(t+1) = Fx(t) + Cv(t)\,,$$
$$y(t) = Hx(t)\,.$$

Here, $v(t) = e(t+1)$.

The associated ARE is

$$P = FPF^T + \lambda^2 CC^T - FPH^T(HPH^T)^{-1}HPF^T\,.$$

The solution turns out to be $P = \lambda^2 CC^T$. (It is easy to show that this is a solution. It is more difficult to show that it is *the* solution required for the estimation problem.) The filter gain is

$$K_{\mathrm{f}} = PH^T(HPH^T)^{-1} = C\,.$$

The Kalman one-step predictor becomes

$$\hat{x}(t+1|t) = (F - FK_{\mathrm{f}}H)\hat{x}(t|t-1) + FK_{\mathrm{f}}y(t)\,.$$

Now:

$$F - FK_fH = \begin{pmatrix} -a_1 & 1 & & \\ \vdots & & \ddots & \\ -a_n & & & 1 \\ 0 & & & 0 \end{pmatrix} - \begin{pmatrix} -a_1 + c_1 \\ \vdots \\ -a_n + c_n \\ 0 \end{pmatrix} (1 \ 0 \ldots 0)$$

$$= \begin{pmatrix} -c_1 & 1 & & \\ \vdots & & \ddots & \\ -c_n & & & 1 \\ 0 & & & 0 \end{pmatrix},$$

and

$$\hat{x}(t+1|t) = \begin{pmatrix} -c_1 & 1 & & \\ \vdots & & \ddots & \\ -c_n & & & 1 \\ 0 & & & 0 \end{pmatrix} \hat{x}(t|t-1)$$

$$+ \begin{pmatrix} c_1 - a_1 \\ \vdots \\ c_n - a_n \\ 0 \end{pmatrix} y(t) \ .$$

Noting that this model is in observable canonical form, it is found that

$$\hat{y}(t|t-1) = \hat{x}_1(t|t-1) = \frac{C(q) - A(q)}{C(q)} y(t) \ .$$

Next, the output innovations are computed as

$$\tilde{y}(t) = y(t) - H\hat{x}(t|t-1) = y(t) - \hat{x}_1(t|t-1)$$
$$= \frac{A(q)}{C(q)} y(t) \ \ [= e(t)] \ ,$$

which is in perfect agreement with what is easily obtained with the polynomial formalism. $\square$

One can give several interpretations of the predictor identity (7.13).

1. The predictor identity can be regarded as a Diophantine equation. Recall that $A(z)$ and $C(z)$ are known and that $F(z)$ and $L(z)$ are to be computed. When the Diophantine equation is reformulated as a set of linear equations in the unknown polynomial equations, this gives the triangular system of equations

$$
\begin{pmatrix}
1 & 0 & | & \\
a_1 & 1 & | & 0 \\
\vdots & & \ddots & | \\
a_n & & 1 & | \\
& \ddots & & a_1 & | & 1 \\
0 & & & \vdots & | & 0 & \ddots \\
& & & a_n & | & & & 1
\end{pmatrix}
\begin{pmatrix}
f_1 \\
\vdots \\
f_{k-1} \\
\hline
\ell_0 \\
\vdots \\
\ell_{n-1}
\end{pmatrix}
=
\begin{pmatrix}
c_1 - a_1 \\
c_2 - a_2 \\
\vdots \\
c_n - a_n \\
0 \\
\vdots \\
0
\end{pmatrix} . \tag{7.21}
$$

It can easily be seen that this system of equations (with $n + k - 1$ equations and the same number of unknowns) always has a unique solution. As the matrix is triangular, it is straightforward to solve the equations successively from top to bottom.

2. It can be seen as a form of polynomial division. Rewrite the predictor identity as

$$
\frac{z^{k-1} C(z)}{A(z)} = F(z) + \frac{L(z)}{A(z)} . \tag{7.22}
$$

The interpretation is then that $z^{k-1}C(z)$ is divided by $A(z)$. The quotient is then the polynomial $F(z)$ of degree $k - 1$ and there is a remainder polynomial $L(z)$.

3. Closely related to the above interpretation is the view of the predictor identity as a truncation of the weighting function. See (7.26) below. In fact, if the transfer function $C(z)/A(z)$ is transformed to a weighting function, one obtains

$$
\frac{C(z)}{A(z)} = h_0 + h_1 z^{-1} + h_2 z^{-2} + \ldots = \sum_{j=0}^{\infty} h_j z^{-j} .
$$

The sum can be truncated after k terms, to give

$$
\begin{aligned}
\sum_{j=0}^{k-1} h_j z^{-j} &= z^{1-k} F(z) , \\
\sum_{j=k}^{\infty} h_j z^{-j} &= z^{1-k} L(z)/A(z) .
\end{aligned} \tag{7.23}
$$

7.2.3 A General Case

Consider the system

$$
y(t) = G(q)u(t) + H(q)e(t) , \tag{7.24}
$$

where $e(t)$ is zero mean complex white noise, $H(\infty) = I$, and $H^{-1}(q)$ is asymptotically stable. This means that $\{e(t)\}$ are the output innovations; see Section 4.3. The best k-step predictor in a mean square sense is sought, that is, a function $f(t)$ of Y^t, such that

$$
V = \mathbf{E} \left[y(t+k) - f(t) \right]^* P \left[y(t+k) - f(t) \right] , \tag{7.25}
$$

where P is a positive definite Hermitian weighting matrix. Assume that all future input values, that is, $u(t+1), \ldots, u(t+k)$, are known at time t.

In order to solve this problem, rewrite the noise filter as

$$H(q) = \sum_{j=0}^{\infty} h_j q^{-j}, \qquad (h_0 = I)$$

and split it into two parts

$$H(q) = \underbrace{[\sum_{j=0}^{k-1} h_j q^{-j}]}_{\triangleq \; q^{-k+1} H_0(q)} + \underbrace{q^{-k}[\sum_{j=0}^{\infty} h_{k+j} q^{-j}]}_{q^{-k} H_1(q)} . \qquad (7.26)$$

Note that $H_0(q)$ so defined will be a polynomial, of degree $k-1$, in q. Rewrite the signal to predict as

$$\begin{aligned}
y(t+k) &= G(q)u(t+k) + H(q)e(t+k) \\
&= G(q)u(t+k) + [q^{-k+1}H_0(q) + q^{-k}H_1(q)]e(t+k) \\
&= G(q)u(t+k) + H_0(q)e(t+1) + H_1(q)e(t) \\
&= G(q)u(t+k) + H_0(q)e(t+1) \\
&\quad + H_1(q)H^{-1}(q)[y(t) - G(q)u(t)] \\
&= [H_0(q)e(t+1)] + [H_1(q)H^{-1}(q)y(t) \\
&\quad + \{H(q)q^k - H_1(q)\}H^{-1}(q)G(q)u(t)] \\
&= [H_0(q)e(t+1)] \\
&\quad + [H_1(q)H^{-1}(q)y(t) + H_0(q)H^{-1}(q)G(q)u(t+1)] \\
&\triangleq \varepsilon(t+1) + z(t) . \qquad (7.27)
\end{aligned}$$

Next, note the following:

- The term $\varepsilon(t+1) = e(t+k) + h_1 e(t+k-1) + \ldots + h_{k-1} e(t+1)$ depends only on *future* values of the white noise $e(t)$. Hence $\varepsilon(t+1)$ is uncorrelated with Y^t.
- The term $z(t)$ is a function of Y^t (and possibly future input variables), and is known at time t.

Thus

$$\begin{aligned}
V &= \mathbf{E}\left[\varepsilon(t+1) + z(t) - f(t)\right]^* P[\varepsilon(t+1) + z(t) - f(t)] \\
&= \mathbf{E}\left[\varepsilon^*(t+1)P\varepsilon(t+1)\right] + \mathbf{E}\left[z(t) - f(t)\right]^* P[z(t) - f(t)] \\
&\geq \mathbf{E}\,\varepsilon^*(t+1)P\varepsilon(t+1) . \qquad (7.28)
\end{aligned}$$

The term $\mathbf{E}\,\varepsilon^*(t+1)P\varepsilon(t+1)$ is something that can in no sense be affected by the choice of predictor $f(t)$. Therefore, the best predictor (in a mean square sense) is the choice $f(t) = z(t)$. The optimal predictor is denoted as follows:

$$\hat{y}(t + k|t) = H_1(q)H^{-1}(q)y(t) + H_0(q)H^{-1}(q)G(q)u(t + 1) . \qquad (7.29)$$

The prediction error is apparently

$$\tilde{y}(t + k) \overset{\triangle}{=} y(t + k) - \hat{y}(t + k|t) = H_0(q)e(t + 1) . \qquad (7.30)$$

and the minimal value of the criterion becomes

$$\begin{aligned} \min V &= \mathbf{E}\, \varepsilon^*(t + 1)P\varepsilon(t + 1) \\ &= \sum_{j=0}^{k-1} \mathbf{E}\, e^*(t + k - j)h_j^* Ph_j e(t + k - j) \\ &= \sum_{j=0}^{k-1} \operatorname{tr}\left[h_j^* Ph_j \mathbf{E}\, e(t)e^*(t) \right] . \end{aligned} \qquad (7.31)$$

Remark For a scalar ARMA process (7.12):

$$H_0(q) = F(q) , \qquad H_1(q) = \frac{qL(q)}{A(q)} ,$$

and

$$H_1(q)H^{-1}(q) = \frac{qL(q)}{A(q)}\frac{A(q)}{C(q)} = \frac{qL(q)}{C(q)} ,$$

which coincides with the previous findings. $\qquad \square$

7.2.4 Prediction of Nonstationary Processes

Some comments on the optimal prediction of nonstationary processes can be given. Consider an ARMA process

$$A(q)y(t) = C(q)e(t) , \qquad (7.32)$$

where $A(z)$ is allowed to have zeros on or outside the unit circle. By spectral factorization applied to the right-hand side of (7.32), it can be assumed that $C(z)$ has all zeros inside the unit circle. The true optimal predictor for such a nonstationary situation is given by the Kalman filter. One could, however, apply the predictor in (7.18), although it would not be optimal in the transient phase. If the process is *perfectly known*, after a transient period the optimal prediction is obtained. The situation is depicted in Figure 7.1.

In the nonstationary case, a small prediction error, $\varepsilon(t + k)$, is still obtained, but it is the difference between two nonstationary processes, and hence is most sensitive to small rounding errors or other imperfections in the calculations. In fact, let the predictor be based on the model in (7.32), while the true process is described by

$$A_0(q)y(t) = C_0(q)e(t) . \qquad (7.33)$$

The prediction error becomes

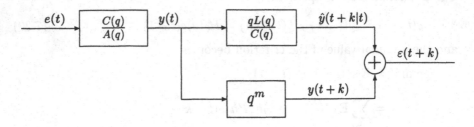

Fig. 7.1. Prediction of an ARMA process

$$\varepsilon(t + k) = y(t + k) - \hat{y}(t + k|t)$$
$$= \left[q^k - \frac{qL(q)}{C(q)} \right] y(t)$$
$$= \frac{qA(q)F(q)}{C(q)} \frac{C_0(q)}{A_0(q)} e(t) . \tag{7.34}$$

In the *ideal* case when $A = A_0$, $C = C_0$, one obtains $\varepsilon(t + k) = qF(q)e(t)$ as before. As soon as the unstable part of A_0 deviates from that of A, the prediction error will have a variance that grows exponentially with time.

7.3 Wiener Filters

The steady-state linear least mean square estimate is derived in this section.

7.3.1 Statement of the Problem

Let $y(t)$ and $s(t)$ be two, possibly complex-valued, correlated and stationary stochastic processes, possibly multivariable. Assume that $y(t)$ is measured and find a causal, asymptotically stable filter $G(q)$ such that $G(q)y(t)$ is a good estimate of $s(t)$. More precisely, it should be the optimal (linear) mean square estimator, that is, it should minimize the criterion

$$V = \mathbf{E} \, \|s(t) - G(q)y(t)\|^2 . \tag{7.35}$$

The analysis of the problem and the derivation of the optimal filter will be done in the frequency domain.

This implies, in particular, that data are assumed to be available since the infinite past $t = -\infty$.

Introduce the estimation error

$$\tilde{s}(t) = s(t) - G(q)y(t) . \tag{7.36}$$

The criterion V (7.35) can be rewritten as

$$
\begin{aligned}
V &= \mathbf{E}\,\|\tilde{s}(t)\|^2 = \mathbf{E}\,\tilde{s}^*(t)\tilde{s}(t) \\
&= \mathbf{E}\,\operatorname{tr}\,\tilde{s}^*(t)\tilde{s}(t) = \mathbf{E}\,\operatorname{tr}\,\tilde{s}(t)\tilde{s}^*(t) \\
&= \operatorname{tr}\,\mathbf{E}\,\tilde{s}(t)\tilde{s}^*(t) = \operatorname{tr}\,r_{\tilde{s}}(0) \\
&= \operatorname{tr}\frac{1}{2\pi i}\oint \phi_{\tilde{s}}(z)\frac{dz}{z}\ .
\end{aligned}
\tag{7.37}
$$

Next, note that

$$
\phi_{\tilde{s}}(z) = \phi_s(z) - G(z)\phi_{ys}(z) - \phi_{sy}(z)G^*(z^{-*}) + G(z)\phi_y(z)G^*(z^{-*})\ .\tag{7.38}
$$

Now let $G(q)$ be the optimal filter and $G_1(q)$ any causal filter. Replace $G(q)$ in (7.35) by $G(q) + \varepsilon G_1(q)$. As a function of ε, V can then be written as

$$
V = V_0 + \varepsilon V_1 + \varepsilon^2 V_2\ .\tag{7.39}
$$

For $G(q)$ to be the *optimal* filter it is required that

$$
V \geq V_0\ \text{ for all }\varepsilon\ ,\tag{7.40}
$$

that is:

$$
\varepsilon V_1 + \varepsilon^2 V_2 \geq 0\ \text{ for all }\varepsilon\ .\tag{7.41}
$$

This equation means that, for any arbitrary and small deviation from the optimal filter, the criterion V will not decrease. Let ε be small enough. Then the first term in (7.41) dominates. Phrased differently, $\varepsilon V_1 \geq 0$. As this result holds for both positive and negative values of ε, it is concluded that $V_1 = 0$. This means that

$$
\begin{aligned}
0 &= \operatorname{tr}\frac{1}{2\pi i}\oint [-G_1(z)\phi_{ys}(z) - \phi_{sy}(z)G_1^*(z^{-*}) \\
&\quad + G(z)\phi_y(z)G_1^*(z^{-*}) + G_1(z)\phi_y(z)G^*(z^{-*})]\frac{dz}{z} \\
&= \operatorname{tr}\frac{1}{2\pi i}\oint [-G_1(z)\phi_{ys}(z) - \phi_{sy}(z)G_1^*(z) \\
&\quad + G(z)\phi_y(z)G_1^*(z) + G_1(z)\phi_y(z)G^*(z)]\frac{dz}{z}\ .
\end{aligned}
\tag{7.42}
$$

This relation must hold for *any* causal and stable filter $G_1(q)$. It can be expected that it will define the optimal filter $G(q)$ uniquely.

The condition in (7.42) can be simplified using some symmetry relations. Using (3.33), (7.38) and (7.42), and for $\operatorname{tr}(AB)$ real-valued, $\operatorname{tr}(AB) = \operatorname{tr}(AB)^* = \operatorname{tr}B^*A^*$, gives

$$
\begin{aligned}
0 &= \operatorname{tr}\frac{1}{2\pi i}\oint [-\phi_{ys}^*(z)G_1^*(z) - \phi_{sy}(z)G_1^*(z) \\
&\quad + G(z)\phi_y(z)G_1^*(z) + G(z)\phi_y^*(z)G_1^*(z)]\frac{dz}{z} \\
&= 2\operatorname{tr}\frac{1}{2\pi i}\oint [G(z)\phi_y(z) - \phi_{sy}(z)]G_1^*(z)\frac{dz}{z}\ .
\end{aligned}
\tag{7.43}
$$

It is possible to give an interpretation and alternative view of (7.43). Consider the original estimation problem. For the optimal filter, the estimation error $\tilde{s}(t)$ should be uncorrelated with all past measurements, $\{y(t-j)\}_{j=0}^{\infty}$. Otherwise there would be another linear combination of the past measurements giving smaller estimation error variance. Compare also the fact that $x - \hat{x}$ and y are uncorrelated; see Section 5.3. Hence

$$\mathbf{E}\,\tilde{s}(t)y^*(t-j) = 0\,, \qquad \text{all } j \geq 0\,, \tag{7.44}$$

or

$$\mathbf{E}\,\tilde{s}(t)[G_1(q)y(t)]^* = 0 \qquad \text{for any stable and causal } G_1(q)\,. \tag{7.45}$$

This can be rewritten as

$$
\begin{aligned}
0 &= \mathbf{E}\,[s(t) - G(q)y(t)][G_1(q)y(t)]^* \\
&= \frac{1}{2\pi\mathrm{i}} \oint [\phi_{sy}(z) - G(z)\phi_y(z)]G_1^*(z)\frac{\mathrm{d}z}{z}\,,
\end{aligned}
\tag{7.46}
$$

which is precisely (7.43), except that the trace operator is omitted.

7.3.2 The Unrealizable Wiener Filter

From (7.43), one easily finds the *unrealizable Wiener filter*. Set

$$G(z)\phi_y(z) = \phi_{sy}(z)\,,$$

giving

$$G(z) = \phi_{sy}(z)\phi_y^{-1}(z)\,. \tag{7.47}$$

The filter is unrealizable since it relies (except in very degenerated cases) on all *future* data points of $y(t)$. Expressed differently: let $G(z)$ be expanded as a Laurent series (with a convergence region covering the unit circle)

$$G(z) = \sum_{j=-\infty}^{\infty} g_j z^{-j}\,. \tag{7.48}$$

A *causal* filter must satisfy $g_j = 0$, all $j < 0$. However, the filter of (7.47) has mostly $g_j \neq 0$ for all j, as is illustrated next.

Example 7.3 As an illustration of the above claim, consider the case of filtering ($k = 0$) a first-order AR process in white noise:

$$s(t) = \frac{1}{q+a}v(t), \qquad y(t) = s(t) + e(t),$$

for which

$$\phi_y(z) = \frac{\lambda_v^2}{(z+a)(z^{-1}+a)} + \lambda_e^2 = \lambda_\varepsilon^2 \frac{(z+c)(z^{-1}+c)}{(z+a)(z^{-1}+a)}$$

holds by spectral factorization, and

$$\phi_{sy}(z) = \phi_s(z) = \frac{\lambda_v^2}{(z+a)(z^{-1}+a)} .$$

Hence, (7.47) implies

$$G(z) = \frac{\lambda_v^2}{\lambda_\varepsilon^2} \; \frac{1}{(z+c)(z^{-1}+c)} = \frac{\lambda_v^2}{\lambda_\varepsilon^2} \sum_{j=-\infty}^{\infty} \frac{(-c)^{|j|}}{1-c^2} z^{-j} ,$$

where the last expression gives the Laurent series. It clearly has all coefficients nonzero. □

Note, though, that when "deriving" (7.47) from (7.43), it was effectively required that (7.43) holds for *any* $G_1(z)$. However, it is only required that (7.43) holds for any *causal and stable* $G_1(z)$. Such an attempt will eventually lead to the optimal *realizable* filter.

7.3.3 The Realizable Wiener Filter

To proceed, assume that the process $y(t)$ has an *innovations representation*

$$\begin{aligned} y(t) &= H(q)e(t) , \\ \mathbf{E}\, e(t)e^*(s) &= \Lambda\delta_{t,s} , \\ H(0) &= I , \end{aligned} \qquad (7.49)$$

with $H(q)$ and $H^{-1}(q)$ asymptotically stable. Then, in particular, on the unit circle

$$\phi_y(z) = H(z)\Lambda H^*(z) . \qquad (7.50)$$

Further, introduce the *causal part* of an analytical function. Let

$$G(z) = \sum_{j=-\infty}^{\infty} g_j z^{-j} , \qquad (7.51)$$

where the series converges in a strip that includes the unit circle. Then the *causal part* of $G(z)$ is defined as

$$[G(z)]_+ = \sum_{j=0}^{\infty} g_j z^{-j} . \qquad (7.52)$$

The *anticausal part* is the complementary part of the sum:

$$[G(z)]_- = \sum_{j=-\infty}^{-1} g_j z^{-j} = G(z) - [G(z)]_+ . \qquad (7.53)$$

It is important to note that the term $g_0 z^{-0}$ in (7.51) appears in the causal part, $[G(z)]_+$.

In order to proceed, the following result will be useful.

Lemma 7.1 *The anticausal part $[G(z)]_-$ of a transfer function $G(z)$ has no poles inside or on the unit circle.*

Proof As the series in (7.51) is such that it converges on the unit circle, it is found that

$$[G(z)]_- = \sum_{j=-\infty}^{-1} g_j z^{-j} = \sum_{k=1}^{\infty} g_{-k} z^k .$$

Therefore, inside and on the unit circle

$$|[G(z)]_-| = |\sum_{k=1}^{\infty} g_{-k} z^k| \le \sum_{k=1}^{\infty} |g_{-k}| < \infty$$

holds, implying that the function $[G(z)]_-$ lacks poles in this region. ∎

Note also that a filter $G(z)$ is causal if and only if

$$G(z) = [G(z)]_+ . \tag{7.54}$$

Using (7.50) and the conventions in (7.52) and (7.53), the optimality condition (7.43) can be formulated as

$$0 = \text{tr} \frac{1}{2\pi i} \oint \{G(z)H(z)\Lambda H^*(z) - \phi_{sy}(z)\}G_1^*(z)\frac{dz}{z}$$

$$= \text{tr} \frac{1}{2\pi i} \oint \{G(z)H(z) - \phi_{sy}(z)\{H^*(z)\}^{-1}\Lambda^{-1}\}\Lambda H^*(z)G_1^*(z)\frac{dz}{z}$$

$$= \text{tr} \frac{1}{2\pi i} \oint \{G(z)H(z) - [\phi_{sy}(z)\{H^*(z)\}^{-1}\Lambda^{-1}]_+$$

$$-[\phi_{sy}(z)\{H^*(z)\}^{-1}\Lambda^{-1}]_-\}\Lambda H^*(z)G_1^*(z)\frac{dz}{z} . \tag{7.55}$$

Next, note that the stability requirements imply that the function $H^*(z)G_1^*(z)$ $= H(z^{-1})G_1(z^{-1})$ does not have any poles inside the unit circle. By construction, the same is true for $[\phi_{sy}(z)\{H^*(z)\}^{-1}\Lambda^{-1}]_-$. The latter function has a zero in $z = 0$. Hence, by the residue theorem:

$$\frac{1}{2\pi i} \oint [\phi_{sy}(z)\{H^*(z)\}^{-1}\Lambda^{-1}]_- \Lambda H^*(z)G_1^*(z)\frac{dz}{z} = 0 . \tag{7.56}$$

The optimal condition of (7.55) is therefore satisfied if

$$\boxed{G(z) = [\phi_{sy}(z)\{H^*(z)\}^{-1}\Lambda^{-1}]_+ H^{-1}(z) . \tag{7.57}}$$

This is the *realizable Wiener filter*. It is clear by its construction that it is a causal and asymptotically stable filter.

7.3.4 Illustration

The Wiener filter will be illustrated in two simple cases. In both cases the model is real-valued.

Example 7.4 Let $y(t)$ be a scalar ARMA process

$$A(q)y(t) = C(q)e(t) , \qquad \mathbf{E}\, e^2(t) = \lambda^2 ,$$

and set $s(t) = y(t - j)$, $j \geq 0$. Then the optimal filter should, of course, be $G(q) = q^{-j}$! The problem is in this case trivial, but is included here for illustration only. In this case

$$H(z) = \frac{C(z)}{A(z)} , \qquad \Lambda = \lambda^2 ,$$

$$\phi_{sy}(z) = z^{-j}\phi_y(z) .$$

The *unrealizable Wiener filter* (7.47) becomes

$$G(z) = z^{-j}\phi_y(z)\phi_y^{-1}(z) = z^{-j} ,$$

which is the expected result.

In order to derive the *realizable Wiener filter*, note that (on the unit circle)

$$\phi_{sy}(z) = z^{-j}\lambda^2 \frac{C(z)}{A(z)}\frac{C^*(z)}{A^*(z)} .$$

Hence, using (7.57), the filter becomes

$$G(z) = \left[z^{-j}\lambda^2 \frac{C(z)}{A(z)}\frac{C^*(z)}{A^*(z)}\frac{A^*(z)}{C^*(z)}\frac{1}{\lambda^2} \right]_+ \frac{A(z)}{C(z)}$$

$$= \left[z^{-j}\frac{C(z)}{A(z)} \right]_+ \frac{A(z)}{C(z)} = z^{-j}\frac{C(z)}{A(z)}\frac{A(z)}{C(z)} = z^{-j} .$$

In both cases the filter will thus be (as it should)

$$\hat{s}(t) = q^{-j}y(t) = y(t - j) . \qquad \qquad \square$$

Example 7.5 Consider the same process as in Example 7.4, but treat the prediction problem

$$s(t) = y(t + k) , \qquad k > 0 .$$

As before, the *unrealizable* filter (7.47) becomes

$$G(z) = z^k\phi_y(z)\phi_y^{-1}(z) = z^k ,$$

meaning that

$$\hat{s}(t) = y(t + k) .$$

Note that it is noncausal, but in this sense it is a perfect estimate since it is without error!

Next, the *realizable* filter is calculated. Similar to the calculations of Example 7.4:

$$G(z) = \left[z^k \lambda^2 \frac{C(z)}{A(z)} \frac{C^*(z)}{A^*(z)} \frac{A^*(z)}{C^*(z)} \frac{1}{\lambda^2} \right]_+ \frac{A(z)}{C(z)}$$

$$= \left[z^k \frac{C(z)}{A(z)} \right]_+ \frac{A(z)}{C(z)} .$$

To proceed, let $A(z)$ and $C(z)$ have degree n, and introduce the polynomial $F(z)$ of degree $k - 1$ and the polynomial $L(z)$ of degree $n - 1$ by the *predictor identity* (7.13):

$$z^{k-1} C(z) \equiv A(z)F(z) + L(z) .$$

This gives

$$G(z) = \left[\frac{zA(z)F(z) + zL(z)}{A(z)} \right]_+ \frac{A(z)}{C(z)} \tag{7.58}$$

$$= \frac{zL(z)}{A(z)} \frac{A(z)}{C(z)} = \frac{zL(z)}{C(z)} . \tag{7.59}$$

The optimal predictor, therefore, has the form

$$\hat{s}(t) = \hat{y}(t + k|t) = \frac{qL(q)}{C(q)} y(t) . \qquad \square$$

7.3.5 Algorithmic Aspects

A convenient procedure for deriving Wiener filters for a given problem formulation can now be described. Another approach will be outlined in Section 7.3.6. The method considered here is based on the following three facts:

1. The optimal filter $G(q)$ satisfies (7.43) for any stable and causal $G_1(q)$.
2. The spectral factorization of the spectrum $\phi_y(z)$ was instrumental in finding the solution.
3. For the optimal Wiener filter (7.57) the integrand of (7.56) becomes

$$[\phi_{sy}(z)\{H^*(z)\}^{-1}]_- H^*(z) G_1^*(z) \frac{1}{z} .$$

It *lacks* poles inside the unit circle.

The derivation procedure is then as follows:

(a) Introduce the spectral factor of $\phi_y(z)$.

(b) Manipulate the integrand of (7.56) so that all stable poles (*i.e.* poles inside the unit circle) are cancelled by zeros.

The procedure may be best understood when illustrated by an example.

Example 7.6 Continue with the prediction of an ARMA process, as dealt with in Examples 7.4 and 7.5. As before (again on the unit circle):

$$\phi_y(z) = \frac{C(z)C^*(z)}{A(z)A^*(z)}\lambda^2 , \qquad \phi_{sy}(z) = z^k\phi_y(z) , \qquad (k > 0) .$$

According to the principle, it is required that the following function has no poles inside the unit circle:

$$f(z) = [G(z)H(z)\Lambda H^*(z) - \phi_{sy}(z)]\frac{1}{z}$$

$$= [G(z) - z^k]\lambda^2 \frac{C(z)C^*(z)}{A(z)A^*(z)} \frac{1}{z} .$$

Now set

$$G(z) = \frac{M_0(z)}{N(z)} , \tag{7.60}$$

where M_0 and N are coprime polynomials. Apparently:

$$f(z) = \frac{M_0(z) - z^kN(z)}{N(z)}\lambda^2 \frac{C(z)C(z^{-1})}{A(z)A(z^{-1})} \frac{1}{z} .$$

As all poles inside the unit circle must be cancelled by zeros, it can be concluded that

  $N(z)$ is a factor of $C(z)$,

  $zA(z)$ is a factor of $M_0(z) - z^kN(z)$.

Now choose

$$N(z) = C(z) , \tag{7.61}$$

which gives

$$M_0(z) - z^kN(z) = zA(z)P(z) \tag{7.62}$$

for some polynomial $P(z)$. Setting $z = 0$ in (7.62) leads to $M_0(0) = 0$, so $M_0(z) = zM(z)$ for some polynomial $M(z)$. Thus from (7.61) and (7.62), after cancelling a factor of z:

$$z^{k-1}C(z) = M(z) - A(z)P(z) . \tag{7.63}$$

This can be recognized as the predictor identity (7.13) after the substitutions

  $M(z) \rightarrow L(z)$,

  $P(z) \rightarrow -F(z)$.

The optimal predictor filter (7.57) becomes

$$G(z) = \frac{zM(z)}{N(z)} = \frac{zM(z)}{C(z)} = \frac{zL(z)}{C(z)} ,$$

which coincides with (7.59).   □

7.3.6 The Causal Part of a Filter, Partial Fraction Decomposition and a Diophantine Equation

When computing the Wiener filter as in (7.57), a systematic method of computing the causal part of a filter is required. As will be shown in this section, this problem is closely tied to partial fraction decomposition and a Diophantine equation.

The following situation will be examined as an archetypical case. Consider evaluation of

$$f(z) = \left[\frac{G(z, z^{-1})}{D(z)F(z^{-1})} \right]_+ ,$$ (7.64)

where

$$G(z, z^{-1}) = \sum_{j=-p}^{\ell} g_j z^j , \qquad p \geq 0, \ \ell \geq 0 ,$$

$$D(z) = z^\nu + d_1 z^{\nu-1} + \ldots d_\nu , \qquad \nu \geq 0 ,$$

$$F(z^{-1}) = z^{-\mu} + f_1 z^{-\mu+1} + \ldots + f_\mu , \qquad \mu \geq 0 .$$

It is no restriction to assume that $p \geq 0$, $\ell \geq 0$. Otherwise one can augment G with some zero coefficients. Assume that $D(z)$ has all zeros strictly inside the unit circle and $z^\mu F(z^{-1})$ all zeros strictly outside.

Now consider the Diophantine equation

$$G(z, z^{-1}) \equiv z^\alpha F(z^{-1})R(z) + z^\beta D(z)L(z^{-1}) ,$$ (7.65)

where α and β are integers, yet to be determined, $R(z)$ is a *polynomial* in z of degree $dR = \nu - \alpha \geq 0$, and $L(z^{-1})$ is a *polynomial* in z^{-1} of degree $dL = \mu + \beta - 1 \geq 0$. From (7.65), one obtains, due to the degree conditions and pole locations:

$$\left[\frac{G(z, z^{-1})}{D(z)F(z^{-1})} \right]_+ = \left[\frac{z^\alpha R(z)}{D(z)} \right]_+ + \left[\frac{z^\beta L(z^{-1})}{F(z^{-1})} \right]_+$$

$$= \frac{z^\alpha R(z)}{D(z)} .$$ (7.66)

Equation (7.66) shows, in essence, that the problem stated can be solved by making a partial fraction decomposition of $G/(DF)$ and retaining the terms with poles inside the unit circle.

It remains to find the appropriate choices of the integers α and β. These numbers must be selected so that (7.65), interpreted as a system of linear equations, has the same number of unknowns as the number of equations. Further, it is of course required that (7.65) has a unique solution. It is postulated that the appropriate choice is

$$
\begin{aligned}
\alpha &= \min(0, \mu - p), \\
\beta &= \max(0, \ell - \nu).
\end{aligned}
\tag{7.67}
$$

As a verification, the number of unknowns in (7.65) becomes

$$
\begin{aligned}
\# \text{ unknowns} &= (dR + 1) + (dL + 1) \\
&= \nu + \mu + 1 - \alpha + \beta \\
&= \nu + \mu + 1 + \max(0, p - \mu) + \max(0, \ell - \nu) \\
&= 1 + \max(\mu, p) + \max(\nu, \ell).
\end{aligned}
\tag{7.68}
$$

This is to be compared with the number of equations, which is determined by the number of powers z^j appearing in (7.65). The highest power is

$$
\begin{aligned}
n_{\max} &= \max(\ell, \alpha + dR, \beta + \nu) \\
&= \max(\ell, \nu, \nu, \ell - \nu + \nu) \\
&= \max(\ell, \nu).
\end{aligned}
$$

The smallest power of z^j turns out to be

$$
\begin{aligned}
n_{\min} &= \min(-p, \alpha - \mu, \beta - dL) \\
&= \min(-p, -\mu, -p, -\mu + 1) \\
&= \min(-p, -\mu) \\
&= -\max(p, \mu).
\end{aligned}
$$

Hence the total number of equations becomes

$$
\begin{aligned}
\# \text{ equations} &= n_{\max} - n_{\min} + 1 \\
&= \max(\ell, \nu) + \max(p, \mu) + 1 \\
&= \# \text{ unknowns}.
\end{aligned}
\tag{7.69}
$$

As for the compatibility of (7.65), it can be seen from the above investigation that all the powers of z^j appear possibly in the left-hand side and definitely in the right-hand side. Hence, there will be no superfluous equations without any unknowns.

Finally, to examine the uniqueness properties of the solution, it is convenient to rewrite (7.65) as a pure polynomial identity, involving only positive powers of z. Multiplying both sides by $z^{-n_{\min}} = z^{\max(p,\mu)}$ gives

$$
\begin{aligned}
z^{\max(p,\mu)} G(z, z^{-1}) &\equiv z^{\alpha + \max(p,\mu)} F(z^{-1}) R(z) \\
&\quad + z^{\beta + \max(p,\mu)} D(z) L(z^{-1}) \\
&\equiv [z^\mu F(z^{-1})] R(z) + [z^{\mu + \beta - 1} L(z^{-1})][z^{1 + \max(p - \mu, 0)} D(z)].
\end{aligned}
\tag{7.70}
$$

However, by construction, the polynomials $z^\mu F(z^{-1})$ and $z^{1+\max(p-\mu,0)}D(z)$ are coprime. Owing to the general properties of Diophantine equations, it can be concluded that equations (7.65) and (7.70) have a unique solution, with the stated polynomial degrees.

The findings are illustrated in an example by reconsidering optimal prediction of an ARMA process.

Example 7.7 Consider the ARMA model

$$A(q)y(t) = C(q)e(t) , \tag{7.71}$$

where

$$A(q) = q^n + a_1 q^{n-1} + \ldots + a_n ,$$
$$C(q) = q^n + c_1 q^{n-1} + \ldots + c_n$$

have all zeros inside the unit circle.

Find the optimal k-steps ahead prediction. Then one must compute

$$f(z) = \left[z^k \frac{C(z)}{A(z)} \right]_+ . \tag{7.72}$$

See Example 7.5. Hence, in this case:

$$\begin{aligned}
G(z, z^{-1}) &= z^k C(z) , \\
D(z) &= A(z) , \\
F(z) &= 1 , \\
p &= 0, \ \ell = k + n, \ \nu = n, \ \mu = 0 .
\end{aligned} \tag{7.73}$$

From (7.67):

$$\begin{aligned}
\alpha &= \min(0, 0) = 0 , \\
\beta &= \max(0, k + n - n) = k ,
\end{aligned}$$

and the polynomial degrees

$$\begin{aligned}
dR &= \nu - \alpha = n , \\
dL &= \beta - 1 = k - 1 .
\end{aligned}$$

The Diophantine equation (7.65) gives

$$z^k C(z) \equiv R(z) + z^k A(z)L(z^{-1}) . \tag{7.74}$$

As $L(z^{-1})$ is of degree $k - 1$, it holds that $z^{k-1}L(z^{-1})$ is a polynomial in z. Therefore, a factor z can be cancelled in (7.74), which implies that

$$R(z) = zR_1(z) ,$$

with $\deg R_1(z) = dR - 1 = n - 1$. Summarizing, one obtains

$$z^{k-1}C(z) \equiv R_1(z) + z^{k-1}A(z)L(z^{-1}) , \tag{7.75}$$

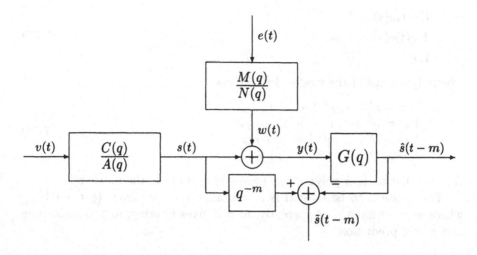

Fig. 7.2. Setup for a polynomial estimation problem

which is recognized as the predictor identity (7.13). The causal part (7.72) is found as (see (7.66))

$$f(z) = \frac{z^0 R(z)}{A(z)} = \frac{z R_1(z)}{A(z)} ,$$

which coincides with the findings in (7.59).   □

7.4 Minimum Variance Filters

7.4.1 Introduction

Many estimation problems for filtering, prediction and smoothing can be solved by polynomial methods. Here, one typical example is analyzed. Consider the measurement of a random signal $s(t)$ in additive noise $w(t)$. This noise source need not be white, but it is assumed to be uncorrelated with the signal $s(t)$. Model $s(t)$ and $w(t)$ as ARMA processes; see Figure 7.2. Thus

$$y(t) = s(t) + w(t) ,$$
$$s(t) = \frac{C(q)}{A(q)} v(t) ,$$
$$w(t) = \frac{M(q)}{N(q)} e(t) ,$$

(7.76)

$$\mathbf{E}\, v(t)v(s) = \lambda_v^2 \delta_{t,s} \, ,$$
$$\mathbf{E}\, e(t)e(s) = \lambda_e^2 \delta_{t,s} \, , \tag{7.77}$$
$$\mathbf{E}\, e(t)v(s) = 0 \, .$$

The polynomials in the model of (7.76) are

$$\begin{aligned}
C(q) &= q^n + c_1 q^{n-1} + \ldots + c_n \, , \\
A(q) &= q^n + a_1 q^{n-1} + \ldots + a_n \, , \\
M(q) &= q^r + m_1 q^{r-1} + \ldots + m_r \, , \\
N(q) &= q^r + n_1 q^{r-1} + \ldots + n_r \, .
\end{aligned} \tag{7.78}$$

Assume that all four polynomials have all zeros inside the unit circle.

The problem to be treated is to estimate $s(t - m)$ from $\{y(t - j)\}_{j=0}^{\infty}$, where m is an integer. (Apparently, $m = 0$ gives filtering, $m > 0$ smoothing and $m < 0$ prediction.)

7.4.2 Solution

The problem is solved using the Wiener filtering techniques of Section 7.3. First, perform a spectral factorization of the output spectrum. Let $|z| = 1$. Then

$$\begin{aligned}
\phi_y(z) &= \lambda_v^2 \frac{C(z)C^*(z)}{A(z)A^*(z)} + \lambda_e^2 \frac{M(z)M^*(z)}{N(z)N^*(z)} \\
&\equiv \lambda_\varepsilon^2 \frac{B(z)B^*(z)}{A(z)N(z)A^*(z)N^*(z)} \, ,
\end{aligned} \tag{7.79}$$

requiring that $B(z)$ is a monic polynomial (that is, it has a leading 1 coefficient) of degree $n + r$, and that it has all zeros inside the unit circle. The polynomial $B(z)$ is therefore uniquely given by the identity

$$\boxed{\begin{aligned}
\lambda_\varepsilon^2 B(z)B^*(z) &\equiv \lambda_v^2 C(z)C^*(z)N(z)N^*(z) \\
&\quad + \lambda_e^2 A(z)A^*(z)M(z)M^*(z) \, .
\end{aligned}} \tag{7.80}$$

Hence, in terms of Section 7.3:

$$H(z) = \frac{B(z)}{A(z)N(z)} \, , \quad \Lambda = \lambda_\varepsilon^2 \, , \quad \phi_{sy}(z) = z^{-m} \lambda_v^2 \frac{C(z)C^*(z)}{A(z)A^*(z)} \, . \tag{7.81}$$

According to Section 7.3 (7.57) the optimal filter becomes

$$\begin{aligned}
G(z) &= \left[z^{-m} \lambda_v^2 \frac{C(z)C^*(z)}{A(z)A^*(z)} \frac{A^*(z)N^*(z)}{B^*(z)} \frac{1}{\lambda_\varepsilon^2} \right]_+ \frac{A(z)N(z)}{B(z)} \\
&= \frac{\lambda_v^2}{\lambda_\varepsilon^2} \left[z^{-m} \frac{C(z)C^*(z)N^*(z)}{A(z)B^*(z)} \right]_+ \frac{A(z)N(z)}{B(z)} \, .
\end{aligned} \tag{7.82}$$

In order to proceed, it is necessary to find the causal part $[\ ]_+$ in (7.82). Apply the technique of Section 7.3.6, and identify the quantities as

$$G(z, z^{-1}) = z^{-m}C(z)C^*(z)N^*(z) ,$$
$$D(z) = A(z) ,$$
$$F(z^{-1}) = B(z^{-1}) , \tag{7.83}$$
$$p = m + n + r, \ell = n - m ,$$
$$\nu = n, \mu = n + r .$$

This approach gives, from (7.67):

$$\alpha = \min(0, n + r - m - n - r) = \min(0, -m) ,$$
$$\beta = \max(0, n - m - n) = \max(0, -m) .$$

The Diophantine equation (7.65) becomes

$$z^{-m}C(z)C^*(z)N^*(z) \equiv z^{\min(0,-m)}B^*(z)R(z)$$
$$+ z^{\max(0,-m)}A(z)L^*(z) , \tag{7.84}$$

where the unknown polynomials have degrees

$$dR = n - \min(0, -m) ,$$
$$dL = n + r - 1 + \max(0, -m) . \tag{7.85}$$

Note that the "-1" that appears in dL has no direct correspondence in dR. The reason is that the direct term $g_0 z^{-0}$ in (7.51) is associated with the causal part of $G(z)$.

The optimal filter is readily found from (7.66). It gives

$$G(z) = \frac{\lambda_v^2}{\lambda_\varepsilon^2} \frac{z^{\min(0,-m)}R(z)}{A(z)} \frac{A(z)N(z)}{B(z)}$$

or

$$\boxed{G(z) = \frac{\lambda_v^2}{\lambda_\varepsilon^2} \frac{z^{\min(0,-m)}R(z)N(z)}{B(z)} .} \tag{7.86}$$

It is worth checking the causality of the filter in (7.86) for illustration. This is done by examining the degrees:

$$\deg[z^{\min(0,-m)}R(z)N(z)] = \min(0, -m) + n - \min(0, -m) + r$$
$$= n + r = \deg[B(z)] .$$

The filter is causal as its numerator and denominator have the same degrees.

7.4.3 The Estimation Error

An expression for the estimation error spectrum is now derived.

For ease of notation, drop the argument z in the derivation; so, in what follows, $S = S(z)$, $S^* = S^*(z)$, etc.

The estimation error can be written as

$$\tilde{s}(t) = s(t-m) - \hat{s}(t-m|t)$$

$$= s(t-m) - Gy(t)$$

$$= q^{-m}\frac{C(q)}{A(q)}v(t) - \frac{\lambda_v^2}{\lambda_\varepsilon^2}q^{\min(0,-m)}\frac{R(q)N(q)}{B(q)}\left[\frac{C(q)}{A(q)}v(t) + \frac{M(q)}{N(q)}e(t)\right]$$

$$= \frac{C(q)}{A(q)B(q)}\left[q^{-m}B(q) - \frac{\lambda_v^2}{\lambda_\varepsilon^2}q^{\min(0,-m)}R(q)N(q)\right]v(t)$$

$$- \frac{\lambda_v^2}{\lambda_\varepsilon^2}q^{\min(0,-m)}\frac{R(q)M(q)}{B(q)}e(t) . \tag{7.87}$$

As $e(t)$ and $v(t)$ are independent, one can write the spectrum of $\tilde{s}(t)$ as (all polynomials having argument z)

$$\phi_{\tilde{s}}(z) = \lambda_v^2\frac{CC^*}{AA^*BB^*}\left[z^mB^* - \frac{\lambda_v^2}{\lambda_\varepsilon^2}z^{-\min(0,-m)}R^*N^*\right]$$

$$\times\left[z^{-m}B - \frac{\lambda_v^2}{\lambda_\varepsilon^2}z^{\min(0,-m)}RN\right] + \frac{\lambda_v^4\lambda_e^2}{\lambda_\varepsilon^4}\frac{RR^*MM^*}{BB^*} . \tag{7.88}$$

This implies that

$$\lambda_\varepsilon^2BB^*AA^*\phi_{\tilde{s}}(z) = \lambda_v^2CC^*\left[\lambda_\varepsilon^2BB^* - \lambda_v^2z^{-m-\min(0,-m)}R^*N^*B\right.$$

$$\left. - \lambda_v^2z^{m+\min(0,-m)}B^*RN + \frac{\lambda_v^4}{\lambda_\varepsilon^2}RR^*NN^*\right]$$

$$+ \frac{\lambda_v^4\lambda_e^2}{\lambda_\varepsilon^2}RR^*MM^*AA^* .$$

Next, recall that (7.84) can be written as

$$z^{-m}CC^*N^* = z^{\min(0,-m)}B^*R + z^{\max(0,-m)}AL^* ,$$

and that (7.80) can be rephrased as

$$\lambda_\varepsilon^2BB^* = \lambda_v^2CC^*NN^* + \lambda_e^2AA^*MM^* .$$

Using these relations repeatedly gives

$$\lambda_\varepsilon^2BB^*AA^*\phi_{\tilde{s}}(z)$$

$$= \lambda_v^2CC^*[\lambda_v^2CC^*NN^* + \lambda_e^2AA^*MM^*]$$

$$- \lambda_v^4z^{-\min(0,-m)}R^*B[z^{\min(0,-m)}B^*R + z^{\max(0,-m)}AL^*]$$

$$- \lambda_v^4z^{\min(0,-m)}RB^*[z^{-\min(0,-m)}BR^* + z^{-\max(0,-m)}A^*L]$$

$$+ \frac{\lambda_v^4}{\lambda_\varepsilon^2}RR^*[\lambda_\varepsilon^2BB^* - \lambda_e^2AA^*MM^*] + \frac{\lambda_v^4\lambda_e^2}{\lambda_\varepsilon^2}RR^*MM^*AA^*$$

$$= \lambda_v^4[z^{\min(0,-m)}B^*R + z^{\max(0,-m)}AL^*]$$

$$\times[z^{-\min(0,-m)}BR^* + z^{-\max(0,-m)}A^*L] + \lambda_v^2\lambda_e^2CC^*AA^*MM^*$$

$$- \lambda_v^4[RR^*BB^* + z^{\max(0,-m)-\min(0,-m)}R^*BAL^*$$

$$+ z^{\min(0,-m)-\max(0,-m)}RB^*A^*L]$$

$$= \lambda_v^2 \lambda_e^2 CC^* AA^* MM^* + \lambda_v^4 AA^* LL^* .$$

Hence

$$\boxed{\phi_{\tilde{s}}(z) = \frac{\lambda_v^2 \lambda_e^2}{\lambda_\varepsilon^2} \frac{CC^* MM^*}{BB^*} + \frac{\lambda_v^4}{\lambda_\varepsilon^2} \frac{LL^*}{BB^*} .} \tag{7.89}$$

Note that the smoothing lag m does not appear explicitly in this equation. It only influences the polynomial L^*. The first term in (7.89) gives a lower bound on the minimal error spectrum that can be obtained. In fact, when $m \to \infty$ (smoothing with a lag tending to infinity), L will tend to zero. The first term in (7.89) corresponds to the minimal loss when the estimate is based on an infinite amount of future data points.

It is possible to give still further interpretations of (7.89). Consider the unrealizable Wiener filter. In the present case,

$$\phi_y(z) = \lambda_\varepsilon^2 \frac{B(z)B^*(z)}{A(z)N(z)A^*(z)N^*(z)} , \tag{7.90}$$

$$\phi_{sy}(z) = \lambda_v^2 z^{-m} \frac{C(z)C^*(z)}{A(z)A^*(z)} . \tag{7.91}$$

Hence the unrealizable Wiener filter becomes

$$G(z) = z^{-m} \frac{\lambda_v^2}{\lambda_\varepsilon^2} \frac{C(z)N(z)C^*(z)N^*(z)}{B(z)B^*(z)} . \tag{7.92}$$

The error spectrum is, in the general case, given by (7.38). Inserting the unrealizable Wiener filter gives

$$\begin{aligned}
\phi_{\tilde{s}}(z) &= \phi_s(z) - \phi_{sy}(z)\phi_y^{-1}(z)\phi_{ys}(z) \\
&= \lambda_v^2 \frac{CC^*}{AA^*} - \frac{\lambda_v^4}{\lambda_\varepsilon^2} \frac{(CC^*)^2}{(AA^*)^2} \frac{AA^* NN^*}{BB^*} \\
&= \frac{\lambda_v^2}{\lambda_\varepsilon^2} \frac{CC^*}{AA^* BB^*} [\lambda_\varepsilon^2 BB^* - \lambda_v^2 CC^* NN^*] \\
&= \frac{\lambda_v^2}{\lambda_\varepsilon^2} \frac{CC^*}{AA^* BB^*} \lambda_e^2 AA^* MM^* \\
&= \frac{\lambda_v^2 \lambda_e^2}{\lambda_\varepsilon^2} \frac{CC^* MM^*}{BB^*} .
\end{aligned} \tag{7.93}$$

This expression is the same as that in the first term of (7.89). Phrased differently, the error spectrum obtained with the unrealizable Wiener filter gives a point-wise lower bound for the error spectrum achievable for a realizable filter.

7.4.4 Extensions

The problem considered in this section can be extended in several ways. The principal approach to solution remains the same. Here are some examples of extensions:

1. Multivariable processes.
2. Estimation of a signal different from, but correlated with, $s(t)$. One may, for example, want to estimate some internal variable in the dynamics, say $C(q)/A(q)v(t)$.
3. Prefiltering of the criterion; that is, the optimal filter should minimize

$$V = \mathbf{E}\left[P(q)\tilde{s}(t)\right]^2 , \tag{7.94}$$

   where $\tilde{s}(t)$ is the estimation error

$$\tilde{s}(t) = s(t-m) - G(q)y(t) , \tag{7.95}$$

   and $P(q)$ is a given rational filter. The idea is, of course, that by using $P(q)$ the frequency contents of the estimation error $\tilde{s}(t)$ can be influenced. For example, if it is required that $\tilde{s}(t)$ should have little energy in a certain frequency band, one should make $P(e^{i\omega})$ large in that band.
4. Drifting signals and disturbances (*i.e.* $A(q)$ or $N(q)$ has zeros on the unit circle). This case can be handled with the approach presented. Compare how the prediction of nonstationary processes was explained in Section 7.2.4.

Another, more indirect, extension is H^∞ filtering. Such filters are based, in analogy with (robust) H^∞ controller design, on not minimizing the variance $\oint \phi_{\tilde{s}}(z)\,dz/z$ of the estimation error. Instead, they are based on a min–max approach. The filter minimizing $\sup_{|z|=1} |\phi_{\tilde{s}}(z)|$ is sought. The filter design treated in this section can be used with a prefiltered criterion to design some classes of such filters.

The third extension above will be considered in some detail. Set the weighting filter to

$$P(q) = \frac{F(q)}{H(q)} , \tag{7.96}$$

where $F(q)$ and $H(q)$ are assumed to be coprime, of the same degree, and have all zeros strictly inside the unit circle.

The filtered estimation error can be rewritten as

$$P(q)\tilde{s}(t) = \frac{F(q)}{H(q)}[s(t-m) - G(q)y(t)]$$

$$= \frac{F(q)C(q)}{H(q)A(q)}v(t-m)$$

$$-G(q)\frac{F(q)}{H(q)}[\frac{C(q)}{A(q)}v(t) + \frac{M(q)}{N(q)}e(t)] . \tag{7.97}$$

It can be seen that this can be interpreted as the estimation error of an original problem with polynomials $\tilde{C}(q)$, $\tilde{A}(q)$, $\tilde{M}(q)$, $\tilde{N}(q)$ as

$$\tilde{C} = FC\,,$$
$$\tilde{A} = HA\,,$$
$$\tilde{M} = FM\,,$$
$$\tilde{N} = HN\,.$$

(7.98)

Proceed as before to obtain the solution. The spectral factorization of (7.80) gives in this case

$$\lambda_{\varepsilon}^2 \tilde{B}(z)\tilde{B}^*(z) \equiv \lambda_v^2 \tilde{C}(z)\tilde{C}^*(z)\tilde{N}(z)\tilde{N}^*(z)$$
$$+\lambda_e^2 \tilde{M}(z)\tilde{M}^*(z)\tilde{A}(z)\tilde{A}^*(z)\,.$$

(7.99)

Using (7.98), it is straightforward to see that

$$\tilde{B}(z) = B(z)F(z)H(z)\,,$$

(7.100)

where $B(z)$ is defined by (7.80). Next consider the Diophantine equation (7.84) to find the optimal filter. The equation becomes

$$z^{-m}FCF^*C^*H^*N^* \equiv z^{\min(0,-m)}B^*F^*H^*\tilde{R}+z^{\max(0,-m)}HA\tilde{L}^*\,.$$

(7.101)

Noting that $F^*(z)H^*(z)$ is a factor of the left-hand side and of the first term of the right-hand side, it can be concluded that it must be a factor of $\tilde{L}^*(z)$ as well. (Owing to the stability constraint, it must be coprime with $H(z)A(z)$.) Thus set

$$\tilde{L}^*(z) = F^*(z)H^*(z)L_0^*(z)\,.$$

(7.102)

Cancelling the factor $F^*(z)H^*(z)$ gives, from (7.101):

$$z^{-m}FCC^*N^* \equiv z^{\min(0,-m)}B^*\tilde{R} + z^{\max(0,-m)}HAL_0^*\,.$$

(7.103)

The optimal filter then becomes (see (7.86))

$$G(z) = \frac{\lambda_v^2}{\lambda_{\varepsilon}^2}\frac{z^{\min(0,-m)}\tilde{R}(z)\tilde{N}(z)}{\tilde{B}(z)}$$

$$= \frac{\lambda_v^2}{\lambda_{\varepsilon}^2}\frac{z^{\min(0,-m)}\tilde{R}(z)N(z)}{B(z)F(z)}\,.$$

(7.104)

7.4.5 Illustrations

For illustration, we now apply the general filter derived in Section 7.4.2 to some simple cases.

Example 7.8 Consider the k-step prediction of an ARMA process:

$$A(q)y(t) = C(q)v(t)\,.$$

(7.105)

Then $M = 1$, $N = 1$, $\lambda_e^2 = 0$, $m = -k$. This gives $B = C$, $\lambda_{\varepsilon}^2 = \lambda_v^2$, from (7.80). The Diophantine equation (7.84) becomes

$$z^k C(z) C^*(z) \equiv z^{\min(0,k)} C^*(z) R(z) + z^{\max(0,k)} A(z) L^*(z)$$
$$\equiv C^*(z) R(z) + z^k A(z) L^*(z) . \tag{7.106}$$

Hence

$$\deg R = n - \min(0, k) = n ,$$
$$\deg L^* = n + 0 - 1 + \max(0, k) = n - 1 + k .$$

As $C^*(z)$ is a factor of the left-hand side and the first term of the right-hand side of (7.106), it must be a factor of $L^*(z)$ as well. Therefore, one can set $L^*(z) = C^*(z) L_0^*(z)$, where $\deg L_0^* = k - 1$. Similarly, z must be a factor of R, so $R = z R_0$, $\deg R_0 = n - 1$. Cancelling the factor $z C^*(z)$ gives

$$z^{k-1} C(z) \equiv R_0(z) + z^{k-1} A(z) L_0^*(z)$$
$$= A(z)[z^{k-1} L_0^*(z)] + R_0(z) ,$$

which is the predictor identity (7.13) that was derived earlier. The optimal predictor becomes, according to (7.86):

$$G(z) = \frac{\lambda_v^2}{\lambda_\varepsilon^2} \frac{z^{\min(0,k)} R(z)}{B(z)} = \frac{z R_0(z)}{C(z)} ,$$

which is consistent with our previous findings.   □

Example 7.9 Consider k-step prediction of an ARMA process observed in *white* noise. Using for a moment the techniques of Chapter 6, it is concluded that

$$\hat{s}(t + k|Y^t) = \mathbf{E}\left[s(t + k)|Y^t\right]$$
$$= \mathbf{E}\left[y(t + k)|Y^t\right] - \mathbf{E}\left[e(t + k)|Y^t\right]$$
$$= \hat{y}(t + k|t) .$$

In this example, $M = N = 1$, $m = -k$. The spectral factorization (7.80) gives

$$\lambda_\varepsilon^2 B(z) B^*(z) \equiv \lambda_v^2 C(z) C^*(z) + \lambda_e^2 A(z) A^*(z) .$$

The Diophantine equation (7.84) becomes

$$z^k C(z) C^*(z) \equiv B^*(z) R(z) + z^k A(z) L^*(z) ,$$

and the optimal predictor is found, from (7.86), to be

$$\hat{s}(t + k|t) = \frac{\lambda_v^2}{\lambda_\varepsilon^2} \frac{R(q)}{B(q)} y(t) .$$

As a further illustration, recall that the measurements $y(t)$ can be expressed as the ARMA model

$$A(q) y(t) = B(q) \varepsilon(t) .$$

According to Example 7.8:

$$\hat{y}(t + k|t) = \frac{qR_0(q)}{B(q)} y(t) ,$$

where

$$z^{k-1}B = R_0 + z^{k-1}AL_0^* .$$

Thus $\hat{s}(t + k|t)$ and $\hat{y}(t + k|t)$ are identical if and only if

$$zR_0(z) = \frac{\lambda_v^2}{\lambda_\varepsilon^2} R(z)$$

$$\iff z^k B - z^k AL_0^* = \frac{\lambda_v^2}{\lambda_\varepsilon^2} \frac{z^k CC^* - z^k AL^*}{B^*}$$

$$\iff \lambda_\varepsilon^2 BB^* - \lambda_\varepsilon^2 B^* AL_0^* = \lambda_v^2 CC^* - \lambda_v^2 AL^*$$

$$\iff \lambda_\varepsilon^2 AA^* - \lambda_\varepsilon^2 B^* AL_0^* = -\lambda_v^2 AL^*$$

$$\iff \lambda_\varepsilon^2 A^* - \lambda_\varepsilon^2 B^* L_0^* = -\lambda_v^2 L^* ,$$

which can be fulfilled by a pertinent polynomial L. Phrased differently, $L^* = \lambda_\varepsilon^2/\lambda_v^2 B^* L_0^* - \lambda_\varepsilon^2/\lambda_v^2 A^*$ is the polynomial of the right structure.

□

Example 7.10 Consider also the smoothing of an ARMA process, observed in white noise. Then

$$M = 1 , \quad N = 1 , \quad m > 0 .$$

The spectral factorization is again

$$\lambda_\varepsilon^2 BB^* \equiv \lambda_v^2 CC^* + \lambda_e^2 AA^* ,$$

and the Diophantine equation (7.84) becomes (deg $R = n+m$, deg $L^* = n-1$)

$$z^{-m}CC^* \equiv z^{-m}B^* R + AL^* .$$

The optimal smoothing is given by (see (7.86))

$$\hat{s}(t - m|t) = \frac{\lambda_v^2}{\lambda_\varepsilon^2} \frac{q^{-m}R(q)}{B(q)} y(t) .$$

□

Example 7.11 Reconsider the smoothing problem of the previous example, but it will now be treated in a time domain format, using also some of the results of Chapter 6. Assume that the noise sequences are Gaussian distributed.

The measured signal can be written as an equivalent ARMA model

$$y(t) = \frac{B(q)}{A(q)} \varepsilon(t) , \qquad \mathbf{E}\, \varepsilon^2(t) = \lambda_\varepsilon^2 , \tag{7.107}$$

where $\varepsilon(t)$ is the output innovation at time t. The optimal smoothing estimate is given by the conditional mean

$$
\begin{aligned}
\hat{s}(t-m|t) &= \mathbf{E}\left[s(t-m)|Y^t\right] \\
&= \mathbf{E}\left[y(t-m) - e(t-m)|y(t), y(t-1), \ldots\right] \\
&= y(t-m) \\
&\quad -\mathbf{E}\left[e(t-m)|\varepsilon(t), \varepsilon(t-1), \ldots \varepsilon(t-m), Y^{t-m-1}\right] \\
&= y(t-m) - \sum_{i=0}^{m}[\mathbf{E}\, e(t-m)\varepsilon(t-i)]\frac{1}{\lambda_\varepsilon^2}\varepsilon(t-i) \, .
\end{aligned}
\tag{7.108}
$$

To proceed, introduce polynomials $F(z)$ and $G(z)$ by

$$
\begin{aligned}
z^m A(z) &= B(z)F(z) + G(z) \, , \\
F(z) &= z^m + f_1 z^{m-1} + \ldots + f_m \, , \\
G(z) &= g_0 z^{n-1} + \ldots + g_n \, .
\end{aligned}
\tag{7.109}
$$

For $i = 0, \ldots, m$, $(f_0 \stackrel{\triangle}{=} 1)$

$$
\begin{aligned}
\mathbf{E}\, e(t-m)\varepsilon(t-i) &= \mathbf{E}\, e(t-m)\frac{A(q)}{B(q)}y(t-i) \\[2mm]
&= \mathbf{E}\, e(t-m)\frac{A(q)}{B(q)}e(t-i) \\[2mm]
&= \mathbf{E}\, e(t-m)\left[\frac{B(q)F(q) + G(q)}{B(q)}q^{-m}e(t-i)\right] \\[2mm]
&= \mathbf{E}\, e(t)\left[F(q)e(t-i) + \frac{G(q)}{B(q)}e(t-i)\right] \\[2mm]
&= \lambda_e^2 f_{m-i}
\end{aligned}
\tag{7.110}
$$

holds. (Note that $\deg G < \deg B$; hence $e(t)$ and $G(q)/B(q)e(t-i)$ are uncorrelated.) Using (7.107) and (7.110) in (7.108), one obtains the following explicit form for the smoothing estimate:

$$
\begin{aligned}
\hat{s}(t-m|t) &= y(t-m) - \sum_{i=0}^{m}\frac{\lambda_e^2}{\lambda_\varepsilon^2}f_{m-i}\frac{A(q)}{B(q)}y(t-i) \\[2mm]
&= \left[q^{-m} - \frac{\lambda_e^2}{\lambda_\varepsilon^2}\sum_{i=0}^{m}f_{m-i}q^{-i}\frac{A(q)}{B(q)}\right]y(t) \\[2mm]
&= \left[q^{-m} - \frac{\lambda_e^2}{\lambda_\varepsilon^2}F(q^{-1})\frac{A(q)}{B(q)}\right]y(t) \, .
\end{aligned}
\tag{7.111}
$$

□

7.5 Robustness Against Modelling Errors

So far, it has been assumed in this book that exact models are available. This is, of course, an idealization. It is of considerable importance to know how sensitive the derived filters and predictors are to modelling errors, and also to derive schemes that are robust (*i.e.* have low sensitivity) to modelling errors. As a starting point, consider a simple example.

Example 7.12 Consider an MA(1) process observed in white noise. It will be of interest to examine the one-step predictor and its prediction error variance.

Assume that the *nominal model* (*i.e.* the model used for computing the predictor) is

$$s(t) = (1 + c_0 q^{-1})v(t) ,$$
$$y(t) = s(t) + e(t) ,$$

(7.112)

where $v(t)$ and $e(t)$ are mutually uncorrelated white noise sequences of variances λ_v^2 and λ_e^2 respectively. Assume first that $\lambda_e^2 = 0$. By using the results of Section 7.2.2 one obtains

$$\hat{y}(t+1|t) = \frac{c_0}{1 + c_0 q^{-1}} y(t) .$$

Assume, further, that the true process is given by

$$y(t) = (1 + c q^{-1})v(t) ,$$

(7.113)

where c may differ from c_0. The prediction error variance becomes

$$
\begin{aligned}
V_{\text{nom}} &= \mathbf{E}\left[y(t+1) - \hat{y}(t+1|t)\right]^2 \\
&= \mathbf{E}\left[(q - \frac{c_0}{1 + c_0 q^{-1}})(1 + c q^{-1})v(t)\right]^2 \\
&= \mathbf{E}\left[\frac{1 + c q^{-1}}{1 + c_0 q^{-1}} v(t+1)\right]^2 \\
&= \lambda_v^2\left[1 + \frac{(c - c_0)^2}{1 - c_0^2}\right] .
\end{aligned}
$$

(7.114)

As a function of c, the error variance V has a minimum for $c = c_0$, which was expected (this means that the nominal model happens to be equal to the true process). If c_0 is grossly in error (deviates considerably from c), the increase in the prediction error variance may be substantial.

Now examine a robustified predictor also, which is less sensitive to deviation of c_0 from c. Assume that the nominal model is

$$y(t) = [1 + (c_0 + \tilde{c})q^{-1}]v(t) ,$$

where $\tilde{c}$ is a random variable of zero mean, variance σ^2, and independent of the noise sequence $\{v(t)\}$. It describes the confidence in the nominal "mean value" c_0. Taking expectation over both $\tilde{c}$ and $v(t)$ gives

$$\mathbf{E}\,y^2(t) = \mathbf{E}\,[v(t) + (c_0 + \tilde{c})v(t-1)]^2$$
$$= \lambda_v^2 + \lambda_v^2(c_0^2 + \sigma^2)\,,$$
$$\mathbf{E}\,y(t)y(t-1) = c_0\lambda_v^2\,,$$
$$\mathbf{E}\,y(t)y(t-\tau) = 0\,, \qquad \tau > 1\,.$$

It is thus found that $y(t)$ has the covariance "structure" of a first-order MA process. As far as its second-order properties are concerned, it can be modelled as

$$y(t) = [1 + \bar{c}q^{-1}]\varepsilon(t)\,, \qquad |\bar{c}| < 1\,.$$

The values of $\bar{c}$ and $\lambda_\varepsilon^2 = \mathbf{E}\,\varepsilon^2(t)$ can be found by spectral factorization. The governing equations become

$$\lambda_v^2[1 + c_0^2 + \sigma^2] = \lambda_\varepsilon^2[1 + \bar{c}^2]\,,$$
$$\lambda_v^2 c_0 = \lambda_\varepsilon^2 \bar{c}\,, \qquad\qquad\qquad (7.115)$$

which uniquely define λ_ε^2 and $\bar{c}$ as functions of λ_v^2, c_0 and σ. Note that, from (7.112), $y(t)$ can be interpreted as consisting of the nominal mean signal model $s(t) = (1 + c_0 q^{-1})v(t)$ and an additive white noise of variance $\lambda_v^2\sigma^2$. The corresponding nominal predictor, taking "the uncertainty parameter" σ into account, is now given by

$$\hat{y}(t+1|t) = \frac{\bar{c}}{1 + \bar{c}q^{-1}}y(t)\,. \qquad\qquad (7.116)$$

It is easy to see that the true prediction error variance now modifies from (7.114) to

$$V_{\mathrm{rob}} = \lambda_v^2\left[1 + \frac{(c - \bar{c})^2}{1 - \bar{c}^2}\right]\,. \qquad\qquad (7.117)$$

The sensitivities of the prediction error variances, with respect to variations of c, for the nominal design (given by (7.114)) and for the robustified design (given by (7.117)), are illustrated in Figure 7.3.

It can be seen that for very small deviations (c being very close to c_0), the nominal design gives the best behaviour and close to optimal value $V = 1$. However, for moderate or large deviations of c from c_0, the robustified design gives the best performance. Its error variance is not as sensitive as the nominal error variance to large errors. □

After this example, which was examined in quite some detail, a brief indication of how the same type of idea can be used in a more general setting is given. The idea can be formulated as follows:

Introduce stochastic descriptions of the parameter uncertainties. Evaluate the output covariance elements. Rewrite the output using a second-order equivalent model transferring the parameter uncertainties to a virtual noise source.

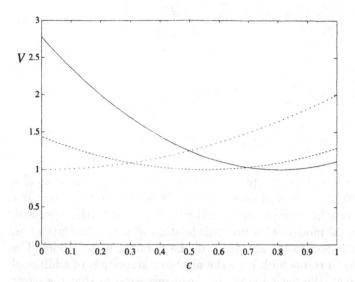

Fig. 7.3. Prediction error variance V as a function of the true c parameter. Nominal design with $c_0 = 0.8$ (*solid line*); robustified design with $c_0 = 0.8$, $\sigma = 0.5$ (*dashed line*); no prediction ($\hat{y} \equiv 0$) (*dash-dotted line*)

Consider the basic setup of Section 7.4.1. Assume that

$$\frac{C(q)}{A(q)} = \frac{C_0(q)}{A_0(q)} + \frac{\tilde{C}(q)}{A_1(q)}, \tag{7.118}$$

$$\frac{M(q)}{N(q)} = \frac{M_0(q)}{N_0(q)} + \frac{\tilde{M}(q)}{N_1(q)}. \tag{7.119}$$

Here, C_0/A_0 and M_0/N_0 are nominal models and A_1 and N_1 are fixed polynomials. Further:

$$\tilde{C}(q) = \tilde{c}_1 q^{n-1} + \ldots + \tilde{c}_n,$$
$$\tilde{M}(q) = \tilde{m}_1 q^{r-1} + \ldots + \tilde{m}_r,$$

account for the uncertainties. Assume that the polynomial coefficients $\{\tilde{c}_i\}$, $\{\tilde{m}_j\}$ have zero mean and given covariances.

The covariance of the perturbed output can be computed. Its spectrum will be ($\overline{\mathbf{E}}$ denotes expectation with respect to the polynomial coefficients $\{\tilde{c}_i\}$, $\{\tilde{m}_j\}$):

$$\phi_y(z) = \overline{\mathbf{E}} \left[\frac{C(z)}{A(z)} \frac{C^*(z^{-*})}{A^*(z^{-*})} \lambda_v^2 + \frac{M(z)}{N(z)} \frac{M^*(z^{-*})}{N^*(z^{-*})} \lambda_e^2 \right]$$

$$= \lambda_v^2 \left[\frac{C_0(z)}{A_0(z)} \frac{C_0^*(z^{-*})}{A_0^*(z^{-*})} + \frac{\overline{\mathbf{E}} \, \tilde{C}(z) \tilde{C}^*(z^{-*})}{A_1(z) A_1^*(z^{-*})} \right]$$

$$+\lambda_e^2 \left[\frac{M_0(z)}{N_0(z)} \frac{M_0^*(z^{-*})}{N_0^*(z^{-*})} + \frac{\overline{E}\tilde{M}(z)\tilde{M}^*(z^{-*})}{N_1(z)N_1^*(z^{-*})} \right] . \tag{7.120}$$

Next, note that

$$\overline{E}\,\tilde{C}(z)\tilde{C}^*(z^{-*}) = \overline{E}\sum_{i=1}^{n}\sum_{j=1}^{n}\tilde{c}_i z^{n-i}\tilde{c}_j(z^{-n+j})^*$$

$$= \sum_{i,j=1}^{n}(\overline{E}\,\tilde{c}_i\tilde{c}_j)z^{n-i}(z^{-n+j})^*$$

has precisely the *same* structure (as a function of z) as the spectrum of a noisy moving average process of order $n-1$. As a consequence, the four terms of (7.120) can be interpreted as follows. The first is the spectrum of the nominal signal model, with no consideration of model uncertainties. Similarly, the third term is the spectrum of the nominal noise model. The second and the fourth terms both have the algebraic structures of additional noise spectra. In fact, the basic model has been rewritten so that the effect of model parameter uncertainties has been transformed into an equivalent noise model (or noise spectrum). By applying spectral factorization, the sum of the last three terms in (7.120) can be written as

$$\lambda_v^2 \left[\frac{\overline{E}\,\tilde{C}(z)\tilde{C}^*(z^{-*})}{A_1(z)A_1^*(z^{-*})} \right] + \lambda_e^2 \left[\frac{M_0(z)}{N_0(z)} \frac{M_0^*(z^{-*})}{N_0^*(z^{-*})} + \frac{\overline{E}\,\tilde{M}(z)\tilde{M}^*(z^{-*})}{N_1(z)N_1^*(z^{-*})} \right]$$

$$\equiv \lambda_\varepsilon^2 \frac{\overline{M}(z)\overline{M}^*(z^{-*})}{\overline{N}(z)\overline{N}^*(z^{-*})} \tag{7.121}$$

for a positive scalar λ_ε^2 and some stable and monic polynomials $\overline{M}$ and $\overline{N}$. Thus it can be seen that estimation problems can be solved using the standard setup of Section 7.4.1 with a modified noise model (replacing M, N, λ_e^2 by $\overline{M}$, $\overline{N}$, λ_ε^2). Note that for this trick to work it has to be assumed that the uncertainties lie in the numerator polynomials only. The polynomials A_1 and N_1 are assumed to be fixed, but they are not required to coincide with A and N respectively.

Exercises

Exercise 7.1 Determine the optimal one-step and two-step predictors for the process

$$y(t) - 0.8y(t-1) = e(t) + 2e(t-1) ,$$

where $e(t)$ is white noise of zero mean and unit variance. Also, determine the prediction error variances.

Exercise 7.2 Modify the expression (7.89) of the error spectrum $\phi_{\tilde{s}}(z)$ when the filter minimizes a prefiltered error, as in (7.94).

Exercise 7.3 Consider the filtering problem of Section 7.4.1 for the scalar case

$$C(q) = q^2 ,$$
$$A(q) = q^2 - 1.5q + 0.7 ,$$
$$M(q) = q^2 - 1.0q + 0.2 ,$$
$$N(q) = q^2 ,$$
$$\lambda_e^2 = \lambda_v^2 = 1 .$$

(a) Compute and plot the signal spectral density $\phi_s(e^{i\omega})$ and the spectral density $\phi_w(e^{i\omega})$ of the disturbance.
(b) Compute the optimal filter $G(q)$ and plot the frequency function for $m = 0, 1, 2, 5, -1, -2, -5$. Also compute and plot the error spectral density $\phi_{\tilde{s}}(e^{i\omega})$.

Exercise 7.4 Reconsider Exercise 7.3 for the filtering case ($m = 0$). Try to design a prefilter of the estimation error so that the unfiltered error becomes less resonant.

Exercise 7.5 Let A and B be square matrices (not necessarily of the same dimension) with all eigenvalues strictly inside the unit circle. Let k be a positive integer and C a matrix. Prove that

$$[z^k(zI - A)^{-1}C(z^{-1}I - B)]_+ = A^k z(zI - A)^{-1}\alpha = z(zI - A)^{-1}A^k\alpha ,$$

where α and β are solutions to the linear system

$$\begin{cases} C = \alpha - A\beta , \\ 0 = -\alpha B + \beta . \end{cases}$$

Exercise 7.6 Consider the dynamic system

$$x(t + 1) = Fx(t) + v(t) ,$$
$$y(t) = Hx(t) + e(t) ,$$

where $v(t)$ and $e(t)$ are mutually white noise sequences of zero mean and covariance matrices R_1 and R_2 respectively. Derive the mean square optimal k-step predictor $\hat{x}(t + k|t)$ using the Wiener filter with the polynomial formalism. Compare the result with what can be derived with the state space formalism.

Hint. The result of Exercise 7.5 is useful.

Exercise 7.7 One might believe from Section 7.4.3 that the error spectrum decreases monotonically as m increases. Examine this conjecture. (Prove it, or find a counterexample.)

Exercise 7.8 Consider the estimation problem defined in Section 7.4. Assume that a prefilter $P(q) = F(q)/H(q)$ is used, and that this prefilter has a sharp resonance at $\omega = \omega_0$. Show, by setting $H(e^{i\omega_0}) = 0$, that the error spectrum obtained achieves the lower bound (7.93) at $\omega = \omega_0$.

Exercise 7.9 Wiener filters give the optimal structure as well as the optimal filter parameters. In most cases it is not possible to give closed form solutions to the optimal filters of a *fixed* structure. Finite impulse response (FIR) filters (filters with all poles in the origin) are exceptions. The reason is that in this case the filtered signal depends *linearly* on the filter parameters. Assume that

$$\hat{s}(t) = G(q)y(t) = \sum_{j=0}^{n} g_j y(t-j),$$

$$\triangleq \varphi(t)g,$$

where

$$\varphi(t) = (y(t) \ldots y(t-n)),$$
$$g = (g_0 \ldots g_n)^T.$$

(a) Show that the filter minimizing the estimation error variance

$$\mathbf{E}\,|s(t) - \hat{s}(t)|^2,$$

with respect to the filter parameter vector g, is given by

$$g = [\mathbf{E}\,\varphi^*(t)\varphi(t)]^{-1}[\mathbf{E}\,s(t)\varphi^*(t)].$$

(b) Consider the particular case of one-step prediction of an ARMA(1,1) process

$$y(t) + ay(t-1) = e(t) + ce(t-1).$$

Derive the one-step optimal predictor using an FIR filter with $n = 0$. Compare with the mean square optimal filter.

(c) Consider the case of k-steps prediction of an ARMA(n_1, n_2) process. When will an FIR filter of order n be identical to the mean square optimal predictor?

Bibliography

In general terms, there is a far richer literature on optimal estimation using state space methods than on polynomial formalism. Some exceptions are

Åström, K.J., Wittenmark, B., 1997. *Computer-Controlled Systems*, third ed. Prentice Hall, Upper Saddle River.

Grimble, M.J., Johnson, M.A., 1988. *Optimal Control and Stochastic Estimation.* John Wiley & Sons, Chichester, UK.

Hunt, K.J. (Ed.), 1993. *Polynomial Methods in Optimal Control and Filtering.* Peter Peregrinus Ltd, Stevenage, UK.

Kučera, V., 1979. *Discrete Linear Control.* John Wiley & Sons, Chichester.

Kučera, V., 1991. *Analysis and Design of Discrete Linear Control Systems.* Prentice Hall International, London.

Optimal prediction of polynomial models as presented here was first developed in

Åström, K.J., 1970. *Introduction to Stochastic Control.* Academic Press, New York.

The algorithmic aspects of the Wiener filter, as outlined in Section 7.3.5, were inspired by

Ahlén, A., M. Sternad, M., 1991. Wiener filter design using polynomial equations. *IEEE Transactions on Signal Processing*, vol SP–39, 2387–2399.

Ahlén, A., Sternad, M., 1993. Optimal filtering problems. In Hunt, K.J., (Ed.), *Polynomial Methods in Optimal Control and Filtering.* Peter Peregrinus Ltd, Stevenage, UK.

Example 7.11 is based on the article

Hagander, P., Wittenmark, B., 1977. A self-tuning filter for fixed-lag smoothing. *IEEE Transactions on Information Theory*, vol. IT–23, 377–384.

8. Illustration of Optimal Linear Estimation

8.1 Introduction

In this chapter, a simple example is analyzed in detail. Both the state space and the polynomial approaches are used.

The system to be considered is a random walk with noisy measurements. The process is modelled as

$$x(t+1) = x(t) + v(t) \,,$$
$$y(t) = x(t) + e(t) \,, \tag{8.1}$$

where $\{v(t)\}$ and $\{e(t)\}$ are mutually independent, Gaussian white noise sequences of zero mean and with variances λ_v^2 and λ_e^2 respectively.

Introduce the parameter β through

$$\beta = \frac{1}{2}[\sqrt{1 + 4\lambda_e^2/\lambda_v^2} - 1] \,. \tag{8.2}$$

Clearly, $\beta \geq 0$, which is a number describing the amount of measurement noise. If λ_e^2/λ_v^2 is small, then $\beta \approx \lambda_e^2/\lambda_v^2$, whereas for large λ_e^2/λ_v^2, $\beta \approx \sqrt{\lambda_e^2/\lambda_v^2}$. The parameter β will be convenient for use in the forthcoming calculations.

Note that (8.2) is equivalent to

$$\lambda_e^2 = \lambda_v^2 \beta(\beta + 1) \,. \tag{8.3}$$

The optimal state estimate $\hat{x}(t|t+m)$ will be examined and its error variance evaluated. All the cases of prediction, filtering and smoothing will be treated. Apparently, $m > 0$ refers to smoothing. For the prediction case ($m < 0$) the notation

$$k \overset{\triangle}{=} -m > 0 \tag{8.4}$$

will be used.

8.2 Spectral Factorization

As stated earlier, spectral factorization plays a central role in coping with the optimal estimation problem. It is the first step in finding the optimal estimates.

For the *state space approach*, the spectral factorization hinges upon the ARE. For the example being studied, the ARE reads

$$P = P + \lambda_v^2 - \frac{P^2}{P + \lambda_e^2} ,$$ (8.5)

giving

$$P^2 - P\lambda_v^2 - \lambda_v^4 \beta(\beta + 1) = 0 ,$$

with the solution (the unfeasible solution is $P = -\beta\lambda_v^2$)

$$P = (1 + \beta)\lambda_v^2 .$$ (8.6)

The Kalman predictor gain becomes

$$K = P/(P + \lambda_e^2) = \frac{1}{1 + \beta} .$$ (8.7)

The innovations form, therefore, becomes

$$\hat{x}(t + 1|t) = \hat{x}(t|t - 1) + K\tilde{y}(t) ,$$
$$y(t) = \hat{x}(t|t - 1) + \tilde{y}(t) ,$$ (8.8)

from which one obtains

$$y(t) = [1 + (qI - 1)^{-1} K]\tilde{y}(t)$$
$$= \frac{q - \frac{\beta}{1+\beta}}{q - 1}\tilde{y}(t) .$$ (8.9)

In particular, this means that $F - KH = \beta/(1 + \beta)$. Note that this quantity lies between zero and one. For small measurement noise, β is small and so is $F - KH$. Similarly, if the noise level is high, β is large and $F - KH$ approaches one.

For the *polynomial approach*, the spectral factorization follows from the treatment given in Section 7.4, with the identification $C(q) = q$, $A(q) = q - 1$, $M = 1$, $N = 1$. Here the noise $v(t)$ has been "relabelled", as otherwise $C(q) = 1$. It is, however, convenient to let $A(z)$ and $C(z)$ be of the same degree.

The spectral factorization (7.80) becomes

$$\lambda_v^2(z)(z^{-1}) + \lambda_e^2(z - 1)(z^{-1} - 1) \equiv \lambda_\varepsilon^2(z + b)(z^{-1} + b) .$$ (8.10)

Equating the powers of z gives the following equations, with b and λ_ε^2 as unknowns:

$$\begin{cases} \lambda_v^2 + 2\lambda_e^2 = \lambda_\varepsilon^2(1 + b^2) , \\ -\lambda_e^2 = \lambda_\varepsilon^2 b . \end{cases}$$

After division, and using (8.3):

$$\frac{b}{1 + b^2} = -\frac{\beta(\beta + 1)}{1 + 2\beta + 2\beta^2} .$$

The solutions are

$$b = -\frac{\beta}{\beta + 1} , \tag{8.11}$$

and $b = -(\beta + 1)/\beta$, which is a false one, since in this case $|b| > 1$.
Further, from (8.10):

$$\lambda_\varepsilon^2 = \lambda_v^2 (1 + \beta)^2 . \tag{8.12}$$

The innovations form is therefore found to be

$$y(t) = \frac{q + b}{q - 1}\tilde{y}(t) = \frac{q - \frac{\beta}{\beta+1}}{q - 1}\tilde{y}(t) , \qquad \mathbf{E}\,\tilde{y}^2(t) = \lambda_v^2 (1 + \beta)^2 . \tag{8.13}$$

This is, of course, the same model as appeared before in (8.6) and (8.9).

8.3 Optimal Prediction

The *state space approach* is used first. The one-step optimal predictor of $x(t)$
is

$$\hat{x}(t + 1|t) = \hat{x}(t|t - 1) + K[y(t) - \hat{x}(t|t - 1)] ,$$

that is:

$$\hat{x}(t + 1|t) = \frac{qK}{q - (1 - K)}y(t) = \frac{q\frac{1}{1+\beta}}{q - \frac{\beta}{1+\beta}}y(t) . \tag{8.14}$$

The prediction error variance is $P = \lambda_v^2 (1 + \beta)$.
  The general k-step predictor is found from (6.40) and is given by

$$\hat{x}(t + k|t) = F^{k-1}\hat{x}(t + 1|t)$$

$$= \hat{x}(t + 1|t) = \frac{q\frac{1}{1+\beta}}{q - \frac{\beta}{1+\beta}}y(t) . \tag{8.15}$$

The prediction error variance can be evaluated from (6.41) and (6.42) as

$$P(t + k|t) = F^{k-1}P(t + 1|t)F^{T^{k-1}} + \sum_{s=0}^{k-2} F^s R_1 F^{T^s}$$

$$= P + (k - 1)R_1$$

$$= \lambda_v^2[1 + \beta + k - 1] = \lambda_v^2[k + \beta] . \tag{8.16}$$

In this case all the predictors ($k = 1, 2, \ldots$) are identical. Further, the prediction error variance grows linearly in k. Both these properties are due to the integrator in the system in (8.1). For a system of the form $x(t+1) = fx(t)+v(t)$, the predictors are related as $\hat{x}(t + k|t) = f^{k-1}\hat{x}(t + 1|t)$, and the prediction error variance grows at an exponential rate as f^{2k}.

Next, use the *polynomial approach*. The Diophantine equation (7.84) will, for k-step prediction, be

$$z^k \equiv (z^{-1} + b)(r_0 z + r_1) + z^k(z-1)(\ell_0 z^{-k} + \ldots + \ell_k), \qquad (8.17)$$

or

$$z^k = (z^{-1} + b)(r_0 z + r_1) + (z-1)(\ell_0 + \ell_1 z + \ldots + \ell_k z^k).$$

Equating the different powers of z gives

$$\begin{aligned}
&\ell_k = 0, \\
&\ell_i = 1, \qquad i = 1, \ldots, k-1, \\
&\ell_0 = r_0 = \tfrac{1}{1+b} = \beta + 1, \\
&r_1 = 0.
\end{aligned} \qquad (8.18)$$

Hence, according to (7.86), the optimal k-steps predictor is

$$\begin{aligned}
\hat{x}(t+k|t) &= \frac{\lambda_v^2}{\lambda_v^2(1+\beta)^2} \frac{qr_0}{q+b} y(t), \\
&= \frac{q\frac{1}{1+\beta}}{q - \frac{\beta}{1+\beta}} y(t),
\end{aligned} \qquad (8.19)$$

which is identical to the previous solution (8.14).

As an alternative way of finding the predictor, note that

$$\hat{x}(t+k|t) = \hat{y}(t+k|t). \qquad (8.20)$$

Noting from (8.9) that $y(t)$ is an ARMA process

$$y(t) - y(t-1) = \tilde{y}(t) - \frac{\beta}{1+\beta}\tilde{y}(t-1), \qquad (8.21)$$

one can use the predictor identity (7.13) to find the predictor:

$$z^{k-1}\left(z - \frac{\beta}{1+\beta}\right) \equiv (z-1)(z^{k-1} + f_1 z^{k-2} + \ldots + f_{k-1}) + \ell_0. \quad (8.22)$$

Equating powers of z gives

$$f_1 = f_2 = \ldots = f_{k-1} = \ell_0 = 1 - \frac{\beta}{1+\beta} = \frac{1}{1+\beta}. \qquad (8.23)$$

The predictor, according to (7.18), (8.20) and (8.22), is

$$\begin{aligned}
\hat{x}(t+k|t) = \hat{y}(t+k|t) &= \frac{q\ell_0}{q - \frac{\beta}{1+\beta}} y(t) \\
&= \frac{q\frac{1}{1+\beta}}{q - \frac{\beta}{1+\beta}} y(t),
\end{aligned} \qquad (8.24)$$

which is identical to the previous finding (*cf.* (8.14)).

The prediction error spectrum can be evaluated from (7.89). It reads in this case

$$\phi_{\tilde{s}}(z) = \frac{\lambda_v^2 \lambda_v^2 \beta(\beta+1)}{\lambda_v^2(\beta+1)^2} \frac{1}{(z+b)(z^{-1}+b)} + \frac{\lambda_v^4}{\lambda_v^2(\beta+1)^2}$$

$$\times \frac{(\beta+1)z^k + z^{k-1} + \ldots + z}{(z+b)}$$

$$\times \frac{(\beta+1)z^{-k} + z^{-(k-1)} + \ldots + z^{-1}}{(z^{-1}+b)} . \tag{8.25}$$

The prediction error variance could be evaluated by integrating (8.25), but a direct approach is easier:

$$\mathbf{E}\left[x(t+k) - \hat{x}(t+k|t)\right] = \mathbf{E}\left[x(t+k) - \frac{q\frac{1}{1+\beta}}{q+b}y(t)\right]^2$$

$$= \mathbf{E}\left[x(t+k) - \frac{q\frac{1}{1+\beta}}{q+b}x(t)\right]^2 + \mathbf{E}\left[\frac{q\frac{1}{1+\beta}}{q+b}e(t)\right]^2$$

$$= \mathbf{E}\left[\{v(t+k) + \ldots + v(t+1)\} + \left\{x(t) - \frac{q\frac{1}{1+\beta}}{q+b}x(t)\right\}\right]^2$$

$$+ \frac{\lambda_e^2}{(1+\beta)^2}\frac{1}{1-b^2}$$

$$= \lambda_v^2 k + \mathbf{E}\left[\frac{q(1-\frac{1}{1+\beta})+b}{q+b}x(t)\right]^2 + \frac{\lambda_v^2\beta}{1+\beta}\frac{(1+\beta)^2}{1+2\beta}$$

$$= \lambda_v^2 k + b^2\mathbf{E}\left[\frac{q-1}{q+b}\frac{q}{q-1}v(t)\right]^2 + \frac{\lambda_v^2\beta(1+\beta)}{1+2\beta}$$

$$= \lambda_v^2 k + \frac{b^2}{1-b^2}\lambda_v^2 + \frac{\lambda_v^2\beta(1+\beta)}{1+2\beta}$$

$$= \lambda_v^2\left[k + \frac{\beta^2}{1+2\beta} + \frac{\beta(1+\beta)}{1+2\beta}\right] = \lambda_v^2(k+\beta) . \tag{8.26}$$

This is, as expected, the same result as derived in (8.16) using the state space approach.

8.4 Optimal Filtering

Using the *state space approach*, one obtains

$$\hat{x}(t|t) = \hat{x}(t|t-1) + K[y(t) - \hat{x}(t|t-1)] .$$

Note that the predictor gain and the filter gain are both equal to K in this case as $F = I$. Hence, from (8.14):

$$\hat{x}(t|t) = (1 - K)\hat{x}(t|t-1) + Ky(t)$$

$$= \left[\frac{K(1-K)}{q-(1-K)} + K\right] y(t) = \frac{Kq}{q-(1-K)} y(t)$$

$$= \frac{q\frac{1}{1+\beta}}{q - \frac{\beta}{1+\beta}} y(t) . \tag{8.27}$$

The filter error variance is

$$\mathbf{E}\left[x(t) - \hat{x}(t|t)\right]^2 = P - \frac{P^2}{P + \lambda_e^2} = \lambda_v^2 \beta . \tag{8.28}$$

It can seen that the previous results (8.15) and (8.16) for the prediction case remain valid in this example if one sets $m = 0$.

For the *polynomial approach*, the Diophantine equation (7.84) becomes

$$1 \equiv (z^{-1} + b)(r_0 z + r_1) + (z - 1)\ell_0 , \tag{8.29}$$

which has the solution

$$r_0 = \frac{1}{1+b} , \qquad r_1 = 0 , \qquad \ell_0 = \frac{-b}{1+b} . \tag{8.30}$$

The filter will be (see (7.86))

$$\hat{x}(t|t) = \frac{\lambda_v^2}{\lambda_e^2} \frac{q(1+\beta)}{q+b} y(t) = \frac{q\frac{1}{1+\beta}}{q - \frac{\beta}{1+\beta}} y(t) , \tag{8.31}$$

which coincides with (8.27).

A direct calculation gives the error variance as follows:

$$\mathbf{E}\left[x(t) - \hat{x}(t|t)\right] = \mathbf{E}\left[x(t) - \frac{q\frac{1}{1+\beta}}{q - \frac{\beta}{1+\beta}}\{x(t) + e(t)\}\right]^2$$

$$= \mathbf{E}\left[\frac{\frac{\beta}{1+\beta}(q-1)}{q - \frac{\beta}{1+\beta}} x(t)\right]^2 + \mathbf{E}\left[\frac{\frac{1}{1+\beta}}{q - \frac{\beta}{1+\beta}} e(t+1)\right]^2$$

$$= \left(\frac{\beta}{1+\beta}\right)^2 \mathbf{E}\left[\frac{1}{q - \frac{\beta}{1+\beta}} v(t)\right]^2 + \frac{\lambda_e^2}{(1+\beta)^2} \frac{1}{1 - (\frac{\beta}{1+\beta})^2}$$

$$= \lambda_v^2 \frac{\beta^2}{1+2\beta} + \lambda_e^2 \frac{1}{1+2\beta} = \frac{\lambda_v^2}{1+2\beta}[\beta^2 + \beta(\beta+1)] = \lambda_v^2 \beta . \tag{8.32}$$

This result coincides with (8.28), again illustrating that the state space and polynomial approaches give identical results.

8.5 Optimal Smoothing

According to the *state space approach* the optimal smoothing estimate is; see (6.54)

$$\hat{x}(t|t+m) = \hat{x}(t|t) + \sum_{s=t+1}^{t+m} K_2(s)[y(s) - \hat{x}(s|s-1)] \,, \tag{8.33}$$

$$K_2(s) = \frac{P}{P+\lambda_e^2}(F - KH)^{s-t} = \frac{1}{1+\beta}\left(\frac{\beta}{1+\beta}\right)^{s-t} . \tag{8.34}$$

The estimation error variance, from (6.106), becomes

$$\mathbf{E}\left[x(t) - \hat{x}(t|t+m)\right]^2 = P - \frac{P^2}{P+\lambda_e^2}\sum_{j=0}^{m}(F - KH)^{2j}$$

$$= \lambda_v^2(1+\beta) - \lambda_v^2\sum_{j=0}^{m}\left(\frac{\beta}{1+\beta}\right)^{2j}$$

$$= \lambda_v^2\left[1 + \beta - \frac{1 - (\beta/(1+\beta))^{2(m+1)}}{1 - (\beta/(1+\beta))^2}\right] . \tag{8.35}$$

In the limiting case, with an infinite smoothing lag:

$$\lim_{m\to\infty}\mathbf{E}\left[x(t) - \hat{x}(t|t+m)\right]^2 = \lambda_v^2\left[1 + \beta - \frac{1}{1 - (\beta/(1+\beta))^2}\right]$$

$$= \lambda_v^2\frac{\beta(\beta+1)}{2\beta+1} . \tag{8.36}$$

This result can also be derived using the analysis of Section 6.7. The Lyapunov equation (6.110) for Q in this case reads

$$Q = \left(\frac{\beta}{1+\beta}\right)^2 Q + \frac{1}{(1+\beta)^2\lambda_v^2} \,,$$

which gives the solution

$$Q = \frac{1}{2\beta+1}\frac{1}{\lambda_v^2} . \tag{8.37}$$

The asymptotic error variance is (cf. (6.111))

$$\lim_{m\to\infty}\mathbf{E}\left[x(t) - \hat{x}(t|t+m)\right]^2 = P - PQP = \frac{\beta(1+\beta)}{2\beta+1}\lambda_v^2 . \tag{8.38}$$

Next, use the *polynomial approach*. The Diophantine equation (7.84) now becomes

$$z^{-m} \equiv z^{-m}(z^{-1} + b)(r_0 z^{m+1} + \ldots + r_{m+1}) + (z - 1)\ell_0 . \tag{8.39}$$

It has the solution

$$r_j = (-b)^{m-j} \,, \qquad j = 1,\ldots,m \,, \qquad r_{m+1} = 0 \,, \tag{8.40}$$

$$r_0 = \frac{(-b)^m}{b+1} \,, \qquad \ell_0 = \frac{(-b)^{m+1}}{b+1} .$$

Following (7.86), the optimal smoother is

$$\hat{x}(t|t+m) = \frac{\lambda_v^2}{\lambda_\varepsilon^2} \frac{q^{-m}R(q)}{q+b} y(t+m)$$

$$= \frac{1}{(\beta+1)^2} \frac{r_0 q + r_1 + r_2 q^{-1} + \ldots + r_{m+1}q^{-m}}{q+b} y(t+m) .$$

(8.41)

It is interesting to evaluate the first part of the estimation error spectrum (7.89). It is easy to see that

$$\phi_{\tilde{s}}(z) \geq \frac{\lambda_e^2 \lambda_v^2}{\lambda_\varepsilon^2} \frac{1}{B(z)B^*(z)} = \frac{\lambda_v^2 \beta(\beta+1)}{(1+\beta)^2} \frac{1}{(z+b)(z^{-1}+b)} .$$

Hence,

$$\mathbf{E}\left[\tilde{s}^2(t)\right] \geq \frac{\lambda_v^2 \beta}{\beta+1} \frac{1}{1-b^2} = \lambda_v^2 \frac{\beta(1+\beta)}{1+2\beta} .$$

Compare with (8.36). The first part of the error spectrum is therefore not only a lower bound, it is also achievable in the limit as $m \to \infty$. This can also be understood in another way. From the solution it follows that $L = \ell_0$ goes to zero exponentially as $m \to \infty$. Hence, the second term of the error spectrum in (7.89) vanishes asymptotically. In fact, the contribution of the second term to the error variance, for finite m, in this example, is

$$\frac{\lambda_v^4}{\lambda_\varepsilon^2} \frac{\ell_0^2}{1-b^2} = \frac{\lambda_v^2}{(1+\beta)^2} \left(\frac{\beta}{1+\beta}\right)^{2m+2} (1+\beta)^2 \frac{1}{1-(\beta/(1+\beta))^2}$$

$$= \lambda_v^2 \frac{\beta^{2m+2}}{(1+\beta)^{2m}} \frac{1}{1+2\beta} ,$$

which is perfectly in accordance with the expression in (8.35) based on the state space approach.

The *alternative polynomial approach*, outlined in Example 7.11, is also worth illustrating. In this case, the polynomial identity (7.109) will be

$$z^m(z-1) = (z+b)(z^m + f_1 z^{m-1} + \ldots + f_m) + g_0 ,$$ (8.42)

giving

$$f_j = -(b+1)(-b)^{j-1} , \qquad (j = 1, \ldots, m) ,$$
$$g_0 = -(b+1)(-b)^m .$$

(8.43)

The smoothing estimate is (set $f_0 = 1$), (see (7.111))

$$\hat{x}(t|t+m) = \left[q^{-m} - \frac{\lambda_e^2}{\lambda_\varepsilon^2}(\sum_{i=0}^{m} f_{m-i}q^{-i})\frac{q-1}{q+b}\right] y(t+m)$$

$$= y(t) - \frac{\beta}{\beta+1}[f_m + f_{m-1}q^{-1}$$

$$+ \ldots + f_1 q^{-m+1} + q^{-m}]\frac{q-1}{1+b} y(t+m) .$$ (8.44)

It is instructive to compare the different forms of the solution. As they are written in different ways, the quantity $w(t) \triangleq (q+b)\hat{x}(t|t+m)$ is now examined.

For the state space approach, using (8.33), (8.34) and (8.27):

$$w(t) = (q+b)\hat{x}(t|t)$$

$$+ \sum_{s=t+1}^{t+m} \frac{1}{1+\beta} \left(\frac{\beta}{1+\beta}\right)^{s-t} (q+b)[y(s) - \hat{x}(s|s-1)]$$

$$= q\frac{1}{1+\beta}y(t)$$

$$+ \sum_{s=t+1}^{t+m} \frac{1}{1+\beta} \left(\frac{\beta}{1+\beta}\right)^{s-t} \left[(q+b)y(s) - q\frac{1}{1+\beta}y(s-1)\right]$$

$$= \frac{1}{1+\beta}y(t+1)$$

$$+ \sum_{s=t+1}^{t+m} \frac{1}{1+\beta} \left(\frac{\beta}{1+\beta}\right)^{s-t} [y(s+1) - y(s)]$$

$$= y(t+1)\left[\frac{1}{1+\beta} - \frac{1}{1+\beta}\frac{\beta}{1+\beta}\right]$$

$$+ y(t+2)\left[\frac{1}{1+\beta}\left(\frac{\beta}{1+\beta}\right) - \frac{1}{1+\beta}\left(\frac{\beta}{1+\beta}\right)^2\right]$$

$$+ \ldots + y(t+m)\left[\frac{1}{1+\beta}\left(\frac{\beta}{1+\beta}\right)^{m-1} - \frac{1}{1+\beta}\left(\frac{\beta}{1+\beta}\right)^m\right]$$

$$+ y(t+m+1)\frac{1}{1+\beta}\left(\frac{\beta}{1+\beta}\right)^m$$

$$= y(t+1)\frac{1}{(1+\beta)^2} + y(t+2)\frac{\beta}{(1+\beta)^3} + \ldots$$

$$+ y(t+m)\frac{\beta^{m-1}}{(1+\beta)^{m+1}} + y(t+m+1)\frac{\beta^m}{(1+\beta)^{m+1}}.$$

For the solution (8.41) using the polynomial approach:

$$w(t) = (q+b)\hat{x}(t|t+m)$$

$$= \frac{1}{(\beta+1)^2}[r_m y(t+1) + r_{m-1}y(t+2)$$

$$+ \ldots + r_1 y(t+m) + r_0 y(t+m+1)]$$

$$= \frac{1}{(1+\beta)^2}\left[y(t+1) + \frac{\beta}{1+\beta}y(t+2)\right]$$

$$+ \ldots + \left(\frac{\beta}{1+\beta}\right)^{m-1} y(t+m) + \frac{\beta^m}{(1+\beta)^{m-1}} y(t+m+1) \Bigg] \; .$$

The solution (8.44) based on the alternative polynomial approach gives

$$w(t) = (q+b)\hat{x}(t|t+m)$$

$$= (q+b)y(t)$$

$$- \frac{\beta}{\beta+1}(q-1)[f_m y(t+m) + f_{m-1} y(t+m-1)$$

$$+ \ldots + f_1 y(t+1) + y(t)]$$

$$= y(t)\left[b + \frac{\beta}{\beta+1}\right] + y(t+1)\left[1 - \frac{\beta}{\beta+1} + f_1 \frac{\beta}{\beta+1}\right]$$

$$+ y(t+2)\frac{\beta}{\beta+1}[f_2 - f_1] + \ldots + y(t+m)\frac{\beta}{\beta+1}[f_m - f_{m-1}]$$

$$- y(t+m+1)\frac{\beta}{\beta+1}f_m$$

$$= y(t+1)\frac{1}{(1+\beta)^2} - y(t+2)\frac{\beta}{(\beta+1)^2}\left[\left(\frac{\beta}{1+\beta}\right)^1 - 1\right] + \ldots$$

$$- y(t+m)\frac{\beta}{(\beta+1)^2}\left[\left(\frac{\beta}{1+\beta}\right)^{m-1} - \left(\frac{\beta}{1+\beta}\right)^{m-2}\right]$$

$$+ y(t+m+1)\frac{\beta}{(\beta+1)^2}\left(\frac{\beta}{1+\beta}\right)^{m-1}$$

$$= y(t+1)\frac{1}{(1+\beta)^2} + y(t+2)\frac{\beta}{(1+\beta)^3} + \ldots$$

$$+ y(t+m)\frac{\beta^{m-1}}{(1+\beta)^{m+1}} + y(t+m+1)\frac{\beta^m}{(1+\beta)^{m+1}} \; .$$

It is thus found, as expected, that all three expressions for the optimal smoothing estimate are in fact equivalent.

8.6 Estimation Error Variance

The optimal prediction, filter and smoothing estimates have now been derived. The variance of the estimation error

$$V_m = \mathbf{E}\left[\hat{x}(t|t+m) - x(t)\right]^2 \tag{8.45}$$

has also been found. See, for example, (8.16), (8.28) and (8.35). Some further illustration of how V_m varies with m and β is given next. For this purpose, the expression for V_m is repeated:

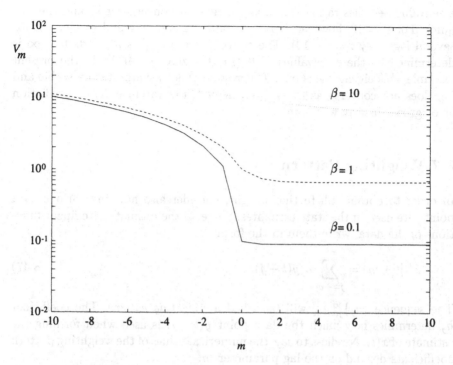

Fig. 8.1. $V_m = \mathbf{E}\left[\hat{x}(t|t+m) - x(t)\right]^2$ as a function of m with β as parameter. Note the logarithmic scale

$$V_m = \begin{cases} \lambda_v^2[-m + \beta] , & m \leq 0 , \\[2mm] \lambda_v^2 \left[\frac{\beta(1+\beta)}{1+2\beta} + \frac{(1+\beta)^2}{1+2\beta}\left(\frac{\beta}{1+\beta}\right)^{2m+2}\right] , & m \geq 0 . \end{cases} \tag{8.46}$$

Clearly, V_m is a decreasing sequence (as m increases). This is also obvious by construction. When more data/information is used, then a more accurate estimate can be found. The variance V_m is plotted as a function of m with β as a parameter in Figure 8.1.

It is clear from Figure 8.1 that V_m is, for fixed m, increasing with increasing β. This is fairly natural. An increased noise level in the measurements should lead to less accurate estimates. One can also see that when the noise level is small or modest (say $\beta < 1$), then the accuracy of the smoothing estimate is only marginally better than the filtering estimate. This means that it does not pay to apply smoothing (compared with filtering) when the data are of good quality (measurement noise has small variance).

In general, V_m converges at an exponential rate both when m tends to $-\infty$ and to ∞. The convergence when m tends to infinity is determined by the eigenvalues of $F - KH$ (or, equivalently, $B(z)$). In the example, this eigenvalue is $\beta/(1 + \beta)$. Note that for small and modest values of λ_e^2/λ_v^2 this eigenvalue becomes rather small, and hence the convergence should then be quick. For $\beta = 10$, however, it pays to apply smoothing and to apply it for several lags (say $m = 5$–10). The convergence of V_m as m tends to $-\infty$ is determined by the eigenvalues of F (*i.e.* the zeros of $A(z)$). In the present example, this eigenvalue is one. The system is not asymptotically stable and V_m does not converge as $m \to -\infty$. In fact, the variance of the prediction error grows linearly with $-m$.

8.7 Weighting Pattern

In order to obtain still further insight, consider also how the various data points are used in the state estimates. Since all the estimates are linear functions of the data, write them in the form

$$\hat{x}(t|t + m) = \sum_{j=-m}^{\infty} h_j y(t - j) \,. \tag{8.47}$$

The sequence $\{h_j\}_{j=-m}^{\infty}$ will be called a *weighting pattern*. The coefficient h_j determines how much the data point $y(t - j)$ is used when forming the estimate of $x(t)$. Needless to say, the numerical values of the weighting pattern coefficients depend on the lag parameter m.

For the case where $m \le 0$, the weighting pattern is easily derived from (8.14) and (8.27). In fact:

$$\hat{x}(t|t + m) = \frac{q\frac{1}{1+\beta}}{q - \frac{\beta}{1+\beta}} y(t + m)$$

$$= \frac{1}{1 + \beta} \sum_{i=0}^{\infty} \left(\frac{\beta}{1 + \beta}\right)^i q^{-i} y(t + m) \,. \tag{8.48}$$

Comparison with (8.47) gives

$$h_j = \begin{cases} \frac{1}{1+\beta} \left(\frac{\beta}{1+\beta}\right)^{j+m} & , j \ge -m \,, \\ 0 \,, & j < -m \,. \end{cases} \tag{8.49}$$

Consider the smoothing case, $m > 0$. Then use the form in (8.41) of the solution and write it as

$$\hat{x}(t|t + m) = \frac{1}{(1 + \beta)^2} \frac{r_0 q + r_1 + \ldots + r_{m+1} q^{-m}}{q + b} y(t + m)$$

$$\triangleq \left[(h_{-m} + h_{-m+1} q^{-1} + \ldots + h_0 q^{-m}) + \frac{\alpha q^{-m}}{q + b} \right] y(t + m) \,. \tag{8.50}$$

By comparing both representations:

$$\frac{1}{(1+\beta)^2}(r_0 q + r_1 + \ldots + r_{m+1} q^{-m})$$

$$= (q + b)(h_{-m} + h_{-m+1} q^{-1} + \ldots + h_0 q^{-m}) + \alpha q^{-m} . \qquad (8.51)$$

Equating the powers of q gives

$$\begin{cases} h_{-m} = \frac{1}{(1+\beta)^2} r_0 = \frac{\beta^m}{(1+\beta)^{m+1}} , \\[2mm] h_{-m+j+1} = \frac{1}{(1+\beta)^2} r_{j+1} - b h_{-m+j} , \qquad j = 0, \ldots, m-1 , \qquad (8.52) \\[2mm] \alpha = -b h_0 . \end{cases}$$

Iterating these equations gives

$$h_{-i} = \frac{1}{(2\beta + 1)} \left(\frac{\beta}{1+\beta} \right)^i \left[1 + \left(\frac{\beta}{1+\beta} \right)^{2m-2i+1} \right],$$

$$i = 0, \ldots, m . \qquad (8.53)$$

We also have that $h_1 = \alpha = -b h_0$, and, more generally, for $j > 0$, $h_j = -b h_{j-1}$, that is:

$$h_j = (-b)^j h_0 = \left(\frac{\beta}{1+\beta} \right)^j \frac{1}{(1+2\beta)} \left[1 + \left(\frac{\beta}{1+\beta} \right)^{2m+1} \right],$$

$$(j \geq 0) . \qquad (8.54)$$

The weighting pattern is displayed in Figures 8.2 and 8.3.

Several properties are illustrated in these figures:

- For small β (little measurement noise) h_0 is large. This means that the measurement $y(t) = x(t) + e(t)$ is given high confidence.
- When m is increasing the weights are decreasing. The explanation is that when more data are available, it is possible to use the new information and rely slightly less on the old information compared with cases with smaller values of m.
- When β is small (the measurement noise has small variance) the weighting pattern is close to its asymptotic $(m \to \infty)$ value already for very small m. In this case, with good quality data, it does not pay to use more than a few lags (if any) in a smoother.
- When m becomes "large", one obtains approximately $h_j = h_{-j}$; that is, the weighting pattern becomes symmetric. This also follows analytically from (8.53) and (8.54). The relation also holds approximately for small and moderate values of m.

8.8 Frequency Characteristics

It is also illustrative to examine the optimal estimates in the frequency domain.

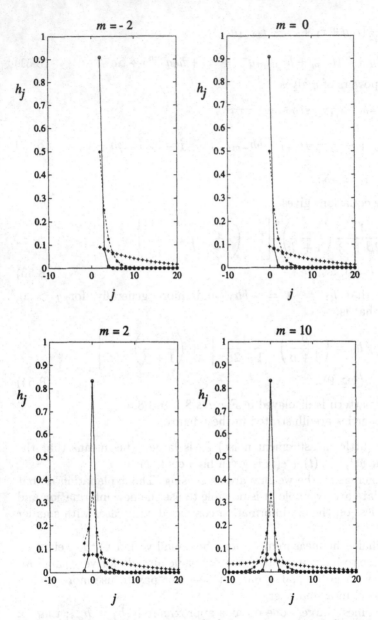

Fig. 8.2. The weighting pattern $\{h_j\}$ for various m and $\beta = 0.1$ (*solid lines*, o), $\beta = 1$ (*dashed lines*, *), $\beta = 10$ (*dotted lines*, +)

Note from (8.19) that the optimal prediction estimate can be written as (recall that $m < 0$ for prediction)

$$\hat{x}(t|t+m) = \frac{q^{\frac{1}{1+\beta}}}{q - \frac{\beta}{1+\beta}} y(t+m)$$

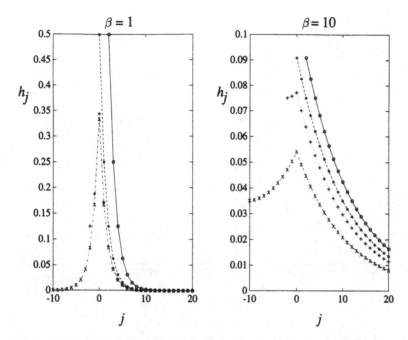

Fig. 8.3. The weighting pattern $\{h_j\}$ for two values of β (left $\beta = 1$, right $\beta = 10$), and $m = -2$ (*solid lines*, o), $m = 0$ (*dashed lines*, *), $m = 2$ (*dotted lines*, +), $m = 10$ (*dash-dotted lines*, x)

$$= \frac{q^{\frac{1}{1+\beta}}}{q - \frac{\beta}{1+\beta}} q^m y(t)$$

$$\triangleq H^{(p)}_{(m)}(q) y(t) . \tag{8.55}$$

The frequency characteristics and the pole–zero configuration of $H^{(p)}_{(m)}(q)$ for $m = -2$ and for various values of β are plotted in Figure 8.4.

The filter clearly has a lowpass character. The static gain is one, which is also clearly seen from (8.55). When the SNR decreases (β increases), the gain and the bandwidth of the filter decrease. This is in perfect agreement with intuition. When the SNR is low, the measurements are less reliable and less confidence should be placed in the high-frequency contents in the signal. Note that the decrease of the phase from 0 to $-360°$ is due to the delay q^{-2}.

Repeat the consideration for the filter case, for which (see (8.31))

$$\hat{x}(t|t) = \frac{q^{\frac{1}{1+\beta}}}{q - \frac{\beta}{1+\beta}} y(t) \triangleq H^{(f)}(q) y(t) . \tag{8.56}$$

The frequency characteristics and the pole–zero configuration are given in Figure 8.5. As is obvious, the only difference from Figure 8.4 is that of the phase curve. Now there is no effect of any additional delay. Instead, there is an intermediate phase drop due to the pole (as there is a lag filter). The drop

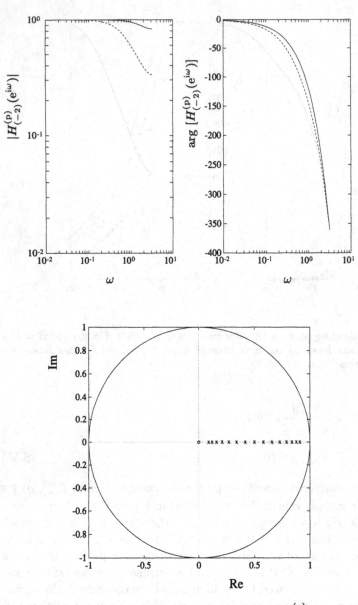

Fig. 8.4. Frequency characteristics (*upper part*) of $H_{(m)}^{(p)}(q)$ for $m = -2$; $\beta = 0.1$ (*solid line*), 1 (*dashed line*), 10 (*dotted line*). Pole–zero configuration of $H_{(m)}^{(p)}(q)$ (except for q^m). The pole migrates from the origin towards $z = 1$ as β increases from 0 to ∞

is more pronounced when the bandwidth is low (the pole is close to $z = 1$, *i.e.* β is large).

Finally, consider the smoothing case. Equation (8.41) gives (now $m > 0$)

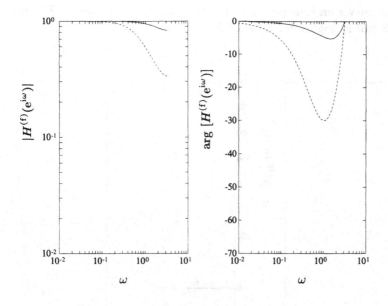

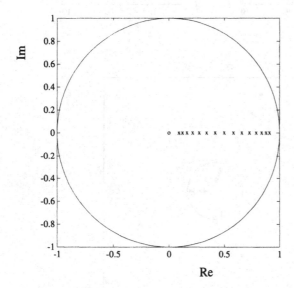

Fig. 8.5. Frequency characteristics (*upper part*) of $H^{(f)}(q)$; $\beta = 0.1$ (*solid line*), 1 (*dashed line*), 10 (*dotted line*). Pole–zero configuration of $H^{(f)}(q)$. The pole migrates from the origin towards $z = 1$ as β increases from 0 to ∞

$$
\hat{x}(t|t+m) = \frac{1}{(\beta+1)^2} \frac{r_0 q + r_1 + \ldots + r_{m+1}q^{-m}}{q - \beta/(1+\beta)} q^m y(t)
$$

$$
\triangleq H^{(s)}_{(m)}(q)y(t) \ . \tag{8.57}
$$

The frequency curves and the pole–zero configuration are displayed in Figure 8.6 for $m = 2$ and in Figure 8.7 for $m = 5$.

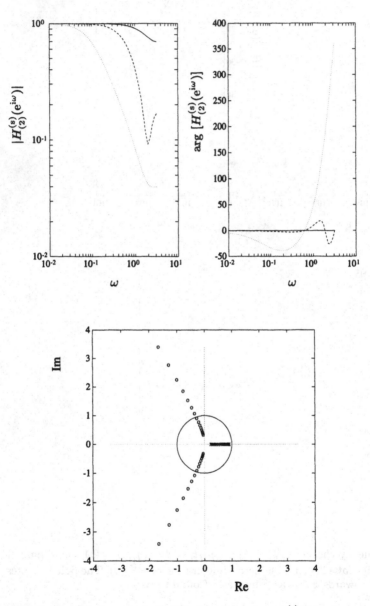

Fig. 8.6. Frequency characteristics (*upper part*) of $H_{(2)}^{(s)}(q)$ for $m = 2$; $\beta = 0.1$ (*solid line*), 1 (*dashed line*), 10 (*dotted line*). Pole–zero configuration of $H_{(2)}^{(s)}(q)$ (except for q^2). The pole migrates from the origin towards $z = 1$ as β increases from 0 to ∞. The zeros move inside the unit circle as β increases

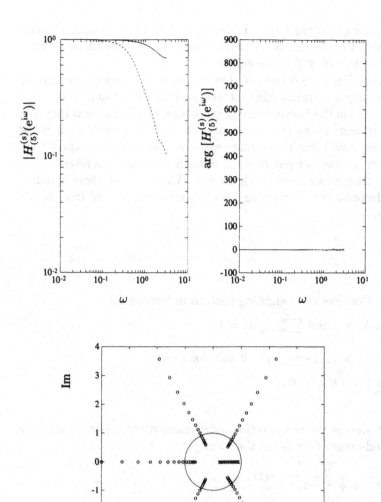

Fig. 8.7. Frequency characteristics (*upper part*) of $H_{(5)}^{(s)}(q)$ for $m = 5$; $\beta = 0.1$ (*solid line*), 1 (*dashed line*), 10 (*dotted line*). Pole–zero configuration of $H_{(5)}^{(s)}(q)$ (except for q^5). The pole migrates from the origin towards $z = 1$ as β increases from 0 to ∞. The zeros move inside the unit circle as β increases

Again, note that the filter has a lowpass character, and a static gain equal to one, and that the gain is reduced for high frequencies when the SNR decreases (β increases). The phase curves sometimes go up to $180m$ degrees (due to the factor q^m in the filter). For small values of β, though, all zeros lie outside the unit circle and the filter becomes nonminimum phase, causing the phase to be close to $0°$ for all frequencies.

It seems from Figures 8.6 and 8.7 that the zero locations of the optimal smoother have only a marginal effect on the performance. Compared with a first-order filter, as in the filtering case (Figure 8.5), it is found that they give marginal adjustments of the frequency curve for high frequencies, where the gain is small anyway. From these diagrams, one can therefore expect that, for this example, it does not pay to use a smoother instead of a filter. This is at least partly true, as was seen in Figure 8.1. When $\beta = 10$, there is indeed a significant decrease in the error variance when m is increased from $m = 0$ to, say, $m = 5$.

Exercises

Exercise 8.1 Consider the weighting pattern in Section 8.7.

(a) Show that $h_j \geq 0$ and $\sum_{j=-m}^{\infty} h_j = 1$.

(b) Show that $h_j > h_{-j}$, $-m \leq j < 0$ and, for fixed j,
$$\lim_{m \to \infty} [h_j - h_{-j}] = 0 .$$

Exercise 8.2 Repeat the computations of Sections 8.4–8.6, but numerically, for some second-order processes of the form
$$s(t) = \frac{1}{1 + a_1 q^{-1} + a_2 q^{-2}} v(t) ,$$
$$y(t) = s(t) + e(t) .$$

Present the results in diagrams such as Figures 8.2–8.7. Study the effect of varying the pole locations as well as the SNR $= \mathbf{E}\, s^2(t)/\mathbf{E}\, e^2(t)$.

Exercise 8.3 Prediction of ARMA processes can be computed in state space form. Consider an MA(1) process
$$y(t) = e(t) + ce(t - 1) , \qquad |c| \leq 1 ,$$

with $e(t)$ being white noise of zero mean and variance λ^2. It can be represented in state space form as
$$x(t + 1) = \begin{pmatrix} 0 & 1 \\ 0 & 0 \end{pmatrix} x(t) + \begin{pmatrix} 1 \\ c \end{pmatrix} v(t) ,$$
$$y(t) = (1 \ \ 0)x(t) .$$

Determine the (time-varying) Kalman filter, $\hat{x}(t + 1|t)$, and compare with the time-invariant predictor derived by the polynomial approach. Examine the Kalman gain $K(t)$. How does it behave when $t_0 \to -\infty$? Give specific attention to the case where $|c| = 1$.

9. Nonlinear Filtering

9.1 Introduction

Some various approaches to state estimation for nonlinear stochastic systems are presented in this chapter. Typically, systems of the form

$$x(t + 1) = f(x(t)) + v(t) ,$$
$$y(t) = h(x(t)) + e(t) ,$$
(9.1)

are treated, where $v(t)$ and $e(t)$ are mutually independent white noise sequences. Recall, from Chapter 5, that the optimal state estimate is given by the conditional mean, $\mathbf{E}\left[x(t)|Y^t\right]$. To compute the optimal state estimate, it is also necessary to compute recursively the conditional pdfs $p(x(t)|Y^t)$. As explained in Section 5.4, in most cases this will require a huge amount of computation, so suboptimal schemes are of practical interest. We also present a numerical scheme based on Monte Carlo simulations for numerically evaluating how the conditional pdfs are propagating. In another section we present a practically important suboptimal algorithm, called the interacting multiple model (IMM) approach. The basic idea is to let the dynamics jump between a fixed number of linear models. Some other nonlinear estimation problems, such as median filters and quantization effects, are also discussed in this chapter.

9.2 Extended Kalman Filters

9.2.1 The Basic Algorithm

The idea of the extended Kalman filter (EKF) is to use the ideas of Kalman filtering for a nonlinear problem. The filter gain is computed by linearizing the nonlinear model. The EKF, in contrast to the Kalman filter for linear systems, is *not* an optimal filter, since its derivation is based on approximations.

Consider the nonlinear model

$$x(t + 1) = f(t, x(t)) + g(t, x(t))v(t) ,$$
$$y(t) = h(t, x(t)) + e(t) ,$$
(9.2)

where $v(t)$ and $e(t)$ are mutually independent Gaussian white noise sequences with zero means and covariances $R_1(t)$ and $R_2(t)$ respectively. The model is expanded in first-order Taylor series around estimates of $x(t)$. The function $f(t, x(t))$ is expanded around the most recent estimate, that is, $\hat{x}(t|t)$. The output function is expanded around the *predicted state* $\hat{x}(t|t-1)$. The reason is that the measurement $y(t)$ has to be compared with its predicted value to obtain a correction term to the filter. The linearizations are thus as follows:

$$f(t, x(t)) \approx f(t, \hat{x}(t|t)) + F(t)(x(t) - \hat{x}(t|t)) ,$$
$$g(t, x(t)) \approx G(t) , \qquad\qquad\qquad\qquad (9.3)$$
$$h(t, x(t)) \approx h(t, \hat{x}(t|t-1)) + H(t)(x(t) - \hat{x}(t|t-1)) ,$$

where

$$F(t) = \left.\frac{\partial f(t, x)}{\partial x}\right|_{x=\hat{x}(t|t)} ,$$
$$G(t) = g(t, x)|_{x=\hat{x}(t|t)} , \qquad\qquad (9.4)$$
$$H(t) = \left.\frac{\partial h(t, x)}{\partial x}\right|_{x=\hat{x}(t|t-1)} .$$

Using the approximation in (9.3), the system becomes linear and can be written as

$$x(t+1) = F(t)x(t) + G(t)v(t) + \tilde{u}(t) ,$$
$$y(t) = H(t)x(t) + e(t) + w(t) ,$$
$$\tilde{u}(t) = f(t, \hat{x}(t|t)) - F(t)\hat{x}(t|t) , \qquad (9.5)$$
$$w(t) = h(t, \hat{x}(t|t-1)) - H(t)\hat{x}(t|t-1) .$$

Now apply the standard Kalman filter to the system of (9.5), treating the terms $\tilde{u}(t)$ and $w(t)$ as *known* functions. Note that the system is time-varying. The state estimates can be written in the following form, which gives the "predicted" estimates $\hat{x}(t+1|t)$ as well as the "filtered" states $\hat{x}(t|t)$ (see Theorem 6.1):

$$\hat{x}(t|t) = \hat{x}(t|t-1) + K(t)[y(t) - H(t)\hat{x}(t|t-1) - w(t)] ,$$
$$\hat{x}(t+1|t) = F(t)\hat{x}(t|t) + \tilde{u}(t) , \qquad\qquad (9.6)$$

$$K(t) = P(t|t-1)H^T(t)[H(t)P(t|t-1)H^T(t) + R_2(t)]^{-1} , \qquad (9.7)$$

$$P(t|t) = P(t|t-1) - P(t|t-1)H^T(t)$$
$$\times [H(t)P(t|t-1)H^T(t) + R_2(t)]^{-1}H(t)P(t|t-1) , \quad (9.8)$$
$$P(t+1|t) = F(t)P(t|t)F^T(t) + G(t)R_1(t)G^T(t) .$$

Using (9.5) the "approximate output innovations" can be rewritten as

$$y(t) - H(t)\hat{x}(t|t-1) - w(t) = y(t) - H(t)\hat{x}(t|t-1) - h(t, \hat{x}(t|t-1))$$
$$+ H(t)\hat{x}(t|t-1)$$
$$= y(t) - h(t, \hat{x}(t|t-1)) . \qquad (9.9)$$

The second part of (9.6) can be rewritten, using (9.5), as

$$\hat{x}(t+1|t) = F(t)\hat{x}(t|t) + \tilde{u}(t) = f(t, \hat{x}(t|t)) . \tag{9.10}$$

Thus (9.6) can be replaced by a more enlightening form. The whole algorithm can be summarized as follows:

$$
\begin{aligned}
H(t) &= \left.\frac{\partial h(t,x)}{\partial x}\right|_{x=\hat{x}(t|t-1)} , \\
K(t) &= P(t|t-1)H^T(t) \\
&\quad \times [H(t)P(t|t-1)H^T(t) + R_2(t)]^{-1} , \\
\hat{x}(t|t) &= \hat{x}(t|t-1) + K(t)[y(t) - h(t, \hat{x}(t|t-1))] , \\
P(t|t) &= P(t|t-1) - K(t)H(t)P(t|t-1) , \\
\hat{x}(t+1|t) &= f(t, \hat{x}(t|t)) , \\
F(t) &= \left.\frac{\partial f(t,x)}{\partial x}\right|_{x=\hat{x}(t|t)} , \\
G(t) &= g(t,x)|_{x=\hat{x}(t|t)} , \\
P(t+1|t) &= F(t)P(t|t)F^T(t) + G(t)R_1(t)G^T(t) .
\end{aligned}
\tag{9.11}
$$

Remark 1 Note that the EKF is *not* an optimal filter. The estimates $\hat{x}(t|t)$, $\hat{x}(t+1|t)$ are *not* conditional means. This is in contrast to the linear case. $\square$

Remark 2 Another contrast to the linear case is that the filter gains $\{K(t)\}$ cannot be *precomputed*. It is apparent that they depend on data (outcome) through (9.4). $\square$

Remark 3 Note that in the general case, there is no guarantee that the filter "converges" or gives an estimate that is close to optimal. $\square$

Remark 4 The filter equations can be modified to account for the effect of an input signal, as in the linear case. Assume that the dynamics is changed to

$$x(t+1) = f(t, x(t), u(t)) + g(t, x(t))v(t) .$$

Then the EKF equations (9.11) remain unchanged with the exception of

$$\hat{x}(t+1|t) = f(t, \hat{x}(t|t), u(t)) . \qquad \square$$

9.2.2 An Iterated Extended Kalman Filter

It is *a priori* not obvious what the relevant linearization point is when computing the filter gain $K(t)$. There are many varieties of the basic EKF. The idea of the version of this section is to iterate over the measurement equation.

The iteration means that the linearization point is changed, and hopefully an improved approximation is achieved.

The algorithm runs as follows. For every t, let i run from 0 to $i_{\max}$, which should be a relatively small integer.

$$K^{(i)}(t) = P(t|t-1)H^{(i)T}(t)$$
$$\times [H^{(i)}(t)P(t|t-1)H^{(i)T}(t) + R_2(t)]^{-1} , \qquad (9.12)$$

$$\hat{x}^{(i)}(t|t) = \hat{x}(t|t-1) + K^{(i)}(t)[y(t) - h(t, \hat{x}(t|t-1))] , \qquad (9.13)$$

where

$$H^{(0)}(t) = \left. \frac{\partial h(t,x)}{\partial x} \right|_{x=\hat{x}(t|t-1)} , \qquad (9.14)$$

$$H^{(i)}(t) = \left. \frac{\partial h(t,x)}{\partial x} \right|_{x=\hat{x}^{(i-1)}(t|t)} \quad (i \geq 1) . \qquad (9.15)$$

Then set

$$\hat{x}(t|t) = \hat{x}^{(i_{\max})}(t|t) , \qquad (9.16)$$

$$P(t|t) = P(t|t-1) - K^{(i_{\max})}(t)H^{(i_{\max})}(t)P(t|t-1) , \qquad (9.17)$$

and compute the prediction estimate as

$$\hat{x}(t+1|t) = f(t, \hat{x}(t|t)) ,$$
$$P(t+1|t) = F(t)P(t|t)F^T(t) + G(t)R_1(t)G^T(t) , \qquad (9.18)$$

where

$$F(t) = \left. \frac{\partial f(t,x)}{\partial x} \right|_{x=\hat{x}(t|t)} , \qquad (9.19)$$

$$G(t) = \left. \frac{\partial g(t,x)}{\partial x} \right|_{x=\hat{x}(t|t)} . \qquad (9.20)$$

After this calculation (for fixed t) a new data point can be measured, t increased and the computation cycle repeated.

9.2.3 A Second-order Extended Kalman Filter

Another variant of the EKF also copes with second-order terms in the Taylor series expansions of $f(x)$ and $h(x)$. One then obtains terms that are quadratic forms in $x(t) - \hat{x}(t)$. Replace such quadratic forms with their "expected values". In this case, set

$$f(t, x(t)) \approx f(t, \hat{x}(t|t)) + F(t)(x(t) - \hat{x}(t|t)) + \overline{f}(t) ,$$
$$g(t, x(t)) \approx G(t) , \qquad (9.21)$$
$$h(t, x(t)) \approx h(t, \hat{x}(t|t-1)) + H(t)(x(t) - \hat{x}(t|t-1)) + \overline{h}(t) ,$$

where $F(t)$, $G(t)$, $H(t)$ are as before; see (9.4). In (9.21) the quantities $\overline{f}(t)$ and $\overline{h}(t)$ are vectors with components

$$(\overline{f}(t))_k = \frac{1}{2}(x(t) - \hat{x}(t|t))^T \left. \frac{\partial^2 f_k}{\partial x^2} \right|_{x=\hat{x}(t|t)} (x - \hat{x}(t|t))$$

$$\approx \frac{1}{2}\text{tr}\left[P(t|t) \left. \frac{\partial^2 f_k}{\partial x^2} \right|_{x=\hat{x}(t|t)} \right] , \tag{9.22}$$

$$(\overline{h}(t))_k = \frac{1}{2}(x(t) - \hat{x}(t|t-1))^T \left. \frac{\partial^2 h_k}{\partial x^2} \right|_{x=\hat{x}(t|t-1)} (x - \hat{x}(t|t-1))$$

$$\approx \frac{1}{2}\text{tr}\left[P(t|t-1) \left. \frac{\partial^2 h_k}{\partial x^2} \right|_{x=\hat{x}(t|t-1)} \right] . \tag{9.23}$$

Comparing with the previous model equations of (9.5), it can be seen that these are still applicable provided the modifications

$$u(t) = f(t, \hat{x}(t|t)) - F(t)\hat{x}(t|t) + \overline{f}(t) ,$$
$$w(t) = h(t, \hat{x}(t|t-1)) - H(t)\hat{x}(t|t-1) + \overline{h}(t) , \tag{9.24}$$

are made. It is possible, therefore, to retain the filter equations in (9.6), (9.7) and (9.8). In this case:

$$\hat{x}(t+1|t) = f(t, \hat{x}(t|t)) + \overline{f}(t) ,$$
$$y(t) - H(t)\hat{x}(t|t-1) - w(t) = y(t) - h(t, \hat{x}(t|t-1)) - \overline{h}(t) . \tag{9.25}$$

To summarize, the second-order algorithm becomes

$$H(t) = \left. \frac{\partial h(t, x)}{\partial x} \right|_{x=\hat{x}(t|t-1)} ,$$

$$K(t) = P(t|t-1)H^T(t)[H(t)P(t|t-1)H^T(t) + R_2(t)]^{-1} ,$$

$$(\overline{h}(t))_k = \frac{1}{2}\text{tr}\left[P(t|t-1) \left. \frac{\partial^2 h_k}{\partial x^2} \right|_{x=\hat{x}(t|t-1)} \right] ,$$

$$\hat{x}(t|t) = \hat{x}(t|t-1) + K(t)[y(t) - h(t, \hat{x}(t|t-1)) - \overline{h}(t)] ,$$

$$(\overline{f}(t))_k = \frac{1}{2}\text{tr}\left[P(t|t) \left. \frac{\partial^2 f_k}{\partial x^2} \right|_{x=\hat{x}(t|t)} \right] , \tag{9.26}$$

$$\hat{x}(t+1|t) = f(t, \hat{x}(t|t)) + \overline{f}(t) ,$$

$$P(t|t) = P(t|t-1) - K(t)H(t)P(t|t-1) ,$$

$$F(t) = \left. \frac{\partial f(t, x)}{\partial x} \right|_{x=\hat{x}(t|t)} ,$$

$$G(t) = g(t, x)|_{x=\hat{x}(t|t)} ,$$

$$P(t+1|t) = F(t)P(t|t)F^T(t) + G(t)R_1(t)G^T(t) .$$

Example 9.1 As an illustration of $\overline{f}(t)$, consider an (artificial) second order system characterized by

$$f(t,x) = \begin{pmatrix} x_1^2 + x_2 \\ x_1 e^{-x_2} \end{pmatrix} \triangleq \begin{pmatrix} f_1(x) \\ f_2(x) \end{pmatrix} .$$

By straightforward differentiation,

$$F = \frac{\partial f}{\partial x} = \begin{pmatrix} \frac{\partial f_1}{\partial x_1} & \frac{\partial f_1}{\partial x_2} \\ \frac{\partial f_2}{\partial x_1} & \frac{\partial f_2}{\partial x_2} \end{pmatrix} = \begin{pmatrix} 2x_1 & 1 \\ e^{-x_2} & -x_1 e^{-x_2} \end{pmatrix} .$$

In order to obtain $F(t)$, this matrix must be evaluated for $x = \hat{x}(t|t)$. In this case

$$\overline{f}(t) = \begin{pmatrix} \overline{f}_1(t) \\ \overline{f}_2(t) \end{pmatrix} ,$$

where

$$\overline{f}_1(t) = \tfrac{1}{2}\mathrm{tr}\left(P(t|t)\frac{\partial^2 f_1}{\partial x^2}\right)_{|x=\hat{x}(t|t)}$$

$$= \tfrac{1}{2}\mathrm{tr}\left(P(t|t)\begin{pmatrix} 2 & 0 \\ 0 & 0 \end{pmatrix}\right) = P_{11}(t|t) ,$$

$$\overline{f}_2(t) = \tfrac{1}{2}\mathrm{tr}\left(P(t|t)\frac{\partial^2 f_2}{\partial x^2}_{|x=\hat{x}(t|t)}\right)$$

$$= \tfrac{1}{2}\mathrm{tr}\left(P(t|t)\begin{pmatrix} 0 & -e^{-x_2} \\ -e^{-x_2} & x_1 e^{-x_2} \end{pmatrix}\right)_{|x=\hat{x}(t|t)}$$

$$= \tfrac{1}{2}e^{-\hat{x}_2(t|t)}[\hat{x}_1(t|t)P_{22}(t|t) - 2P_{12}(t|t)] . \qquad \square$$

9.2.4 An Example

As a simple illustration of the differences between the variants of extended Kalman filtering and optimal filtering, consider the simple nonlinear system

$$\begin{aligned} x(t+1) &= x^2(t), \qquad t = 0 , \\ y(t) &= x(t) + e(t) . \end{aligned} \tag{9.27}$$

The initial state $x(0)$ is assumed to be uniformly distributed over $[-a, a]$. The measurement noise $e(0)$ is assumed to be uniformly distributed over $[-b, b]$, and to be independent of $x(0)$.

In order to simplify the notation, set

$$x = x(0), \qquad z = x(1), \qquad y = y(0), \qquad e = e(0).$$

The problem to be examined is the estimation of x and z from y.

It is readily derived that x and e have zero means and variances $a^2/3$ and $b^2/3$ respectively.

Standard EKF

When applying the EKF, the appropriate initial values are

$$\hat{x}(0|-1) = Ex(0) = 0 , \qquad P(0|-1) = \text{var}[x(0)] = a^2/3 .$$

Straightforward use of (9.6)–(9.8) gives

$$K(0) = \frac{a^2/3}{a^2/3 + b^2/3} ,$$

$$\hat{x} = \frac{a^2}{a^2 + b^2} y , \tag{9.28}$$

$$P(0|0) = \frac{a^2 b^2/3}{a^2 + b^2} , \tag{9.29}$$

$$\hat{z} = \hat{x}^2 = \frac{a^4}{(a^2 + b^2)^2} y^2 . \tag{9.30}$$

Second-order EKF

When applying the algorithm (9.26), the previous estimate $\hat{x}$, as well as $P(0|0)$, is obtained. The prediction differs, however. In this case:

$$\overline{f}(0) = \frac{1}{2} P(0|0) 2 = \frac{a^2 b^2/3}{a^2 + b^2} .$$

Hence, the predicted state estimate is now

$$\hat{z} = \hat{x}^2 + \overline{f}(0) = \frac{a^4}{(a^2 + b^2)^2} y^2 + \frac{a^2 b^2/3}{a^2 + b^2} . \tag{9.31}$$

Optimal estimates

According to the development in Section 5.2, the mean square optimal estimates of x and z are the conditional means

$$\hat{x} = \mathbf{E}\left[x|y\right] = \int x p(x|y) \, dx , \tag{9.32}$$

$$\hat{z} = \mathbf{E}\left[x^2|y\right] = \int x^2 p(x|y) \, dx . \tag{9.33}$$

In order to evaluate these expectations, the conditional pdf $p(x|y)$ must be found. For this, recall that

$$p(x|y) = \frac{p(x, y)}{p(y)} , \tag{9.34}$$

and that $p(x, y)$ is uniformly distributed in the area displayed in Figure 9.1. The corners of the area have the coordinates $(a, a+b)$, $(a, a-b)$, $(-a, -a-b)$ and $(-a, -a+b)$.

In order to find the conditional pdf $p(x|y)$, one can then use (9.34). Let y be fixed. Apparently, $p(x|y)$ will be uniformly distributed over an interval.

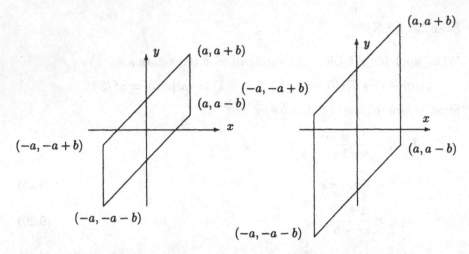

Fig. 9.1. The joint pdf $p(x, y)$ is uniformly distributed in the areas shown. Left: $a \geq b$. Right: $a \leq b$. The coordinates of the corners are displayed

Note that $p(y)$, the denominator in (9.34), acts solely as a scaling factor to make $p(x|y)$ a pdf.

Proceeding in this way, the following result is obtained:

Case 1 $a \geq b$

$$-a - b \leq y \leq -a + b \Rightarrow p(x|y) \quad \text{uniform over} \quad (-a, y + b)$$
$$-a + b \leq y \leq a - b \Rightarrow p(x|y) \quad \text{uniform over} \quad (y - b, y + b) \qquad (9.35)$$
$$a - b \leq y \leq a + b \Rightarrow p(x|y) \quad \text{uniform over} \quad (y - b, a)$$

Case 2 $a \leq b$

$$-a - b \leq y \leq a - b \Rightarrow p(x|y) \quad \text{uniform over} \quad (-a, y + b)$$
$$a - b \leq y \leq -a + b \Rightarrow p(x|y) \quad \text{uniform over} \quad (-a, a) \qquad (9.36)$$
$$-a + b \leq y \leq a + b \Rightarrow p(x|y) \quad \text{uniform over} \quad (y - b, a)$$

Using (9.35) and (9.36), the conditional means (9.32) and (9.33) can be evaluated. The case where $a \geq b$, $-a - b \leq y \leq -a + b$ is considered in some detail. Then, $p(x|y)$ is uniform over an interval of length $y + b - (-a) = y + a + b$. Hence, in this particular case,

$$p(x|y) = \begin{cases} 1/(y + a + b), & -a \leq x \leq y + b, \\ 0, & \text{elsewhere}, \end{cases}$$

and therefore,

$$\hat{x} = \int_{-a}^{y+b} x p(x|y)\, dx = \frac{1}{y + a + b} \frac{1}{2} [(y + b)^2 - (-a)^2]$$

$$= \frac{y+b-a}{2} , \tag{9.37}$$

$$\hat{z} = \int_{-a}^{y+b} x^2 p(x|y)\,dx = \frac{1}{y+a+b}\frac{1}{3}[(y+b)^3 - (-a)^3]$$

$$= \frac{1}{3}[(y+b)^2 - a(y+b) + a^2] . \tag{9.38}$$

Proceeding similarly for the other intervals, the following results are obtained:

Case 1 $a \geq b$

$$\hat{x} = \begin{cases} [y+b-a]/2 , & -a-b \leq y \leq -a+b , \\ y , & -a+b \leq y \leq a-b , \\ [y+a-b]/2 , & a-b \leq y \leq a+b , \end{cases} \tag{9.39}$$

$$\hat{z} = \begin{cases} [(y+b)^2 - a(y+b) + a^2]/3 , & -a-b \leq y \leq -a+b , \\ y^2 + b^2/3 , & -a+b \leq y \leq a-b , \\ [(y-b)^2 + a(y-b) + a^2]/3 , & a-b \leq y \leq a+b . \end{cases} \tag{9.40}$$

Case 2 $a \leq b$

$$\hat{x} = \begin{cases} [y+b-a]/2 , & -a-b \leq y \leq a-b , \\ 0 , & a-b \leq y \leq -a+b , \\ [y-b+a]/2 , & -a+b \leq y \leq a+b , \end{cases} \tag{9.41}$$

$$\hat{z} = \begin{cases} [(y+b)^2 - a(y+b) + a^2]/3 , & -a-b \leq y \leq a-b , \\ a^2/3 , & a-b \leq y \leq -a+b , \\ [(y-b)^2 + a(y-b) + a^2]/3 , & -a+b \leq y \leq a+b . \end{cases} \tag{9.42}$$

The estimates have now been derived. They are displayed, as functions of the observed variable y, in Figure 9.2. Observe that in the optimal estimates the a priori information is used in particular so that the constraints $|\hat{x}| \leq a$, $0 \leq \hat{z} \leq a^2$, are automatically satisfied.

It can be seen from the diagram that the estimates of x look relatively similar, whereas the differences are more apparent for the estimates of z.

To evaluate the statistical performance of the estimates would require straightforward, but occasionally very tedious, calculations to compute the expected mean square errors (MSEs), $\mathbf{E}[\hat{x} - x]^2$ and $\mathbf{E}[\hat{z} - z]^2$, in the various cases. (See Exercise 9.13.) Instead, results of Monte Carlo simulations are shown in Table 9.1. In each case, $N = 10\,000$ trials were used and the MSE evaluated as $\frac{1}{N}\sum_{i=1}^{N}(\hat{x}_i - x_i)^2$ and $\frac{1}{N}\sum_{i=1}^{N}(\hat{z}_i - z_i)^2$ respectively. The parameter a was chosen as 1, and can be viewed as a simple scaling factor. The numerical results give the same picture as does Figure 9.2. The quality of the estimates $\hat{x}$ does not differ very much. The EKF estimate of z is clearly inferior to the other estimates of z. The mean square optimal estimates give the smallest MSEs.

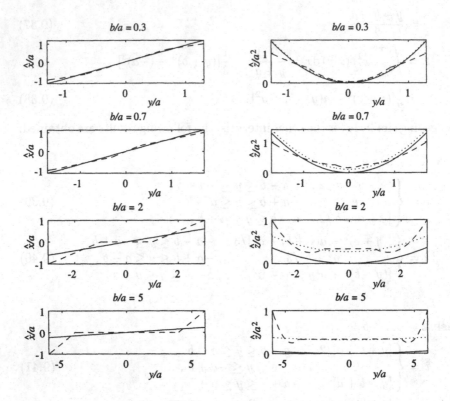

Fig. 9.2. *Left*: Comparisons of the EKF estimate of x (9.28); (*solid line*), and the optimal estimate (9.39) and (9.41); (*dashed line*). *Right*: Comparisons of the EKF estimate of z, (9.30); (*solid line*), the second-order EKF estimate (9.31); (*dotted line*) and the optimal estimate (9.40) and (9.42); (*dashed line*)

Table 9.1. Performance evaluation by Monte Carlo simulations of estimators

Method	b	0.3	0.7	2	5
EKF	MSE($\hat{x}$)	0.027	0.108	0.27	0.32
	MSE($\hat{z}$)	0.031	0.085	0.16	0.19
Second-order EKF	MSE($\hat{x}$)	0.027	0.108	0.27	0.32
	MSE($\hat{z}$)	0.030	0.075	0.087	0.088
Optimal estimate	MSE($\hat{x}$)	0.025	0.105	0.25	0.30
	MSE($\hat{z}$)	0.024	0.064	0.081	0.085

9.3 Gaussian Sum Estimators

It was seen previously, in Section 5.2, that the optimal state estimate is often given by the conditional mean. To compute it, one needs to calculate the conditional pdf $p(x(t)|Y^t)$ recursively, which can often be a formidable task; see Section 5.4. Linear systems with Gaussian disturbances are the clear exception. *Gaussian sum estimators* are constructed as a reasonable way of approximating the recursive computation of the conditional pdfs. The basic underlying idea is to approximate an arbitrary pdf, say $p(x)$, by a weighted sum of Gaussian distributions, such as

$$p(x) \approx \sum_{i=1}^{m} \alpha_i \gamma(x; m_i, P_i) , \tag{9.43}$$

where $\alpha_i \geq 0$, $\sum_{i=1}^{m} \alpha_i = 1$ and $\gamma(x; m, P)$ is the pdf of an n-dimensional $N(m, P)$ random vector

$$\gamma(x; m, P) = \frac{1}{(2\pi)^{n/2} (\det P)^{1/2}} \exp\left(-\frac{1}{2}(x - m)^T P^{-1}(x - m)\right) . \tag{9.44}$$

In fact, it can be shown that by selecting m, $\{\alpha_i\}$, $\{m_i\}$ and $\{P_i\}$ appropriately, the approximation in (9.43) can be made arbitrarily good.

Now apply this idea to the computation of the conditional pdfs for a nonlinear system of the form

$$x(t + 1) = f(x(t)) + g(x(t))v(t) ,$$
$$y(t) = h(x(t)) + e(t) , \tag{9.45}$$
$$\mathbf{E}\, v(t)v^T(t) = R_1 , \quad \mathbf{E}\, e(t)e^T(t) = R_2 ,$$

where $v(t)$ and $e(t)$ are mutually independent white noise sequences of zero mean. Assume that the approximation

$$p(x(t)|Y^t) \approx \sum_{i=1}^{m} \alpha_i(t)\gamma(x(t); m_i(t), P_i(t)) \tag{9.46}$$

is available. Then the conditional mean and its covariance matrix can easily be found using Lemma 2.1. This gives

$$\hat{x}(t|t) = \int x(t)p(x(t)|Y^t)\, dx(t)$$
$$= \sum_{i=1}^{m} \alpha_i(t) \int x(t)\gamma(x(t); m_i(t), P_i(t))\, dx(t)$$
$$= \sum_{i=1}^{m} \alpha_i(t)m_i(t) , \tag{9.47}$$

$$P(t|t) = \mathbf{E}\,[x(t) - \hat{x}(t|t)][x(t) - \hat{x}(t|t)]^T$$
$$= \sum_{i=1}^{m} \alpha_i(t) \int [x(t) - \hat{x}(t|t)][x(t) - \hat{x}(t|t)]^T$$
$$\times \gamma(x(t); m_i(t), P_i(t))\, dx(t)$$
$$= \sum_{i=1}^{m} \alpha_i(t)[P_i(t) + (m_i(t) - \hat{x}(t|t))(m_i(t) - \hat{x}(t|t))^T] . \tag{9.48}$$

The updating consists of m parallel EKFs to find $\{m_i(t), P_i(t)\}$.

Consider first *the measurement update*, which concerns how to update

$$p(x(t)|Y^{t-1}) \approx \sum_{i=1}^{m} \alpha_i(t-1)\gamma(x(t); \overline{m}_i(t), \overline{P}_i(t)) \tag{9.49}$$

to

$$p(x(t)|Y^t) \approx \sum_{i=1}^{m} \alpha_i(t)\gamma(x(t); m_i(t), P_i(t)) . \tag{9.50}$$

Proceed by applying an EKF for each term in the sum. Note that if the system is linear and there is just one term, then the standard Kalman filter is used. As in Sections 6.4 and 9.2.1, this gives the updating

$$
\begin{aligned}
m_i(t) &= \overline{m}_i(t) + K_i(t)[y(t) - h(\overline{m}_i(t))] , \\
S_i(t) &= H_i(t)\overline{P}_i(t)H_i^T(t) + R_2 , \\
P_i(t) &= \overline{P}_i(t) - \overline{P}_i(t)H_i^T(t)S_i^{-1}(t)H_i(t)\overline{P}_i(t) , \\
K_i(t) &= \overline{P}_i(t)H_i^T(t)S_i^{-1}(t) , \\
H_i(t) &= \left.\frac{\partial h}{\partial x}\right|_{x=\overline{m}_i(t)} .
\end{aligned}
\tag{9.51}
$$

The weights of the individual pdfs are updated as

$$\alpha_i(t) = \frac{\alpha_i(t-1)\gamma(y(t); h(\overline{m}_i(t)), S_i(t))}{\sum_{i=1}^{m} \alpha_i(t-1)\gamma(y(t); h(\overline{m}_i(t)), S_i(t))} . \tag{9.52}$$

This means that $\alpha_i(t)$ will be influenced by how well the corresponding pdf can explain the prediction error $y(t) - h(\overline{m}_i(t))$.

Next deal with the *time update*, which concerns how to update

$$p(x(t)|Y^t) \approx \sum_{i=1}^{m} \alpha_i(t)\gamma(x(t); m_i(t), P_i(t)) \tag{9.53}$$

to

$$p(x(t+1)|Y^t) \approx \sum_{i=1}^{m} \alpha_i(t)\gamma(x(t+1); \overline{m}_i(t+1), \overline{P}_i(t+1)) . \tag{9.54}$$

Apply m parallel EKFs again:

$$
\begin{aligned}
\overline{m}_i(t+1) &= f(m_i(t)) , \\
\overline{P}_i(t+1) &= F_i(t)P_i(t)F_i(t) + G_i(t)R_1G_i^T(t) , \\
F_i(t) &= \left.\frac{\partial f(x)}{\partial x}\right|_{x=m_i(t)} , \\
G_i(t) &= g(x)|_{x=m_i(t)} , \\
\hat{x}(t+1|t) &= \sum_{i=1}^{m} \alpha_i(t)\overline{m}_i(t+1) ,
\end{aligned}
\tag{9.55}
$$

$$P(t+1|t) = \mathbf{E}\left[x(t+1) - \hat{x}(t+1|t)\right]\left[x(t+1) - \hat{x}(t+1|t)\right]^T$$

$$= \sum_{i=1}^{m} \alpha_i(t)[\overline{P}_i(t+1)$$

$$+ (\overline{m}_i(t+1) - \hat{x}(t+1|t))(\overline{m}_i(t+1) - \hat{x}(t+1|t))^T].$$

Note that the weights $\{\alpha_i(t)\}$ are the same in (9.53) and (9.54).

The Gaussian sum estimator has now been described. In most practical cases, it is required that m, the number of filters used, is chosen large, so that the "individual" covariance matrices $P_i(t)$ and $\overline{P}_i(t)$ can be small.

9.4 The Multiple Model Approach

9.4.1 Introduction

The multiple model approach accounts for techniques, where the underlying dynamics are linear, but can follow one of several linear models. Furthermore, it is not known *a priori*, which of these linear models applies. It is also possible that the system is switching over time between the different linear models. The technique has been used frequently within navigation, tracking and surveillance of aircraft traffic, with both civilian and military applications. The different linear models can then correspond to different types of manoeuvre, such as

- constant speed and altitude
- acceleration in one direction
- a turn.

9.4.2 Fixed Models

We consider first the case when there are r different and constant modes. The different modes, or linear models of the usual state-space types, are denoted

$$\{M_j\}_{j=1}^{r}. \tag{9.56}$$

The associated linear model M_j can be described as

$$x_{(j)}(t+1) = F_j x_{(j)}(t) + v_{(j)}(t),$$

$$y(t) = H_j x_{(j)}(t) + e_{(j)}(t), \qquad j = 1, \ldots, r. \tag{9.57}$$

$$\mathbf{E}\, v_{(j)}(t)v_{(j)}^T(t) = R_{1,j}, \qquad \mathbf{E}\, e_{(j)}(t)e_{(j)}^T(t) = R_{2,j}.$$

Note that the model order (the dimension of the state vector $x_{(j)}(t)$) can be allowed to differ between the different models.

We assume that the prior probability that the system is in mode M_j is known, and that this is expressed as

$$\mu_j(0) \overset{\triangle}{=} P(M_j|Y^0), \qquad j = 1, \ldots, r . \tag{9.58}$$

Consider the case when the mode is fixed over time. This means that j is constant, but unknown. The basic idea in the multiple model approach is then to run r different linear Kalman filters, each corresponding to a separate model, in parallel. In addition, based on the observations, the likelihood for the system to follow each model is assessed, and a weighted sum of the *a posteriori* pdfs of the individual filters can be applied for finding the most appropriate model, and state estimate.

The *a posteriori* probability that model M_j is correct, given data up to time t, can be evaluated using Bayes' rule. It is recursively given by

$$\begin{aligned}
\mu_j(t) &= P(M_j|Y^t) = P(M_j|y(t), Y^{t-1}) \\
&= \frac{p(y(t)|Y^{t-1}, M_j)P(M_j|Y^{t-1})}{p(y(t)|Y^{t-1})} \\
&= \frac{p(y(t)|Y^{t-1}, M_j)}{\sum_{i=1}^r p(y(t)|Y^{t-1}, M_i)\mu_i(t-1)}\mu_j(t-1), \quad j = 1, \ldots, r .
\end{aligned} \tag{9.59}$$

We note that in (9.59) the quantity in the numerator, $p(y(t)|Y^{t-1}, M_j)$, is the likelihood that the output at time t follows the linear model M_j. This is given by the Gaussian distribution for the innovation at time t:

$$p(y(t)|Y^{t-1}, M_j) = p(\tilde{y}_{(j)}(t)) = \gamma(\tilde{y}_{(j)}(t); 0, \Lambda_j(t)) . \tag{9.60}$$

Here, $\tilde{y}_{(j)}(t)$ denotes the innovation at time t using model M_j, $\Lambda_j(t)$ is its covariance matrix, and $\gamma(\cdot; \cdot, \cdot)$ is the pdf of a Gaussian distributed random vector.

Each separate mode ($j = 1, \ldots, r$) gives a *mode-conditioned state estimate* $\hat{x}_{(j)}(t|t)$.

The latest mode probabilities $\mu_j(t)$ can be used to combine the mode-conditioned estimates and covariances. This is possible only in the case that the different linear models are all of the same order. In fact, under the assumptions:

- the r linear models are of the same order,
- the true dynamics are among one of the r models,
- the true dynamics are constant over time, and do not jump between the models,

the pdf of the state will be a Gaussian mixture (that is, a Gaussian sum)

$$P(x(t)|Y^t) = \sum_{j=1}^r \mu_j(t)\gamma(x(t); \hat{x}_{(j)}(t|t), P_j(t|t)) , \tag{9.61}$$

where estimation using model M_j gives the state estimate $\hat{x}_{(j)}(t|t)$ and the associated covariance matrix $P_j(t|t)$.

From the distribution (9.61) one can readily compute the optimal state estimate as

$$\hat{x}(t|t) = \sum_{j=1}^{r} \mu_j(t)\hat{x}_{(j)}(t|t) , \qquad (9.62)$$

which has the covariance matrix

$$P(t|t) = \sum_{j=1}^{r} \mu_j(t) \left[P_j(t|t) + [\hat{x}_{(j)}(t|t) - \hat{x}(t|t)][\hat{x}_{(j)}(t|t) - \hat{x}(t|t)]^T \right] . \quad (9.63)$$

If the set of linear models includes the true dynamics, and no jump occurs, then the probability of the correct mode will converge to unity as time increases. This means that for a large time span we will, with a probability approaching one, know which of the linear models is correct.

9.4.3 Switching Models

We consider now the case when there can be switches between the different models. More precisely, we assume that the jump process between the different models is a Markov chain, with known transition probabilities

$$p_{ij} = P(M(t) = M_j | M(t-1) = M_i) . \qquad (9.64)$$

We assume that these transition probabilities are time-invariant and independent of the model states $\{x_{(j)}(t)\}$. This leads to the switches forming a homogeneous Markov chain. In practice, the matrix P, defined in (9.64), often has diagonal elements that are close to one, meaning that the probability is high that there is no switch. The matrix P is an example of a stochastic matrix: all its elements are positive, and the row sums are all equal to one.

The total system, with the r models and the switching between them, is an example of a hidden Markov model. The state of the Markov chain is hidden, as it is not directly measurable. What can be observed is only the output variable $y(t)$.

The conditional probability of the state $x(t)$ given the observations Y^t is still a Gaussian mixture. However, noting that at each time instant the Markov chain can take r different values, and there can be a switch at each time step, we find that over the interval $[1, t]$ there are in fact r^t different possible paths for the switching Markov chain. Hence we have in this case

$$p(x(t)|Y^t) = \sum_{k=1}^{r^t} p(x(t)|M^{(k)}, Y^t) P(M^{(k)}|Y^t) , \qquad (9.65)$$

where we let $M^{(k)}$ denote the whole time history of the switching models.

The mean value of the conditional distribution in (9.65) will be the optimal state estimate. However, owing to the exponentially increasing number of terms it is not practical to use, and suboptimal approximations are needed. The IMM algorithm to be presented in Section 9.4.4, is based on using only r filters operating in parallel.

9.4.4 Interacting Multiple Models Algorithm

A basic idea of the IMM algorithm is to limit the amount of possible combination of models. Assume here that the state vectors of all the r models have the same dimension. At each time step, only r different models are used. Before the time update, the current estimates are mixed with weightings (probabilities) that are also updated. For each of the parallel r filters, the conditional pdf is assumed to be a mixture of r Gaussian pdfs, and they are then approximated via moment matching to a single Gaussian pdf (with details given later on).

The r different models will be subject to switches, and we adopt the time-varying notation $\{M_j(t)\}_{j=1}^r$ to account for this.

Using Bayes' rule, we get

$$
p(x(t)|Y^t) = \sum_{j=1}^r p(x(t)|M_j(t), Y^t) p(M_j(t)|Y^t)
$$

$$
= \sum_{j=1}^r p(x(t)|M_j(t), y(t), Y^{t-1}) \mu_j(t) . \tag{9.66}
$$

Further, we have

$$
p(x(t)|M_j(t), y(t), Y^{t-1}) = \frac{p(y(t)|M_j(t), x(t))}{p(y(t)|M_j(t), Y^{t-1})}
$$

$$
\times p(x(t)|M_j(t), Y^{t-1}) . \tag{9.67}
$$

The last factor of (9.67) is evaluated, with some approximation, using the expression for the total probability:

$$
p(x(t)|M_j(t), Y^{t-1})
$$

$$
= \sum_{i=1}^r p(x(t)|M_j(t), M_i(t-1), Y^{t-1}) P(M_i(t-1)|M_j(t), Y^{t-1})
$$

$$
\approx \sum_{i=1}^r p(x(t)|M_j(t), M_i(t-1), \hat{x}_{(i)}(t-1|t-1), P_i(t-1|t-1))
$$

$$
\times \mu_{i|j}(t-1|t-1) . \tag{9.68}
$$

Here, $\mu_{i|j}(t-1|t-1)$ denotes the probability of a switch from model $M_i(t-1)$ to $M_j(t)$, and the state estimate and its covariance matrix for model $M_i(t-1)$ are denoted by $\hat{x}_{(i)}(t-1|t-1)$ and $P_i(t-1|t-1)$, respectively.

Apparently, (9.68) is a mixture of r Gaussian pdfs, (assuming that each state estimate is Gaussian). We next replace the sum by one single Gaussian, using moment matching (further details are given later, see Step 2).

$$
p(x(t)|M_j(t), Y^{t-1}) = \gamma(x(t); \hat{x}_{(j)}^0(t-1|t-1), P_j^0(t-1|t-1)) . \tag{9.69}
$$

The whole IMM algorithm runs us follows. We describe below the updating of the filter quantities from time $t-1$ to time t.

1. **Calculation of the mixing probabilities** The probability that we
   have a switch from $M_i(t-1)$ to $M_j(t)$, conditioned on the observations
   Y^{t-1} is

$$\mu_{i|j}(t-1|t-1) \triangleq P(M_i(t-1)|M_j(t),Y^{t-1})$$

$$= \frac{1}{\bar{c}_j}P(M_j(t)|M_i(t-1),Y^{t-1})P(M_i(t-1)|Y^{t-1})$$

$$= \frac{1}{\bar{c}_j}p_{ij}\mu_i(t-1) . \qquad (9.70)$$

Here the left-hand side signifies the *mixing probabilities*, and the denominator $\bar{c}_j$ is just the normalizing constant,

$$\bar{c}_j = \sum_{i=1}^{r}p_{ij}\mu_i(t-1) , \qquad j=1,\dots,r . \qquad (9.71)$$

2. **Mixing** Starting with the state estimates $\{\hat{x}_{(i)}(t-1|t-1)\}_{i=1}^{r}$, one com-
   putes the mixed "initial condition" for the filter $M_j(t)$ using (9.68)

$$\hat{x}_{(j)}^0(t-1|t-1) = \sum_{i=1}^{r}\mu_{i|j}(t-1|t-1)\hat{x}_{(i)}(t-1|t-1) ,$$

$$j=1,\dots,r . \qquad (9.72)$$

The associated covariance matrix reads

$$P_j^0(t-1|t-1) = \sum_{i=1}^{r}\mu_{i|j}(t-1|t-1)\Big\{P(t-1|t-1)$$

$$+[\hat{x}_{(i)}(t-1|t-1) - \hat{x}_{(j)}^0(t-1|t-1)]$$

$$\times[\hat{x}_{(i)}(t-1|t-1) - \hat{x}_{(j)}^0(t-1|t-1)]^T\Big\} ,$$

$$j=1,\dots,r . \qquad (9.73)$$

3. **Mode-matched filters** The state estimate (9.72) and the covariance
   matrix (9.73) are used as inputs to the filter $M_j(t)$. With the new out-
   put measurement $y(t)$ available, the new filter state estimate $\hat{x}_{(j)}(t|t)$ is
   computed using the standard Kalman update. Further, an updated fil-
   ter covariance matrix $P_j(t|t)$ is computed as well. These calculations are
   carried out for the r different filters $(j=1,\dots,r)$. The details are as
   follows. The "initial values" are $\hat{x}_{(j)}^0(t-1|t-1)$ and $P_j^0(t-1|t-1)$. The
   predictor of the state vector is

$$\hat{x}_{(j)}(t|t-1) = F_j\hat{x}_{(j)}^0(t-1|t-1) + Gu(t-1) , \qquad (9.74)$$

and of the output

$$\hat{y}_{(j)}(t|t-1) = H_j\hat{x}_{(j)}(t|t-1) . \qquad (9.75)$$

The associated update of the Riccati equation becomes

$$P_j(t|t-1) = F_j P_j^0(t-1|t-1) F_j^T + R_{1,j} \,. \tag{9.76}$$

The filter estimate of the state vector is

$$\hat{x}_{(j)}(t|t) = \hat{x}_{(j)}(t|t-1) + \tilde{K}_j \left(y(t) - H_j \hat{x}_{(j)}(t|t-1) \right) \tag{9.77}$$

where the filter gain is

$$\tilde{K}_j(t) = P_j(t|t-1) H_j^T \left(H_j P_j(t|t-1) H_j^T + R_{2,j} \right)^{-1} . \tag{9.78}$$

Finally, the covariance matrix of the filter error is

$$P_j(t|t) = P_j(t|t-1) - \tilde{K}_j(t) H_j P_j(t|t-1) \,. \tag{9.79}$$

In addition, the likelihood that the filter $M_j(t)$ is applicable is expressed using the Gaussian pdf implied by (9.72) and (9.73), as

$$S_j(t) = p(y(t)|M_j(t), Y^{t-1}) \,. \tag{9.80}$$

Using the model M_j and the assumption of Gaussian distributed data, $y(t)$ conditioned on Y^{t-1} has mean value $\hat{y}_{(j)}(t|t-1)$ and covariance matrix $H_j P_j(t|t-1) H_j^T + R_{2,j}$. Hence

$$S_j = \gamma(y(t); H_j \hat{x}_{(j)}(t|t-1), H_j P_j(t|t-1) H_j^T + R_{2,j}) \,. \tag{9.81}$$

4. **Mode probability update** Here we update the probabilities $\{\mu_j(t)\}_{j=1}^r$ as follows:

$$
\begin{aligned}
\mu_j(t) &\triangleq P(M_j(t)|Y^t) \\
&= \frac{1}{c} p(y(t)|M_j(t), Y^{t-1}) P(M_j(t)|Y^{t-1}) \\
&= \frac{1}{c} S_j(t) \sum_{i=1}^r P(M_j(t)|M_i(t-1), Y^{t-1}) P(M_i(t-1)|Y^{t-1}) \\
&= \frac{1}{c} S_j(t) \sum_{i=1}^r p_{ij} \mu_i(t-1) \,, \qquad j = 1, \dots, r \,.
\end{aligned}
\tag{9.82}
$$

or, in brief:

$$\mu_j(t) = \frac{1}{c} S_j(t) \bar{c}_j \,, \qquad j = 1, \dots, r \,. \tag{9.83}$$

Above, $\bar{c}_j$ is given in (9.71), and the normalizing denominator factor c is given by

$$c = \sum_{i=1}^r S_j(t) \bar{c}_j \,. \tag{9.84}$$

5. **Estimate and covariance combination** It is enough to carry out the above calculations per each time step. When explicit state estimates and covariance matrices are needed as well, the following weighted combinations can be computed:

$$\hat{x}(t|t) = \sum_{j=1}^{r} \mu_j(t)\hat{x}_{(j)}(t|t) , \tag{9.85}$$

$$P(t|t) = \sum_{j=1}^{r} \mu_j(t) \left\{ P_j(t|t) \right.$$

$$\left. + [\hat{x}_{(j)}(t|t) - \hat{x}(t|t)][\hat{x}_{(j)}(t|t) - \hat{x}(t|t)]^T \right\} . \tag{9.86}$$

The computations for processing the IMM algorithm in one time step are displayed in Figure 9.3.

The IMM algorithm is next illustrated in a simple example.

Example 9.2 To make the example simple, but still illustrative, we consider the case of shifting between two first-order systems with different time constants and different static gains. The two models are

$$M_1 : \begin{cases} x(t+1) = 0.95x(t) + 0.05u(t) + 0.1v(t) \\ \quad y(t) = x(t) + 0.5e(t) \end{cases} \tag{9.87}$$

$$M_2 : \begin{cases} x(t+1) = 0.5x(t) + 2u(t) + 0.3v(t) \\ \quad y(t) = x(t) + 0.5e(t) \end{cases} \tag{9.88}$$

where $v(t)$ and $e(t)$ are uncorrelated white noise sequences of zero mean and unit variance. The system starts in M_1 with zero initial conditions. The input is a square wave of unit amplitude and period 80 sampling intervals. There is a switch to M_2 at $t = 100$, back to M_1 at $t = 175$, and again to M_2 at $t = 311$. The simulated data are shown in Figure 9.4.

Some different estimators of the state $x(t)$ were tested. In all cases, the initial estimate was set to $\hat{x}(0|-1) = 5$, with a given accuracy of $P(0) = 1$. As a reference, we first consider a Kalman filter that utilizes full information about which of the two models is operating. This would give a lower bound of what performance can be achieved. The results obtained using this type of Kalman filter are displayed in Figure 9.5. It can be seen that the estimate follows the actual state quite accurately.

Next, we test Kalman filters using only model M_1 or model M_2. The results are shown in Figures 9.6 and 9.7. Not surprisingly, the filters perform well when the model is valid, and detoriate strongly in periods when a wrong model is used.

Finally, we tried the IMM algorithm. In this case the initial probabilities for the two models were each set as 0.5. Further, the transition matrix for shifts between the models was set to

$$\begin{pmatrix} 0.98 & 0.02 \\ 0.02 & 0.98 \end{pmatrix}$$

meaning that at each time point the probability of a model change is 0.02. The state estimates obtained are displayed in Figure 9.8. It turns out that the

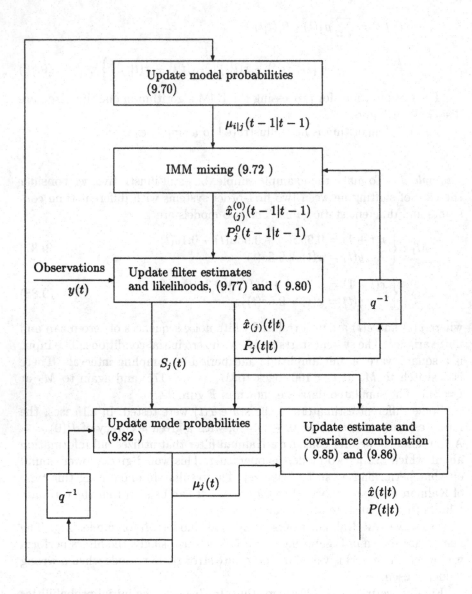

Fig. 9.3. Flowchart of the IMM algorithm

performance is much superior to the use of one single model. It has a quality close to that shown in Figure 9.5, despite the fact that here it is not known which of the two models is in operation. In Figure 9.9 we show the estimated

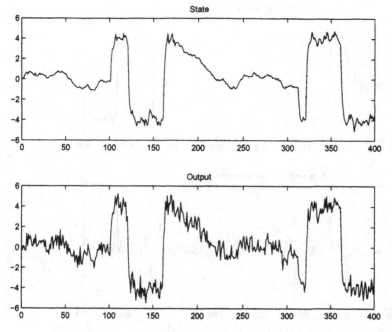

Fig. 9.4. Generated state variable and output signal, Example 9.2

probability for model M_1. Over most of the experiment, it has been possible

9.5 Monte Carlo Methods for Propagating the Conditional Probability Density Functions

It was found in Chapter 5 that the conditional pdfs, which are needed to evaluate the optimal state estimates in the general case, are propagating according to (see (5.24))

$$p(x(t+1)|Y^{t+1}) = \frac{p(y(t+1)|x(t+1))}{p(y(t+1)|Y^t)}$$

$$\times \int p(x(t+1)|x(t))p(x(t)|Y^t)\,dx(t) . \qquad (9.89)$$

Further, the conditional mean filter estimate will be

$$\hat{x}(t) \triangleq \hat{x}(t|Y^t) = \int x(t)p(x(t)|Y^t)\,dx(t) , \qquad (9.90)$$

and it has the (conditional, or *a posteriori*) covariance matrix

$$P(t) = \mathbf{E}\left[\hat{x}(t) - x(t)\right]\left[\hat{x}(t) - x(t)\right]^T$$

$$= \int [\hat{x}(t) - x(t)][\hat{x}(t) - x(t)]^T p(x(t)|Y^t)\,dx(t) . \qquad (9.91)$$

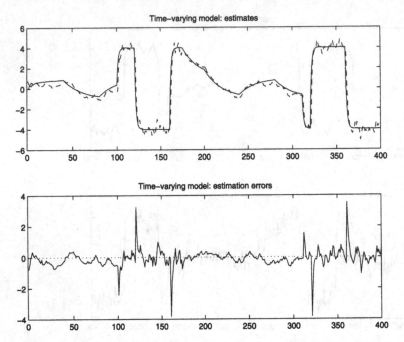

Fig. 9.5. State estimate using a Kalman filter with information of the current model, Example 9.2: (*upper part, solid line*) state estimate, (*upper part, dashed line*) true state, (*lower part*) estimation error

Assume now that the dynamics of the system are given as in (9.1):

$$x(t+1) = f(x(t)) + v(t) ,$$
$$y(t) = h(x(t)) + e(t) .$$
$$(9.92)$$

The Monte Carlo based methods for evaluating the conditional pdfs and the optimal estimate (9.90) are based on the assumptions:

- A prior estimation of $x(0|Y^0)$ is known (or assumed to be given).
- The conditional pdf $p(y(t)|x(t))$ has a known functional form. This corresponds to knowing the pdf of the measurement noise $e(t)$.
- The pdf of the process noise $p(v(t))$ is known, and is available for sampling.

The basic algorithm runs as follows. Set $t = 0$.

1. Assume that the conditional pdf $p(x(t)|Y^{t-1})$ is known, and that we have been able to generate M independent samples, $\{x_t^i\}_{i=1}^M$, from it. The number of samples M is taken as a relatively large number (say, at least $M = 10^3$–10^4). This means that we approximate the conditional pdf as

$$p(x(t)|Y^{t-1}) \approx \frac{1}{M} \sum_{i=1}^{M} \delta(x(t) - x_t^i) .$$
$$(9.93)$$

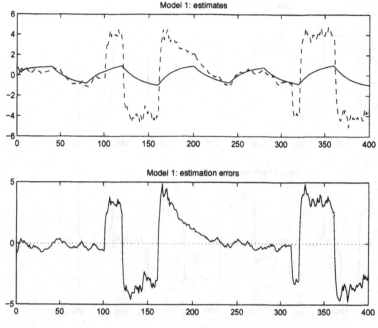

Fig. 9.6. State estimate using a Kalman filter with model 1 as a fixed model, Example 9.2: (*upper part, solid line*) state estimate, (*upper part, dashed line*) true state, (*lower part*) estimation error

Set $M = rN$, with r an integer (take at least $r \geq 10$).

2. Next we compare and assess these values using the available output at time t, that is $y(t)$, and assign a normalized weight to each sample x_t^i:

$$w_i = \frac{p(y(t)|x_t^i)}{\sum_{i=1}^{M} p(y(t)|x_t^i)} , \qquad i = 1, \ldots , M . \tag{9.94}$$

3. Generate a new set $\{x_t^{i*}\}_{i=1}^{N}$ by resampling with replacement N times from the discrete set $\{x_t^j\}_{j=1}^{M}$, where

$$P(x_t^{i*} = x_t^j) = w_j . \tag{9.95}$$

The resampling is actually done as follows.

For $i = 1 \ldots N$: take u_i as a sample from a uniform distribution over the interval $(0, 1]$. Find an M such that

$$\sum_{j=0}^{M-1} w_j < u_i \leq \sum_{j=0}^{M} w_j , \tag{9.96}$$

where the convention $w_0 = 0$ is used. Then take

$$x_t^{i*} = x_t^M . \tag{9.97}$$

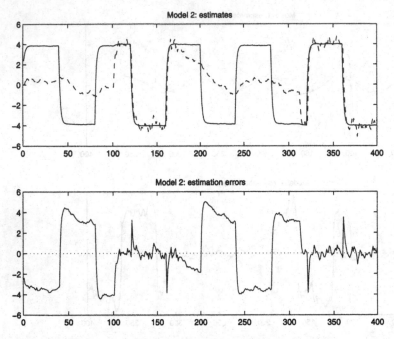

Fig. 9.7. State estimate using a Kalman filter with model 2 as a fixed model, Example 9.2: (*upper part, solid line*) state estimate, (*upper part, dashed line*) true state, (*lower part*) estimation error

4. Predict each of the resampled states independently, r times. Thus, generate $\{x_{t+1}^j\}_{j=1}^M$ as

$$x_{t+1}^{(i-1)r+k} \sim p(x(t+1)|x(t) = x_t^{i*}),$$
$$i = 1, \ldots, N, \quad k = 1, \ldots, r. \tag{9.98}$$

5. Replace t by $t+1$ and go to Step 2.

In addition to the above propagation of the conditional pdfs, we find that the conditional mean and its covariance matrix are then readily approximated using (9.93) as

$$\hat{x}(t) = \mathbf{E}\, x(t) p(x(t)|Y^{t-1})\, \mathrm{d}x(t)$$

$$= \int \frac{1}{N} \sum_{i=1}^N x(t) w_i \delta(x(t) - x_t^i)\, \mathrm{d}x(t)$$

$$= \frac{1}{N} \sum_{i=1}^N x_t^i w_i. \tag{9.99}$$

Similarly:

$$P(t) = \frac{1}{N} \sum_{i=1}^N w_i [x_t^i - \hat{x}(t)][x_t^i - \hat{x}(t)]^T. \tag{9.100}$$

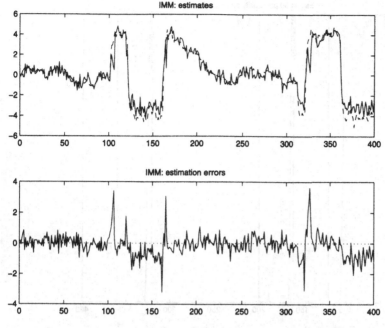

Fig. 9.8. State estimate using the IMM algorithm, Example 9.2: (*upper part, solid line*) state estimate, (*upper part, dashed line*) true state, (*lower part*) estimation error

9.6 Quantized Measurements

In many cases, data are quantized in amplitude. This can, for example, be due to analogue to digital conversion. The quantized data may have slightly different properties from the original one. In this section, the influence of amplitude quantization on the variance is examined briefly, as this is an important dispersion measure.

Let x be a continuously distributed random variables and let x_Q be its quantized value, in the following way:

$$x_Q = j\,\Delta, \quad \text{if } j\,\Delta - \frac{\Delta}{2} \le x < j\,\Delta + \frac{\Delta}{2}. \tag{9.101}$$

A *simplified approach* would be to regard x_Q as a measurement of x, disturbed by a noise that is uniformly distributed over the interval $(-\Delta/2, \Delta/2)$. This would mean

$$x_Q = x + e. \tag{9.102}$$

Assuming that x and e are independent:

$$\begin{aligned} \text{var}(x_Q) &= \text{var}(x) + \text{var}(e) \\ &= \text{var}(x) + \Delta^2/12. \end{aligned} \tag{9.103}$$

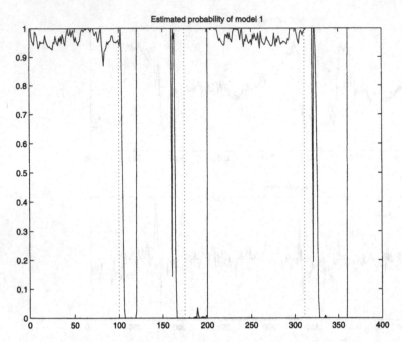

Fig. 9.9. Estimate of the probability of model M_1 being in operation, Example 9.2. The dotted lines show the time points when there was a shift between M_1 and M_2

This is the so-called *Sheppard correction.*

To examine the validity of the Sheppard correction, a detailed analysis invoking the characteristic function $\varphi_x(\omega)$ of x is needed. The function $\varphi_x(\omega)$ is the Fourier transform of the pdf of x:

$$\varphi_x(\omega) = \int_{-\infty}^{\infty} e^{i\omega x}\, p(x)\, dx = \mathbf{E}\left[e^{i\omega x}\right]. \tag{9.104}$$

The details of the analysis are given in Section 9.A.1. Explicit correction terms for (9.103) are derived. In particular, it is shown that the Sheppard correction is exact if the characteristic function has a bounded support so that (for any small ε)

$$\varphi_x(\omega) \equiv 0, \qquad |\omega| \ge \frac{2\pi}{\Delta} - \varepsilon.$$

Note that this also gives an upper bound for the quantization level Δ.

9.7 Median Filters

9.7.1 Introduction

Median filters form a special class of simple *nonlinear* filters. The idea is simple. Let

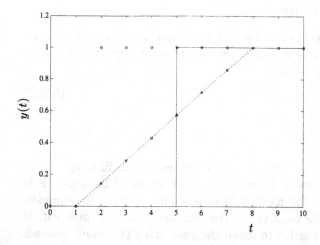

Fig. 9.10. Step responses for median filter (*solid line*) and mean filter (*dashed line*), $n = 7$ ($m = 3$). The step (*dotted line*) is applied at time $t = 2$

$$n = 2m + 1 \tag{9.105}$$

be an odd integer. Then the output $y(t)$ of the filter is the median of the last n input values: $u(t - n + 1), u(t - n + 2), \ldots, u(t)$. In the following, some properties of median filters are derived. It is pertinent to compare this with the arithmetic mean value. Taking the mean corresponds to a *linear* filter of the form

$$H(q) = \frac{1}{n}[1 + q^{-1} + \ldots + q^{-n+1}]. \tag{9.106}$$

This filter is said to be of FIR type, and all its poles lie at the origin.

9.7.2 Step Response

Both the median filter and the mean filter give a phase lag. However, they differ significantly when applied to a step signal. The step shape is retained for the median filter, whereas the mean filter yields a more smoothed result (in fact a ramp). These properties are illustrated in Figure 9.10.

It is easy to verify that the output of the median filter rises from zero to one with a lag of m steps. In contrast, the output of the mean filter increases linearly from zero to one during n steps. It can sometimes be of interest to filter out noise but to retain the original shape of the input signal. This can be the case in image processing. For such situations, median filters may be of interest. It is worth noting that, of course, other forms of step response can be obtained with other filters. For example, linear filters with poles not at the origin (so-called infinite impulse response (IIR) filters) give a response that tends to unity exponentially.

9.7.3 Response to Sinusoids

By considering sinusoids as the input signal, it is possible to form a frequency function, although the concept is not really relevant for nonlinear filters. Assume that the input signal is

$$u(t) = \sin \omega t , \qquad (9.107)$$

and that

$$\omega n < 1 ,$$

which means that a minor part of one period is used for the filtering.

When all the data points lie between two peak values of the input, it is obvious that the effect of the median filter is a pure delay of m samples. When the data include the peak value, the situation is a little different. It will, of course, never be possible to reach the peak value (*i.e.* one). Instead, the filter output settles at a fixed value for some time.

To investigate this maximum value, it is convenient to consider the input

$$u(t) = \cos \omega t , \qquad t = 0, \pm 1, \ldots \qquad (9.108)$$

It is straightforward to find the maximum value of the output. Consider first the case when m is an even integer. Then

$$|y(t)| \leq \cos \omega \frac{m}{2} . \qquad (9.109)$$

Moreover, for the input (9.108), equality holds in (9.109) for $t = 0, \pm 1, \ldots$, $\pm m/2$. The output thus reaches its extreme value for a period of $m + 1$ samples. Similarly, for the case when m is an odd integer:

$$|y(t)| \leq \cos \omega \frac{m+1}{2} , \qquad (9.110)$$

and this extreme value is reached for $t = 0, \pm 1, \ldots , \pm \frac{m+1}{2}$. The output achieves its extreme for $m + 2$ time points.

The effect of the median filter on a sine wave is illustrated in Figure 9.11.

For an arithmetic mean filter, the frequency function can easily be calculated, which becomes

$$H(e^{i\omega}) = \frac{1}{n} \frac{1 - e^{-i\omega n}}{1 - e^{i\omega}} = \frac{1}{n} \frac{e^{i\omega n/2} - e^{-i\omega n/2}}{e^{i\omega/2} - e^{-i\omega/2}} e^{-i\omega(n-1)/2}$$

$$= \frac{1}{n} \frac{\sin \frac{\omega n}{2}}{\sin \frac{\omega}{2}} e^{-i\omega(n-1)/2} . \qquad (9.111)$$

Hence there is a phase shift corresponding to a delay of $(n - 1)/2 = m$ samples. This is the same shift as is obtained for the median filters when step functions or the flanks of a sinusoid are considered. For the arithmetic mean filter, the amplitude is reduced by the factor

$$|H(e^{i\omega})| = \frac{1}{n} \frac{\sin \frac{\omega n}{2}}{\sin \frac{\omega}{2}} . \qquad (9.112)$$

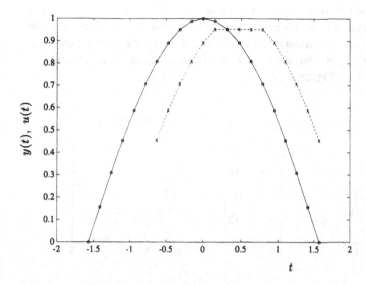

Fig. 9.11. Effect of the median filter on a sine wave: original sine wave $u(t)$ (*solid line*, o), filtered sine wave $y(t)$ (*dashed line*, x), $n = 7$ ($m = 3$)

It is of some interest to note that this gain is somewhat smaller than the extreme values (9.109), (9.110), at least for low frequencies and n not too large. In fact, for small ω

$$|H(e^{i\omega})| \approx \frac{1}{n} \frac{\frac{\omega n}{2} - \frac{1}{6}\left(\frac{\omega n}{2}\right)^3}{\frac{\omega}{2} - \frac{1}{6}\left(\frac{\omega}{2}\right)^3}$$

$$= \frac{1 - \frac{1}{24}\omega^2 n^2}{1 - \frac{1}{24}\omega^2} \approx 1 - \frac{1}{24}\omega^2(n^2 - 1).$$

In contrast, the bound in (9.109) becomes

$$\cos\frac{\omega m}{2} \approx 1 - \frac{1}{2}\left(\frac{\omega m}{2}\right)^2 = 1 - \frac{1}{32}\omega^2(n-1)^2,$$

and (9.110) gives

$$\cos\frac{\omega(m+1)}{2} \approx 1 - \frac{1}{2}\left(\frac{\omega(m+1)}{2}\right)^2 = 1 - \frac{1}{32}\omega^2(n+1)^2,$$

which implies the above statement.

9.7.4 Effect on Noise

In order to study this situation, assume that the input data are zero mean white noise. Then use Lemma 9.1 in Section 9.A.2, where the pdf for the median is derived. For further reference, the pdfs for the maximum and minimum values are also derived.

One might expect that, in the present case, the median will not differ very much from the mean value. To examine whether this can be a reasonable approximation, resort to numerical evaluations for the case of a Gaussian distribution. Then $u(t)$ is assumed to consist of independent $N(0, 1)$ variables. The pdfs are shown in Figure 9.12.

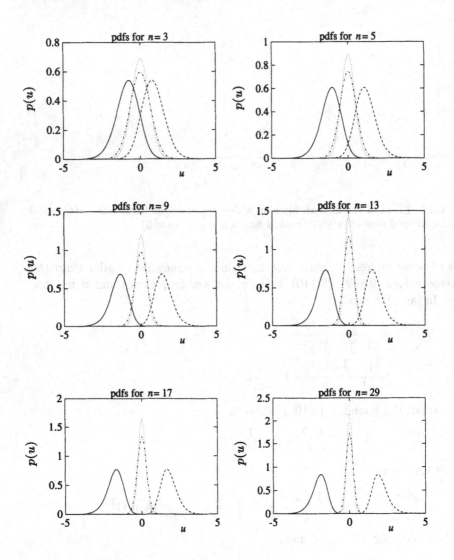

Fig. 9.12. Pdfs for various values of n. Minimum of $\{u(t)\}$ (*solid line*); maximum of $\{u(t)\}$ (*dashed line*); mean of $\{u(t)\}$ (*dotted line*); median of $\{u(t)\}$ (*dash−dotted line*)

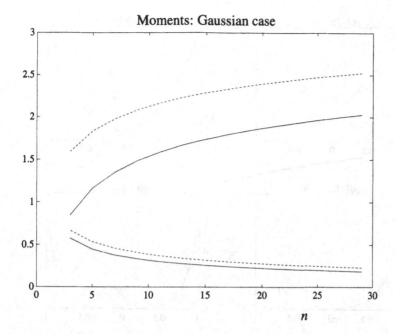

Moments: Gaussian case

Fig. 9.13. Moments, for the Gaussian case, as a function of n. *Lower curves*: standard deviation of the mean (*solid line*); standard deviation of the median (*dashed line*). *Upper curves*: mean of maximum value (*solid line*); mean plus one standard deviation of maximum value (*dashed line*); mean minus one standard deviation of maximum value (*dashed–dotted line*)

It can be seen from Figure 9.12 that the mean and the median do not differ very much. The pdfs in both cases are "well centred" around zero. It is also interesting to note that the maximum and minimum values of the pdfs change very slowly with increasing n.

Some mean and standard deviations were also computed numerically. These are displayed in Figure 9.13.

As a further illustration, the above calculations were repeated with a uniform distribution of the inputs. These have the pdf

$$f(u) = \begin{cases} 1/2 \,, \ |u| < 1 \,, \\ 0 \,, \quad \text{elsewhere} \,. \end{cases} \tag{9.113}$$

The results are given in Figures 9.14 and 9.15.

In this case, there is a more pronounced difference between the mean and the median variables. Compare Figures 9.13 and 9.15.

The median filter is well suited to handling data with "impulsive" noise. It is often applied in image processing, where one aim may be to sharpen contours in noisy images. An important class of such "impulsive noise" is the Bernoulli–Gaussian process, which can be modelled as

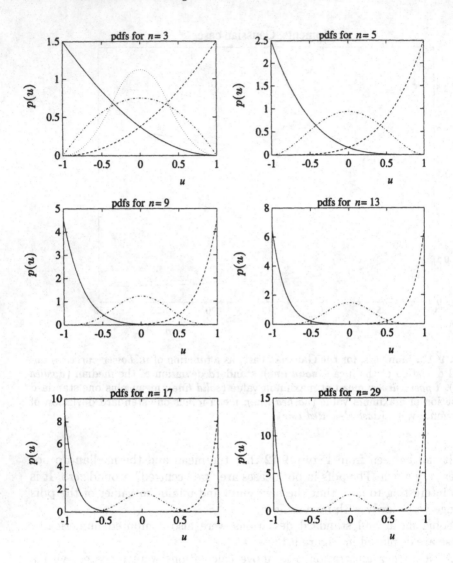

Fig. 9.14. Pdfs for various values of n (uniform distribution). Minimum of $\{u(t)\}$ (*solid line*); maximum of $\{u(t)\}$ (*dashed line*); mean of $\{u(t)\}$ (*dotted line*); median of $\{u(t)\}$ (*dash–dotted line*)

$$
\begin{aligned}
y(t) &= q(t)e(t) \,, \\
e(t) &\quad \text{Gaussian}, \\
q(t) &= \begin{cases} 1 & \text{with prob } \lambda \,, \\ 0 & \text{with prob } 1 - \lambda \,, \end{cases}
\end{aligned}
\tag{9.114}
$$

$e(t)$ and $q(t)$ being independent, and λ normally small.

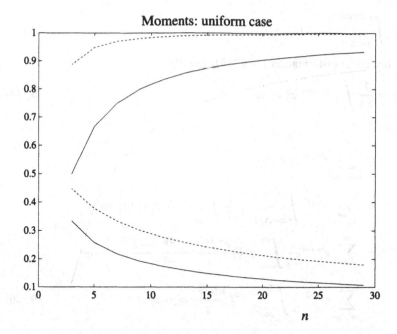

Fig. 9.15. Moments, for the uniform distribution, as a function of n. *Lower curves*: Standard deviation of the mean (*solid line*); standard deviation of the median (*dashed line*). *Upper curves*: mean of maximum value (*solid line*); mean plus one standard deviation of maximum value (*dashed line*); mean minus one standard deviation of maximum value (*dashed-dotted line*)

9.A Appendix. Auxiliary results

9.A.1 Analysis of the Sheppard Correction

Recall that the characteristic function $\varphi_x(\omega)$ is given by

$$\varphi(\omega) = \mathbf{E}\left[e^{i\omega x}\right] = \int_{-\infty}^{\infty} e^{i\omega x}\, p(x)\, dx \, . \tag{9.115}$$

By expanding both sides of (9.115) in a Taylor series, it is easily found that

$$\begin{aligned} \varphi(0) &= 1 \, , \\ \mathbf{E}\,x &= -i\varphi_x'(0) \, , \\ \mathbf{E}\,x^2 &= -\varphi_x''(0) \, . \end{aligned} \tag{9.116}$$

Next, the relations between the characteristic functions of x and x_Q are examined. The pdf of x_Q is discrete and is defined as

$$p_{x_Q}(y) = \begin{cases} p_j \, , & y = j\,\Delta, \qquad j = 0, \pm 1, \pm 2, \dots \, , \\ 0 \, , & \text{elsewhere} \, . \end{cases} \tag{9.117}$$

where

$$p_j = \int_{j\Delta-\Delta/2}^{j\Delta+\Delta/2} p(x)\,\mathrm{d}x \ . \tag{9.118}$$

Using the inverse transformation to (9.115), namely

$$p(x) = \frac{1}{2\pi} \int_{-\infty}^{\infty} e^{-i\omega x}\,\varphi(\omega)\,\mathrm{d}\omega \ , \tag{9.119}$$

one obtains

$$
\begin{aligned}
\varphi_Q(\omega) &= \sum_{j=-\infty}^{\infty} e^{i\omega j\Delta}\, p_j \\
&= \sum_{j=-\infty}^{\infty} e^{i\omega j\Delta} \int_{j\Delta-\Delta/2}^{j\Delta+\Delta/2} \left[\frac{1}{2\pi} \int_{-\infty}^{\infty} e^{-iyx}\,\varphi(y)\,\mathrm{d}y \right]\,\mathrm{d}x \\
&= \sum_{j=-\infty}^{\infty} e^{i\omega j\Delta} \frac{1}{2\pi} \int_{-\infty}^{\infty} \varphi(y) \frac{e^{-ij\Delta y}[e^{-i\Delta y/2} - e^{i\Delta y/2}]}{-iy}\,\mathrm{d}y \\
&= \sum_{j=-\infty}^{\infty} e^{ij\omega\Delta} \frac{1}{\pi} \int_{-\infty}^{\infty} e^{-ij\Delta y} \frac{\sin\frac{\Delta y}{2}}{y}\,\varphi(y)\,\mathrm{d}y \ . \tag{9.120}
\end{aligned}
$$

Now set

$$F(\omega) \triangleq \sum_{k=-\infty}^{\infty} \frac{\sin(\frac{\Delta\omega}{2}+\pi k)}{\frac{\Delta\omega}{2}+k\pi}\, \varphi\left(\omega + k\frac{2\pi}{\Delta}\right) \ . \tag{9.121}$$

The purpose is to show that $\varphi_Q(\omega) = F(\omega)$. As a first step it can be seen that $\varphi_Q(\omega)$ and $F(\omega)$ are both periodic, with period $2\pi/\Delta$, since

$$\varphi_Q\left(\omega + \frac{2\pi}{\Delta}\right) = \varphi_Q(\omega) \ ,$$

$$F\left(\omega + \frac{2\pi}{\Delta}\right) = \sum_{k=-\infty}^{\infty} \frac{\sin(\frac{\Delta\omega}{2}+\pi+\pi k)}{\frac{\Delta\omega}{2}+\pi+k\pi}\, \varphi\left(\omega + \frac{2\pi}{\Delta} + k\frac{2\pi}{\Delta}\right) = F(\omega) \ .$$

Hence $\varphi_Q(\omega)$ and $F(\omega)$ can be developed in Fourier series:

$$
\begin{aligned}
\varphi_Q(\omega) &= \sum_{n=0}^{\infty} c_n\, e^{in\omega\Delta} \ , \\
F(\omega) &= \sum_{n=0}^{\infty} f_n\, e^{in\omega\Delta} \ ,
\end{aligned}
\tag{9.122}
$$

where

$$
\begin{aligned}
c_n &= \frac{\Delta}{2\pi} \int_0^{2\pi/\Delta} \varphi_Q(\omega)\, e^{-in\omega\Delta}\,\mathrm{d}\omega \ , \\
f_n &= \frac{\Delta}{2\pi} \int_0^{2\pi/\Delta} F(\omega)\, e^{-in\omega\Delta}\,\mathrm{d}\omega \ .
\end{aligned}
\tag{9.123}
$$

To prove that $\varphi_Q(\omega) = F(\omega)$, it can be shown equivalently that $c_n = f_n$, for all n. From the above expression (9.120) for $\varphi_Q(\omega)$:

$$c_n = \frac{1}{\pi} \int_{-\infty}^{\infty} e^{-in\Delta y} \frac{\sin(\frac{\Delta y}{2})}{y}\,\varphi(y)\,\mathrm{d}y \ ,$$

whereas (using the substitution $y = \omega + k\frac{2\pi}{\Delta}$)

$$f_n = \frac{\Delta}{2\pi} \int_0^{2\pi/\Delta} e^{-in\omega\Delta} \sum_{k=-\infty}^{\infty} \frac{\sin(\frac{\Delta\omega}{2} + \pi k)}{\frac{\Delta\omega}{2} + k\pi} \varphi\left(\omega + k\frac{2\pi}{\Delta}\right) d\omega$$

$$= \frac{\Delta}{2\pi} \sum_{k=-\infty}^{\infty} \int_{k\frac{2\pi}{\Delta}}^{k\frac{2\pi}{\Delta} + \frac{2\pi}{\Delta}} e^{-in(y - k\frac{2\pi}{\Delta})\Delta} \frac{\sin(\frac{\Delta y}{2})}{\frac{\Delta y}{2}} \varphi(y) \, dy$$

$$= \frac{1}{\pi} \sum_{k=-\infty}^{\infty} \int_{k\frac{2\pi}{\Delta}}^{(k+1)\frac{2\pi}{\Delta}} e^{-iny\Delta} \frac{\sin\frac{\Delta y}{2}}{y} \varphi(y) \, dy$$

$$= \frac{1}{\pi} \int_{-\infty}^{\infty} e^{-iny\Delta} \frac{\sin\frac{\Delta y}{2}}{y} \varphi(y) \, dy$$

$$= c_n \, .$$

To summarize so far, this proves that

$$\varphi_Q(\omega) = \sum_{k=0}^{\infty} \frac{\sin(\frac{\Delta\omega}{2} + \pi k)}{\frac{\Delta\omega}{2} + k\pi} \varphi\left(\omega + k\frac{2\pi}{\Delta}\right) \tag{9.124}$$

$$\triangleq \sum_{k=0}^{\infty} f_k(\omega)\varphi\left(\omega + k\frac{2\pi}{\Delta}\right) \, , \tag{9.125}$$

where

$$f_k(\omega) = \frac{\sin(\frac{\Delta\omega}{2} + \pi k)}{\frac{\Delta\omega}{2} + k\pi} = \frac{(-1)^k \sin\frac{\Delta\omega}{2}}{\frac{\Delta\omega}{2} + k\pi} \, . \tag{9.126}$$

Straightforward differentiation of (9.125) gives

$$\varphi_Q'(\omega) = \sum_k \left[f_k'(\omega)\varphi\left(\omega + k\frac{2\pi}{\Delta}\right) + f_k(\omega)\varphi'\left(\omega + k\frac{2\pi}{\Delta}\right) \right] \, , \tag{9.127}$$

$$\varphi_Q''(\omega) = \sum_k \left[f_k''(\omega)\varphi\left(\omega + k\frac{2\pi}{\Delta}\right) + 2f_k'(\omega)\varphi'\left(\omega + k\frac{2\pi}{\Delta}\right) \right.$$

$$\left. + f_k(\omega)\varphi''\left(\omega + k\frac{2\pi}{\Delta}\right) \right] \, . \tag{9.128}$$

The function $f_0(\omega)$ can be expanded as

$$f_0(\omega) = \frac{\sin\frac{\Delta\omega}{2}}{\frac{\Delta\omega}{2}} \approx 1 - \frac{1}{6}\left(\frac{\Delta\omega}{2}\right)^2 + \dots \, ,$$

showing that

$$f_0(0) = 1 \, , \qquad f_0'(0) = 0 \, , \qquad f_0''(0) = -\frac{\Delta^2}{12} \, . \tag{9.129}$$

From (9.126) it is easily seen that

$$f_k(0) = 0 \qquad \text{for } k \neq 0 \, . \tag{9.130}$$

Using (9.116), one now obtains

$$\mathrm{var}(x_Q) = -\varphi_Q''(0)$$

$$= -f_0''(0) - \varphi''(0) - \sum_{k>0} \left[f_k''(0)\varphi\left(k\frac{2\pi}{\Delta}\right) + 2f_k'(0)\varphi'\left(k\frac{2\pi}{\Delta}\right) \right]$$

$$= \mathrm{var}(x) + \frac{\Delta^2}{12} - \sum_{k>0} \left[f_k''(0)\varphi\left(k\frac{2\pi}{\Delta}\right) + 2f_k'(0)\varphi'\left(k\frac{2\pi}{\Delta}\right) \right] . \quad (9.131)$$

The sum $(\sum_{k>0} \cdots)$ in (9.131) is the modification that would be needed for the Sheppard correction (9.103) to be exact.

There is an important special case. Assume that $\varphi(\omega)$ has *bounded support* so that, for any small ε:

$$\varphi(\omega) \equiv 0 , \quad \text{when } |\omega| \geq \frac{2\pi}{\Delta} - \varepsilon . \quad (9.132)$$

Then the modification sum cancels and the Sheppard correction holds true exactly. Note that (9.132) gives an *upper* bound for the quantization level Δ.

9.A.2 Some Probability Density Functions

Lemma 9.1 *Let* $x_1, \ldots , x_{2n+1}$ *be iid random variables, with mean* m, *variance* σ^2, *pdf* $f(x)$, *and distribution function*

$$F(x) = \int_{-\infty}^{x} f(y)\,\mathrm{d}y .$$

Set

$$z_1 = \min_i \{x_i\} ,$$

$$z_2 = \max_i \{x_i\} , \quad (9.133)$$

$$z_3 = \text{median of } \{x_i\} .$$

The pdfs of these variables are

$$f_1(z) = (2n+1)f(z)[1 - F(z)]^{2n} , \quad (9.134)$$

$$f_2(z) = (2n+1)f(z)[F(z)]^{2n} , \quad (9.135)$$

$$f_3(z) = \frac{(2n+1)!}{n!n!} f(z)[F(z)]^n [1 - F(z)]^n . \quad (9.136)$$

Proof By straightforward calculation for z_1:

$$F_1(z) = P(\min\{x_i\} < z) = 1 - P(\min\{x_i\} > z) = 1 - P(x_i > z \;\forall i)$$

$$= 1 - [\prod_{i=1}^{2n+1} P(x_i > z)] = 1 - [1 - P(x_i < z)]^{2n+1}$$

$$= 1 - [1 - F(z)]^{2n+1} .$$

Differentiation gives

$$f_1(z) = \frac{\mathrm{d}}{\mathrm{d}z} F_1(z) = (2n+1)f(z)[1 - F(z)]^{2n} ,$$

which is (9.134). Similarly, for z_2:

$$F_2(z) = P(\max\{x_i\} < z) = P(x_i < z \text{ all } i)$$

$$= \prod_{i=1}^{2n+1} F(z) = [F(z)]^{2n+1} ,$$

$$f_2(z) = \frac{\mathrm{d}}{\mathrm{d}z} F_2(z) = (2n+1)f(z)[F(z)]^{2n} .$$

To derive $f_3(z)$, note that

$$f_3(z)\,\mathrm{d}z = P(z \le z_3 < z + \mathrm{d}z)$$
$$= P(n \text{ of } \{x_i\} < z, \quad n \text{ of } \{x_i\} > z + \mathrm{d}z, \quad 1 \text{ of } \{x_i\} \in (z, z + \mathrm{d}z))$$
$$= \frac{(2n+1)!}{n!n!} [F(z)]^n [1 - F(z)]^n f(z)\,\mathrm{d}z ,$$

from which (9.136) follows.   ∎

Exercises

Exercise 9.1 Consider the following simplified model of a surveillance radar tracking an aircraft and estimating its position and velocity:

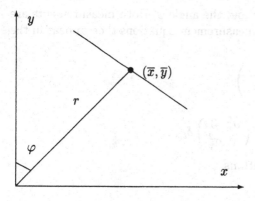

Consider two dimensions only. The position of the aircraft is $(\bar{x}, \bar{y})$ in Cartesian coordinates. The following model is to be employed for the filter design. Set

$$x = \begin{pmatrix} \bar{x} \\ \dot{\bar{x}} \\ \bar{y} \\ \dot{\bar{y}} \end{pmatrix} .$$

Then set

$$
\dot{x} = \begin{pmatrix} 0 & 1 & 0 & 0 \\ 0 & 0 & 0 & 0 \\ 0 & 0 & 0 & 1 \\ 0 & 0 & 0 & 0 \end{pmatrix} x + \begin{pmatrix} 0 & 0 \\ 1 & 0 \\ 0 & 0 \\ 0 & 1 \end{pmatrix} \begin{pmatrix} x_e \\ y_e \end{pmatrix} ,
$$

$$
E \begin{pmatrix} x_e(t) \\ y_e(s) \end{pmatrix} \left(x_e(s) \; y_e(s) \right) = \begin{pmatrix} r_x & 0 \\ 0 & r_y \end{pmatrix} \delta(t-s) .
$$

where the noise variances r_x and r_y are set to zero, the state model describes a movement with constant velocity. Inclusion of the noise variances may facilitate the estimation problem when the aircraft makes some manoeuvres.

As measurements must be taken in discrete time, a sampled form of the model will be needed. For the sampling interval h the model above becomes

$$
x(t+h) = \begin{pmatrix} 1 & h & 0 & 0 \\ 0 & 1 & 0 & 0 \\ 0 & 0 & 1 & h \\ 0 & 0 & 0 & 1 \end{pmatrix} x(t) + v(t) ,
$$

where

$$
E\, v(t) v^T(t) = R_{\mathrm{d}} = \begin{pmatrix} r_x h^3/3 & r_x h^2/2 & 0 & 0 \\ r_x h^2/2 & r_x h & 0 & 0 \\ 0 & 0 & r_y h^3/3 & r_y h^2/2 \\ 0 & 0 & r_y h^2/2 & r_y h \end{pmatrix} .
$$

The radar measures the distance r and the angle φ. Both measurements are contaminated by white noise. The measurement equations then appear in the following nonlinear form:

$$
\begin{pmatrix} r \\ \varphi \end{pmatrix} = \begin{pmatrix} \sqrt{\bar{x}^2 + \bar{y}^2} \\ \mathrm{atan}(\bar{x}/\bar{y}) \end{pmatrix} + \begin{pmatrix} e_r \\ e_\varphi \end{pmatrix} ,
$$

$$
E \begin{pmatrix} e_r(t) \\ e_\varphi(t) \end{pmatrix} \left(e_r(s) \; e_\varphi(s) \right) = \begin{pmatrix} \sigma_r^2 & 0 \\ 0 & \sigma_\varphi^2 \end{pmatrix} \delta_{t,s} .
$$

Consider the following filter algorithms.

F1 : A standard EKF.

F2 : "Solve" the measurement equation and write it in the form

$$
\begin{pmatrix} z_1 \\ z_2 \end{pmatrix} = \begin{pmatrix} \bar{x} \\ \bar{y} \end{pmatrix} + \begin{pmatrix} e_1 \\ e_2 \end{pmatrix} ,
$$

where e has a *state dependent* covariance matrix. Apply an EKF to this model.

F3 : The iterated EKF.

F4 : The second-order EKF.

The user choices in the filters are the noise variances r_x and r_y, and the initial values $\hat{x}(0| - 1)$, $P(0| - 1)$. Assume the measurement inaccuracies σ_r^2 and σ_φ^2 to be known.

(a) Examine the behaviour of the filters for the case of an aircraft flying with constant velocity. Try the numerical values

$$x(0) = \begin{pmatrix} -5000 \text{ m} \\ 200 \text{ m/s} \\ 1000 \text{ m} \\ 0 \text{ m/s} \end{pmatrix} ,$$

$r_x = 0$, $r_y = 0$, $\sigma_r = 50$ m, $\sigma_\varphi = 0.01$ rad, $h = 0.5$ s, and generate 100 data points. What is the effect of the user choices? You may present your results in diagrams displaying

(i) $\bar{y}(t)$ versus $\bar{x}(t)$, $\hat{\bar{y}}(t)$ versus $\hat{\bar{x}}(t)$, etc.

(ii) $\bar{x}(t)$ versus t, $\hat{\bar{x}}(t)$ versus t, etc.

(iii) $\bar{y}(t)$ versus t, $\hat{\bar{y}}(t)$ versus t, etc.

(b) Repeat the examination for the case where the aircraft is making a turn. The trajectory is assumed to satisfy

$$\bar{x}(t) = \bar{x}_0 + v_x t + a_x t^2/2 ,$$
$$\bar{y}(t) = \bar{y}_0 + v_y t + a_y t^2/2 .$$

Try the numerical values $\bar{x}_0 = -5000$ m, $\bar{y}_0 = 1000$ m, $v_x = 300$ m/s, $a_x = -1$ m/s^2, $v_y = -30$ m/s, $a_y = 0.5$ m/s^2, 200 data points.

Exercise 9.2 Consider the nonlinear system

$$x(t + 1) = \frac{1}{1 + x(t)} + v(t) ,$$
$$y(t) = x(t) + e(t) ,$$

where $v(t)$ and $e(t)$ are mutually independent Gaussian white noise sequences with zero mean and variances r_1 and r_2 respectively. Examine by simulations an EKF and a Gaussian sum filter for estimating the state for the output measurements. Try, for example, the numerical values $r_1 = 1/3$, $r_2 = 1$, $t \leq 100$, $\hat{x}(0) = 0$, $\text{cov}[\hat{x}(t)] = 10 \times r_2$. For the Gaussian sum filter, let it start with a non-Gaussian distribution.

Exercise 9.3 Consider the following simplified model of a servo where the output $\theta(t)$ is an angle, which is measured with resolvers giving $\sin(\theta(t))$ and $\cos(\theta(t))$:

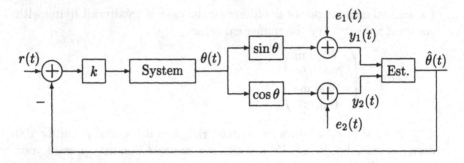

The dynamics of the system are in continuous time $G(s) = 1/(s(s+1))$, which after sampling gives $H(q) = (b_1 q + b_2)/(q^2 + a_1 q + a_2)$ with (h is the sampling interval)

$$a_1 = -1 - e^{-h},$$
$$a_2 = e^{-h},$$
$$b_1 = h - 1 + e^{-h},$$
$$b_2 = -he^{-h} + 1 - e^{-h}.$$

The measurement noise sources, $e_1(t)$ and $e_2(t)$, are assumed to be white, mutually independent, of zero mean and variance r_e. Examine numerically how the closed loop system behaves when the reference signal is a square wave. Try the numerical values $h = 0.3$, $k = 0.4$, $r_e = 0.05$ and full period of $r(t)$ 80–100. Try different values of the square wave amplitude. (Values around π are of particular interest!) The block labelled "Est." is an estimator whose output $\hat{\theta}(t)$ should resemble the true output $\theta(t)$. Consider the following cases:

(a) The idealized case with $\hat{\theta}(t) \equiv \theta(t)$.
(b) The simple estimate

$$\hat{\theta}(t) = \text{atan}\left(\frac{y_1(t)}{y_2(t)}\right).$$

If the computations are done in Matlab use the command **atan2**, which gives the angle in the interval $(-\pi, \pi)$.
(c) An EKF using $y_1(t)$ and $y_2(t)$ as measurements.

Examine, for example, the behaviour of the system by plotting $r(t)$, $\theta(t)$ and $\hat{\theta}(t)$.

Exercise 9.4 Consider a uniform distribution over the interval $(-1, 1)$:

$$p(x) = \begin{cases} \frac{1}{2}, & |x| \leq 1, \\ 0, & \text{elsewhere}. \end{cases}$$

Try to approximate $p(x)$ by a weighted sum of Gaussian pdfs. Use your own approach for selecting the parameters, or use the following one. Let $x_1 \ldots x_m$ be equidistantly spaced in $(-1, 1)$ with $x_{i+1} - x_i = \Delta$. Choose a number k and take the approximation as

$$P_A(x) = \sum_{i=1}^{m} \frac{1}{m} \gamma(x; x_i, k\Delta^2).$$

Exercise 9.5 Consider the estimation of a constant from noisy measurements, modelled as

$$x(t+1) = x(t),$$
$$y(t) = x(t) + e(t),$$

where $\{e(t)\}$ is a white noise sequence of random variables, uniformly distributed over $(-1, 1)$. Examine by simulation the following approaches for recursively estimating the state. Also illustrate the uncertainty (say standard deviations) of the approaches:

(a) the standard Kalman filter.
(b) the conditional mean $\mathbf{E}\left[x(t)|Y^t\right]$.
(c) a Gaussian sum estimate.

Exercise 9.6 Examine, by means of numerical computations, the validity of the Sheppard correction

$$\text{var}(x_Q) \approx \text{var}(x) + \Delta^2/12$$

for Gaussian distributed data and for uniformly distributed data.

Exercise 9.7 Let $u(t)$ be white noise with mean value m. Assume that the pdf of $u(t)$ is symmetric around m.

(a) Show that the expected value of the median of $\{u(t)\}_{t=1}^{2n+1}$ is m.
(b) Let f_1 [f_2] be the pdf of the maximum [minimum] value of $\{u(t)\}_{t=1}^{2n+1}$. Show that $f_1(m+z) = f_2(m-z)$.

Exercise 9.8 Consider a signal $s(t)$ of the form

$$s(t) = \begin{cases} 0, & 0 \leq t \leq 50, \\ 1, & 51 \leq t \leq 100, \\ 0, & 101 \leq t \leq 150, \\ (t-150)/50, & 151 \leq t \leq 200, \\ 1, & 201 \leq t \leq 350. \end{cases}$$

Assume that $s(t)$ is observed with white Gaussian noise of unit variance. Simulate the signal and the measurements and try to recover $s(t)$ from the measurements using various filters:

(a) a median filter,
(b) a linear IIR filter,
(c) a mean filter,
(d) a Kalman filter (try to set up a state space model that can be relevant for describing the properties of the signal; you may also allow its noise characteristics in terms of the matrices R_1, R_{12}, R_2 to be time-varying).

Test various filter parameters. Examine to some extent what *a priori* information is useful when tuning the filters.

Exercise 9.9 EKFs can be used in order to derive recursive parameter estimation algorithms. Consider a linear model

$$\bar{x}(t+1) = A(\theta)\bar{x}(t) + B(\theta)u(t) + v(t) ,$$

$$y(t) = C(\theta)\bar{x}(t) + e(t) ,$$

$$\mathbf{E}\begin{pmatrix} v(t) \\ e(t) \end{pmatrix} \left(v^T(s)\ e^T(s) \right) = \begin{pmatrix} R_1(\theta) & 0 \\ 0 & R_2(\theta) \end{pmatrix} \delta_{t,s} ,$$

where θ is a parameter vector.

(a) Show that an EKF, applied to a model for

$$x(t) = \begin{pmatrix} \bar{x}(t) \\ \theta \end{pmatrix} ,$$

can be used to derive a recursive identification algorithm for estimating the parameter vector θ.

(b) Derive explicitly such an EKF for estimating the parameter a in the AR(1) model

$$y(t) + ay(y-1) = v(t) .$$

The noise variance $\lambda^2 = \mathbf{E}\, v^2(t)$ can be assumed to be known.

Exercise 9.10 Consider the problem of estimating a constant from noisy measurements. Model the situation as

$$\begin{aligned} x(t+1) &= x(t) , \\ y(t) &= x(t) + e(t) . \end{aligned} \qquad t = 1, 2, \ldots, N .$$

(a) Assume that $\hat{x}(0)$, $\{e(t)\}$ are independent Gaussian random variables with zero mean and $\mathbf{E}\,\hat{x}^2(0) = r_0$, $\mathbf{E}\,e^2(t) = r_2$. Determine the conditional mean $\hat{x}(N|N)$. What happens when there is no *a priori* information, that is $r_0 \to \infty$?

(b) Consider the case as above and let $r_0 \to \infty$. Let $\hat{x}(N|N)$ be the estimate of the constant. What is the (unconditional) variance of that estimate?

(c) Assume that there is no *a priori* information and that the $\{e(t)\}$ are independently and uniformly distributed in the interval $(-a, a)$. Determine the conditional pdf of $(x(N)|Y^N)$. What is the conditional mean $E[x(N)|Y^N]$?

Hint. Show first that $p(x(t)|Y^t)$ is a uniform distribution.

(d) Let the conditional mean derived in (a) be denoted by $\hat{x}_a$ and that derived in (c) be denoted by $\hat{x}_c$. Show that these two estimates differ in general. Evaluate their (unconditional) variances under the assumption that the $\{e(t)\}$ are uniformly distributed in $(-a, a)$.

Exercise 9.11 Consider the following (very simplified) model for the re-entry of a satellite into the atmosphere:

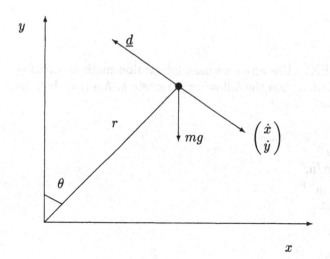

The satellite is affected by gravity and by the atmosphere. The latter gives a contribution to the velocity and can be written as

$$d = \frac{1}{2}\rho(y)c_d(\dot{x}^2 + \dot{y}^2)\begin{pmatrix} \dfrac{-\dot{x}}{\sqrt{\dot{x}^2+\dot{y}^2}} \\ \dfrac{-\dot{y}}{\sqrt{\dot{x}^2+\dot{y}^2}} \end{pmatrix},$$

where ρ is the density of the atmosphere, c_d is a ballistic constant and $(\dot{x}\ \dot{y})^T$ is the velocity vector of the satellite. The density is assumed to vary with the height as (the barometer formula)

$$\rho(y) = \rho_0\, e^{-ky},$$

where k is a constant and ρ_0 is the density of air at sea level. The gravity force is $(0 \ -mg)^T$, where m is the mass of the satellite. Newton's laws give the equation

$$\ddot{x} = -\frac{1}{2m}\rho_0 c_d \, e^{-ky}(\dot{x}^2 + \dot{y}^2)^{1/2}\dot{x} + a_x \,,$$

$$\ddot{y} = -\frac{1}{2m}\rho_0 c_d \, e^{-ky}(\dot{x}^2 + \dot{y}^2)^{1/2}\dot{y} - g + a_y \,,$$

where a_x and a_y are variations in the accelerations due to unmodelled effects. Measurements of the satellite's position are assumed to be made by a tracking radar, giving distance and elevation as (see the diagram above)

$$r = \sqrt{x^2 + y^2} + \Delta R \,,$$
$$\theta = \tan^{-1}(x/y) + \Delta\theta \,,$$

where ΔR and $\Delta\theta$ can be assumed to be uncorrelated sensor noise.

Construct and simulate an EKF with discrete-time measurements, estimating the velocity and position of the satellite, using the state vector

$$\overline{x} = \begin{pmatrix} x \\ \dot{x} \\ y \\ \dot{y} \end{pmatrix} \,.$$

Also test the iterated EKF. Use any pertinent integration method for solving the differential equation. Use the following parameters. Assume that the sensor noise is Gaussian.

$$m = 10^3 \text{ kg} \,,$$
$$g = 10 \text{ m/s}^2 \,,$$
$$\rho_0 = 1.3 \text{ kg/m}^3 \,,$$
$$k = 10^{-4} \text{ m}^{-1} \,,$$
$$c_d = 0.6 \text{ m}^2 \,,$$
$$\mathbf{E}\, a_x^2 = (10 \text{ m/s}^2)^2 \,,$$
$$\mathbf{E}\, a_y^2 = (10 \text{ m/s}^2)^2 \,,$$
$$\mathbf{E}\,(\Delta R)^2 = (200 \text{ m})^2 \,,$$
$$\mathbf{E}\,(\Delta\theta)^2 = (2 \times 10^{-3} \text{ rad})^2 .$$

Let the sampling frequency be $f_s = 20$ Hz. Choose the following initial conditions:

$$x(0) = -75 \text{ km} \,,$$
$$y(0) = 45 \text{ km} \,,$$
$$\dot{x}(0) = 5000 \text{ m/s} \,,$$
$$\dot{y}(0) = -2000 \text{ m/s} \,,$$

$$\hat{x}(0|-1) = 0 \text{ km},$$
$$\hat{y}(0|-1) = 25 \text{ km},$$
$$\dot{\hat{x}}(0|-1) = 0 \text{ m/s},$$
$$\dot{\hat{y}}(0|-1) = -2500 \text{ m/s},$$
$$\mathbf{E}\,\Delta\hat{x}(0|-1)^2 = (50 \text{ km})^2,$$
$$\mathbf{E}\,\Delta\hat{y}(0|-1)^2 = (25 \text{ km})^2,$$
$$\mathbf{E}\,\Delta\dot{\hat{x}}(0|-1)^2 = (5000 \text{ m/s})^2,$$
$$\mathbf{E}\,\Delta\dot{\hat{y}}(0|-1)^2 = (2500 \text{ m/s})^2.$$

Simulate the system until $y < 5000$ m. Also study what happens if c_d is increased or a_x, a_y are increased (say by of factor 2–4). Illustrate the simulation results and study the accuracy of the filter by plotting the estimation error versus time, and the square roots of the diagonal elements of $P(t)$. Also try to apply the filter without using any information about the velocity, except that the maximal velocity at re-entry is 7 km/s. Hence, in this case, modify the following initial values:

$$\dot{\hat{x}}(0|-1) = -100 \text{ m/s},$$
$$\dot{\hat{y}}(0|-1) = -100 \text{ m/s},$$
$$\mathbf{E}\,\Delta\dot{\hat{x}}(0|-1)^2 = (7000 \text{ m/s})^2,$$
$$\mathbf{E}\,\Delta\dot{\hat{y}}(0|-1)^2 = (7000 \text{ m/s})^2.$$

Remark The initial estimates of the velocities are chosen differently from zero in the last case, as

$$\dot{\hat{x}}(0|-1)^2 + \dot{\hat{y}}(0|-1)^2 > 0$$

is a necessary condition for F to exist. ☐

Exercise 9.12 Consider one-step prediction of the following simple time-varying process:

$$y(t+1) = a(t)y(t) + v(t),$$

where $v(t)$ is white noise of zero mean and unit variance. The parameter $a(t)$ shifts between two levels, a_0 and a_1. The probability that $a(t)$ differs from $a(t-1)$ is p (for all t), and shifts at different time instants are independent.

Simulate the system and compute some estimates $\hat{y}(t+1|t)$; see below. Evaluate and compare the behaviour by computing and plotting

$$V(t) = \sum_{s=1}^{t} \varepsilon^2(s) = \sum_{s=1}^{t} [y(s) - \hat{y}(s|s-1)]^2.$$

Try the numerical values $a_0 = 0.49$, $a_1 = 0.99$, $p = 0.03$ and use a data set of 500 points.

(a) Assume that $a(t)$ is known. Then $\hat{y}(s|s-1) = a(s-1)y(s-1)$.

(b) A simple robustified estimator would be to use the mean value of $a(t)$ to obtain a time-invariant estimator, that is:

$$\hat{y}(t+1|t) = \bar{a}y(t) ,$$

$$\bar{a} = \frac{a_0 + a_1}{2} .$$

(c) A more sophisticated estimator can be designed using the ideas of Section 7.5. Replace $a(t)$ by $\bar{a} + \tilde{a}$, where $\tilde{a}$ is a "small" variable. Then make the approximation (linearization in $\tilde{a}$)

$$y(t) = \frac{q}{q - (\bar{a} + \tilde{a})} v(t)$$

$$\approx \frac{q}{q - \bar{a}} v(t) + \frac{q\tilde{a}}{(q - \bar{a})^2} v(t) .$$

Now regard $\tilde{a}$ as a random variable of zero mean and variance $\frac{1}{4}(a_0 - a_1)^2$. Further, examining

$$z(t) \stackrel{\triangle}{=} (q - \bar{a})^2 y(t) = q(q - \bar{a})v(t) + q\tilde{a}v(t) ,$$

it is found that $z(t)$ has a covariance function as an MA(1) process, as

$$\mathbf{E}\, z^2(t) = 1 + \bar{a}^2 + \frac{1}{4}(a_0 - a_1)^2 ,$$

$$\mathbf{E}\, z(t)z(t-1) = -\bar{a} ,$$

$$\mathbf{E}\, z(t)z(t-\tau) = 0 , \qquad \tau \geq 2 .$$

By spectral factorization one can then derive an "equivalent" model:

$$y(t) = \frac{q+c}{(q-\bar{a})^2} \epsilon(t) , \qquad \mathbf{E}\, \epsilon^2(t) = \lambda^2 ,$$

for which the optimal predictor is

$$\hat{y}(t+1|t) = \frac{(c + 2\bar{a})q - \bar{a}^2}{q + c} y(t) .$$

(d) A further possibility is to use a nonlinear extended model

$$\bar{x}(t) = \begin{pmatrix} x(t) \\ a(t) \end{pmatrix} ,$$

$$\bar{x}(t+1) = \begin{pmatrix} a(t)x(t) \\ \alpha a(t) + m \end{pmatrix} + \begin{pmatrix} v_1(t) \\ v_2(t) \end{pmatrix} ,$$

and apply an EKF. Show also that pertinent values α, m, $\lambda_{v_2}^2$ of such a model are

$$\alpha = 1 - 2p ,$$
$$m = p(a_0 + a_1) ,$$
$$\lambda_{v_2}^2 = p(1 - p)(a_1 - a_0)^2 .$$

Exercise 9.13 Consider the example analyzed in Section 9.2.4. Derive the theoretical MSEs in the following cases.

(a) $\mathrm{MSE}(\hat{x})$ when the EKF is used.
(b) $\mathrm{MSE}(\hat{z})$ when the EKF is used.
(c) $\mathrm{MSE}(\hat{z})$ when the second-order EKF is used.
(d) $\mathrm{MSE}(\hat{x})$ when the optimal estimator is used.

Compare with the results obtained by Monte Carlo simulations.

Exercise 9.14 Let $u(t)$, $t = 1 \ldots 2n + 1 = N$ be independent uniformly distributed random variables:

$$f(u) = \begin{cases} 1/2 \, , \ |u| \leq 1 \, , \\ 0 \, , \quad \text{elsewhere} \, . \end{cases}$$

(a) Show that the median of $\{u(t)\}$ has variance $\frac{1}{2n+3}$.

   Hint. It holds that

   $$\frac{(2n+1)!}{2^{2n}n!n!} \sum_{i=0}^{n} \binom{n}{i} \frac{(-1)^i}{3+2i} = \frac{1}{2n+3} \, .$$

(b) Let y be the maximum value of $\{u(t)\}$. Show that

   $$\mathbf{E}\, y = \frac{N-1}{N+1} \, ,$$

   $$\text{var}(y) = \frac{4N}{(N+1)^2(N+2)} \, .$$

Exercise 9.15 This problem illustrates an alternative approach to the non-linear filtering problem, namely the use of a partial differential equation to describe the evolution of the pdf of the state probability vector in between measurement instances. Heuristically, this method looks on the pdf as a representation of a flow of probability with the constraint that the integral of the pdf over its support always equals one, and hence probability cannot be destroyed. The above fact lends itself to arguments similar to those used when deriving partial differential equations, *e.g.* for heat flow with continuity equation arguments. In the probabilistic case, however, the mathematics is difficult, and here only the major result is reproduced. This is the celebrated Fokker–Planck equation or the forward Kolmogorov equation. The formulation is as follows.

Consider the n-dimensional stochastic differential equation

$$dx(t) = f(x(t)) \, dt + G(x(t)) \, d\beta(t) \, , \qquad \mathbf{E}\, d\beta(t) \, d\beta^T(t) = Q \, dt \, ,$$
$$x(t_0) = x_0 \, ,$$

where $d\beta(t)$ is a Brownian motion process independent of the initial state $x_0(t)$. Then the pdf $p(x,t)$ propagates according to the Fokker–Planck equation

$$\frac{\partial p(x,t)}{\partial t} = -\sum_{i=1}^{n} \frac{\partial\left(p(x,t)f_i(x)\right)}{\partial x_i} + \frac{1}{2}\sum_{i,j=1}^{n} \frac{\partial^2\left(p(x,t)(G(x)QG(x)^T)_{i,j}\right)}{\partial x_i \partial x_j},$$

$$p(x,0) = p_0(x).$$

Now consider the system

$$\dot{x} = \begin{pmatrix} 0 & 1 \\ 0 & 0 \end{pmatrix} x$$

$$y(kh) = \Delta_{\text{round}}\left(\frac{(1\ 0)\,x(kh)}{\Delta}\right)$$

which describes e.g. constant altitude velocity flight, where the altitude is reported as mode C altitude returns, quantized in steps of $\Delta = 100$ f. For the state variables all uncertainties are due to the unknown intitial values. The measurements are assumed to occur in discrete-time with sampling interval h. The "round" operator gives the integer value closest to the argument. The measurement is to be processed and differentiated so that the velocity determined is accurate enough to allow prediction, say 2 min ahead. Such predictions are used in collision avoidance systems for air traffic control systems. In this case the prediction error is not Gaussian, and it turns out that Kalman filters do not give high enough accuracy for the problem in question. For this reason an algorithm that accounts optimally for the quantization is to be derived here.

(a) Write the Fokker–Planck equation for the constant altitude velocity problem.

(b) Solve the Fokker–Planck equation with the initial condition at time t_0

$$p(x_1, x_2, t_0) = p_0(x_1, x_2)$$

and show that the solution has the form

$$p(x_1, x_2, t) = p_0(x_1 - x_2(t - t_0), x_2).$$

Hint. Introduce new variables according to

$$u = x_1 - x_2(t - t_0),$$
$$t' = t - t_0.$$

Make a change of variables in the Fokker–Planck equation, solve it and transform back.

(c) Assume that the initial pdf is uniform with support of a convex polygon with n corners. Prove also that the pdf after propagation is uniform, with support of a convex polygon with n corners.

Hint. Uniformity follows from (b). To prove convexity take two points that are within the final polygon. Use the propagation result of (b) to transform the line between the points to the initial time, where convexity holds. Use the fact that linearity preserves convexity and the transformation to draw conclusions about convexity after propagation.

(d) Use a geometric argument to explain why the pdf after update with a quantized measurement at time $t = kh$ preserves uniformity as well as support by a convex polygon, albeit possibly with a different number of corners.

Hint. Draw a figure indicating the restriction of the measurement when acting upon the propagated polygon.

(e) Derive the optimal state estimate by a computation of the conditional mean.

Hint. Divide the polygon into a set of disjoint triangles, and compute the contribution to the mean from its triangle. In particular, for a triangle with one corner in the origin, and the other two apprearing in arbitrary positions, show that the area of the triangle is

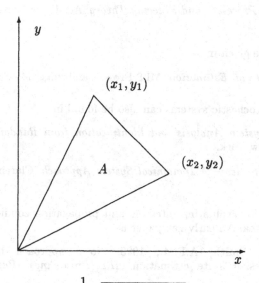

$$A = \frac{1}{2}\sqrt{(x_1 y_2 - x_2 y_1)^2} \, ,$$

and that the centre of mass is

$$\int_A \begin{pmatrix} x \\ y \end{pmatrix} dA = A \frac{1}{3} \begin{pmatrix} x_1 + x_2 \\ y_1 + y_2 \end{pmatrix} .$$

(f) Compute the covariance of the optimal state estimate using techniques similar to that of part (e).

(g) The conclusion of the previous parts of the exercise is that the conditional pdf given the quantized measurements is uniform, and is nonzero precisely on a convex polygon. Hence an algorithm for optimal state estimation can be constructed by propagation and update of only the corners of a convex polygon (as a convex polygon is completely determined by its corners). The number of corners may of course change in the update step. Write Matlab code for such an algorithm and compare it with the Kalman filter using the scenarios in the paper by Sviestins and Wigren (2001).

Remark The reference mentioned above describes many aspects of the topic of this problem. The estimation method is used practically. It is in operation at Swedish major air traffic control centres.

Bibliography

Extended Kalman filtering is a classical subject. Two major sources are

Anderson, B.D.O., Moore, J.B., 1979. *Optimal Filtering*. Prentice Hall, Englewood Cliffs, NJ.

Jazwinski, A.H., 1970. *Stochastic Processes and Filtering Theory*. Academic Press, New York.

whereas more practical aspects are given in

Gelb, A. (Ed.), 1975. *Applied Optimal Estimation*. MIT Press, Cambridge, MA.

Various aspects of nonlinear stochastic systems can also be found in

Bendat, J.S., 1990. *Nonlinear System Analysis and Identification from Random Data*. John Wiley & Sons, New York.

Tong, H., 1990. *Nonlinear Time Series – A Dynamical System Approach*. Clarendon Press, Oxford, UK.

Monte Carlo based methods for evaluating integrals and propagating conditional pdfs is an active research area. An early key paper is

Gordon, N.J., Salmond, D.J., Smith, A.F.M., 1993. Novel approach to nonlinear/non-Gaussian Bayesian state estimation. *IEE Proceedings, Part F*, vol 140, 107–113.

See also the book

Doucet, A., de Freitas, N., Gordon, N. (Eds), 2001. *Sequential Monte Carlo Methods in Practice*, Springer-Verlag, New York.

The multiple model approach (described in Section 9.4) is treated in detail in

Bar-Shalom. Y., Li, X.-R., 1993. *Estimation and Tracking. Principles, Techniques and Software.* Artech House, Norwood, MA.

Blackman, S., Popoli, R., 1999. *Design and Analysis of Modern Tracking Systems.* Artech House, Norwood, MA.

A detailed background to Exercise 9.15 is given in the paper

Sviestins, E., Wigren, T., 2001. Nonlinear techniques for mode C climb/descent rate estimation in ATC systems. *IEEE Transactions on Control Systems Technology*, vol 9, 163–174.

10. Introduction to Optimal Stochastic Control

10.1 Introduction

Some optimal control problems for stochastic systems are studied in this chapter. As demonstrated in Section 10.2, the randomness of the system can be introduced in various ways, which leads to different types of solution. A basic principle, dynamic programming, for deriving and describing optimal stochastic controllers is presented in Section 10.4. As this principle, although of fundamental importance for the theory, requires a huge amount of computation for even very simple examples, we also need to consider suboptimal schemes. Some aspects of such controllers are provided in Section 10.5.

10.2 Some Simple Examples

10.2.1 Introduction

In this section, some typical properties of optimal control for stochastic systems are studied by means of simple examples. The systems are all varieties of the first-order model

$$x(t+1) = ax(t) + bu(t) , \qquad t = 0,\dots,N-1 ,$$
$$y(t) = x(t) .$$

(10.1)

Stochastic aspects can be attributed in several ways, such as

- process noise,
- measurement noise,
- a and/or b being random parameters.

The criterion to optimize is the variance of the final state

$$J = \mathbf{E}\, x^2(N) .$$

(10.2)

10.2.2 Deterministic System

Consider the system of (10.1). Assume that a and b are known, and that $x(t)$ is exactly observable. This is a fully deterministic problem and the expectation operator in the criterion of (10.2) can be skipped. Obviously:

$$x^2(N) = [ax(N-1) + bu(N-1)]^2 .$$

Hence the absolute minimum $J = 0$ is achieved by the choice

$$u(N-1) = -\frac{a}{b}x(N-1) , \qquad (10.3)$$

$$u(0), \ldots u(N-2) \quad \text{arbitrary.}$$

To comply with forthcoming cases, study in particular the (time-invariant) dead-beat regulator

$$u(t) = -\frac{a}{b}x(t) , \qquad t = 0, \ldots , N-1 , \qquad (10.4)$$

which clearly satisfies (10.3). In the absence of any disturbances, it forces the state to zero at time $t = 1$.

10.2.3 Random Time Constant

Now consider the system

$$x(t+1) = a(t)x(t) + bu(t) ,$$
$$y(t) = x(t) , \qquad (10.5)$$
$$x(0) \quad \text{given} ,$$

where $\{a(t)\}$ is a sequence of iid random numbers with mean m and variance σ^2. Assume that m and σ^2 are known. The criterion is still given by (10.2).

To evaluate the criterion $J = \mathbf{E}\, x^2(N)$, use the following general result (see Lemma 2.2):

$$\mathbf{E}\, x = \mathbf{E}_y\{\mathbf{E}\,[x|y]\} , \qquad (10.6)$$

where $\mathbf{E}_y$ denotes expectation with respect to y. This trick will be used repeatedly in this chapter when deriving optimal controllers.

It is thus possible to rewrite the criterion as

$$J = \mathbf{E}\, x^2(N) = \mathbf{E}\,[\mathbf{E}\,(x^2(N)|X^{N-1}, U^{N-1})] , \qquad (10.7)$$

where the outer expectation is with respect to

$$X^{N-1} = \{x(0), \ldots x(N-1)\} .$$

As $u(t)$ will be a causal feedback from the states, it follows that $U^{t-1} = \{u(0) \ldots u(t-1)\}$ must be a function of X^{t-1} for $t = 1, \ldots , N$. Now it is easy to see that

$$\mathbf{E}\,[x^2(N)|X^{N-1}, U^{N-1}] = \mathbf{E}\,[a(N-1)x(N-1) + bu(N-1)]^2$$
$$= [mx(N-1) + bu(N-1)]^2 + \sigma^2 x^2(N-1) . \qquad (10.8)$$

(Recall that, owing to the problem formulation, $a(N-1)$ is independent of X^{N-1}.) The optimal value of $u(N-1)$ is therefore

$$u(N-1) = -\frac{m}{b}x(N-1) \, , \tag{10.9}$$

which gives

$$J = \mathbf{E}\,\sigma^2 x^2(N-1) = \sigma^2 \mathbf{E}\,x^2(N-1) \, . \tag{10.10}$$

Comparing (10.10) and (10.2), one can repeat the procedure successively to determine all input values. The optimal control law becomes

$$u(t) = -\frac{m}{b}x(t) \quad t = 0, \dots , N-1 \, , \tag{10.11}$$

and

$$J = (\sigma^2)^N x^2(0) \, . \tag{10.12}$$

The optimal control law of (10.11) is an example of the *certainty equivalence principle*. If, for the system of (10.5), the stochastic variables (in this case $a(t)$) are replaced by their expected values, a deterministic system is obtained. The certainty equivalence principle is based on the hypothesis that the solution to the deterministic system so obtained is in fact also an optimal solution for the original, stochastic system. It has thus been found that the certainty equivalence principle holds in this example. Another case where it is applicable is linear quadratic Gaussian control, as will be shown in Section 11.2. Later, cases where it does *not* apply will also be presented.

10.2.4 Noisy Observations

Let the system now be

$$\begin{aligned} x(t+1) &= ax(t) + bu(t) \, , \\ y(t) &= x(t) + e(t) \, , \end{aligned} \tag{10.13}$$

with a and b known, $e(t)$ being white Gaussian noise of zero mean and known variance r_2. In this case, the control law is required to be a causal function of the available measurements,

$$u(t) = f(Y^t) \, . \tag{10.14}$$

Introduce the conditional mean of the states as

$$m(t) = \mathbf{E}\,[x(t)|Y^t] \, , \tag{10.15}$$

and let their precision be expressed as

$$\sigma^2(t) = \mathrm{var}[x(t)|Y^t] = \mathbf{E}\,[\{x(t) - m(t)\}^2|Y^t] \, . \tag{10.16}$$

It is now possible to write

$$J = \mathbf{E}\,x^2(N) = \mathbf{E}\,[\mathbf{E}\,[x^2(N)|Y^{N-1}]] \, , \tag{10.17}$$

where the outer expectation is with respect to Y^{N-1}.

$$\mathbf{E}\left[x^2(N)|Y^{N-1}\right] = \mathbf{E}\left[\{ax(N-1) + bu(N-1)\}^2|Y^{N-1}\right]$$
$$= [am(N-1) + bu(N-1)]^2 + a^2\sigma^2(N-1) \quad (10.18)$$

holds. Hence it is found, as in Section 10.2.2, that the optimal input at time $t = N - 1$ is

$$u(N-1) = -\frac{a}{b}m(N-1). \quad (10.19)$$

In order to complete the derivation of the regulator, the conditional mean $m(t)$ and its variance $\sigma^2(t)$ must be found. These quantities are given precisely by the Kalman filter (see Section 6.3).

10.2.5 Process Noise

Let the system be

$$x(t+1) = ax(t) + bu(t) + v(t), \quad (10.20)$$
$$y(t) = x(t),$$

with $x(0)$ given, a and b known, $v(t)$ white noise with zero mean and known variance r_1. As in previous cases:

$$J = \mathbf{E}\left[\mathbf{E}\left[x^2(N)|X^{N-1}\right]\right],$$
$$\mathbf{E}\left[x^2(N)|X^{N-1}\right] = \mathbf{E}\left[\{ax(N-1) + bu(N-1) + v(N-1)\}^2|X^{N-1}\right]$$
$$= [ax(N-1) + bu(N-1)]^2 + r_1. \quad (10.21)$$

The optimal control at time $t = N - 1$ becomes

$$u(N-1) = -\frac{a}{b}x(N-1) \quad (10.22)$$

as in (10.4).

10.2.6 Unknown Time Constants and Measurement Noise

This case is a combination of the situations given in Sections 10.2.3 and 10.2.4. Consider the system

$$x(t+1) = ax(t) + bu(t), \quad (10.23)$$
$$y(t) = x(t) + e(t),$$

where a is a random variable with known mean $\mathbf{E}\,a = m$ and variance $\mathbf{E}\,(a-m)^2 = \sigma^2$, b is known, and $e(t)$ is white noise of zero mean and variance r_2.

As there is measurement noise in the system, it is not possible to determine a exactly from measurements.

One might conjecture, from Sections 10.2.3 and 10.2.4, that the optimal regulator would be

$$u(N-1) = -\frac{\hat{m}(N-1)}{b}\hat{x}(N-1), \quad (10.24)$$

where

$$\hat{m}(N-1) = \mathbf{E}\left[a|Y^{N-1}\right],$$
$$\hat{x}(N-1) = \mathbf{E}\left[x(N-1)|Y^{N-1}\right]. \tag{10.25}$$

The regulator (10.24) would be obtained by applying the certainty equivalence principle. As it turns out, the control (10.24) is *not* optimal.

Rewrite the criterion as

$$J = \mathbf{E}\, x^2(N)$$
$$= \mathbf{E}\left[\mathbf{E}\left[x^2(N)|Y^{N-1}\right]\right]. \tag{10.26}$$

In this case:

$$\mathbf{E}\left[x^2(N)|Y^{N-1}\right] = \mathbf{E}\left[\{ax(N-1)+bu(N-1)\}^2|Y^{N-1}\right]$$
$$= [\widehat{ax}(N-1)+bu(N-1)]^2 + \Sigma(N-1), \tag{10.27}$$

where

$$\widehat{ax}(N-1) = \mathbf{E}\left[ax(N-1)|Y^{N-1}\right],$$
$$\Sigma(N-1) = \mathrm{var}[ax(N-1)|Y^{N-1}]. \tag{10.28}$$

The optimal control is apparently

$$u(N-1) = -\frac{1}{b}\widehat{ax}(N-1). \tag{10.29}$$

However, in general:

$$\widehat{ax}(N-1) = \mathbf{E}\left[ax(N-1)|Y^{N-1}\right].$$
$$\neq \hat{m}(N-1)\hat{x}(N-1)$$
$$= \mathbf{E}\left[a|Y^{N-1}\right]\mathbf{E}\left[x(N-1)|Y^{N-1}\right] \tag{10.30}$$

holds, which illustrates that the certainty equivalence principle *does not* give optimal control in this case. See also Exercise 10.1.

10.2.7 Unknown Gain

Consider now the system

$$x(t+1) = ax(t) + bu(t) + v(t),$$
$$y(t) = x(t), \tag{10.31}$$

where $x(0)$ is given, a is known, b is a random variable with nonzero mean and finite variance, $v(t)$ is white Gaussian noise of zero mean and variance r_1.

If one were to try the certainty equivalence principle, it would be relevant to consider the deterministic system

$$x(t+1) = ax(t) + \beta(t)u(t), \tag{10.32}$$

where

$$\beta(t) = \mathbf{E}\left[b|Y^t\right] = \mathbf{E}\left[b|X^t\right].$$ (10.33)

Also introduce

$$p(N-1) = \operatorname{var}[b|X^{N-1}]$$
$$= \mathbf{E}\left[\{b - \beta(N-1)\}^2|X^{N-1}\right].$$ (10.34)

As derived earlier, in Section 10.2.2, the optimal control law of (10.32) would then be

$$u(t) = -\frac{a}{\beta(t)}x(t).$$ (10.35)

This will *not* be the optimal regulator for the system of (10.31). Phrased differently, the certainty equivalence principle *does not* give optimality for this case. In order to prove this assertion, first evaluate the criterion for the regulator (10.35). The contribution from the last stage turns out to be

$$\mathbf{E}\left[x^2(N)|X^{N-1}\right]$$
$$= \mathbf{E}\left[\left\{ax(N-1) - \frac{ba}{\beta(N-1)}x(N-1) + v(N-1)\right\}^2|X^{N-1}\right]$$
$$= p(N-1)\left\{\frac{ax(N-1)}{\beta(N-1)}\right\}^2 + r_1.$$ (10.36)

Now try another regulator, namely:

$$u(t) = -a\frac{\beta(t)}{\beta^2(t) + p(t)}x(t).$$ (10.37)

Compared with (10.35), the regulator gain has been reduced so as to take the uncertainty in $\beta(t)$ into account. When the estimate $\beta(t)$ is uncertain (*i.e.* $p(t)$ is large), the gain is reduced significantly. For the regulator of (10.37), the contribution to the loss function will be

$$\mathbf{E}\left[x^2(N)|X^{N-1}]\right]$$
$$= \mathbf{E}\left[\left\{ax(N-1)\right.\right.$$

$$\left.\left. - \frac{ba\beta(N-1)}{\beta^2(N-1) + p(N-1)}x(N-1) + v(N-1)\right\}^2|X^{N-1}\right]$$
$$= \left\{ax(N-1) - \frac{a\beta^2(N-1)}{\beta^2(N-1) + p(N-1)}x(N-1)\right\}^2$$
$$+ p(N-1)\left\{\frac{a\beta(N-1)}{\beta^2(N-1) + p(N-1)}\right\}^2 x^2(N-1) + r_1$$
$$= p(N-1)\frac{\{ax(N-1)\}^2}{\beta^2(N-1) + p(N-1)} + r_1.$$ (10.38)

This is apparently smaller than the corresponding expression in (10.36), which shows that the regulator (10.35) cannot be optimal.

Note that the regulator of (10.37) is sometimes called a *cautious controller*. One can, in fact, show that it is the optimal regulator for the system of (10.31) in the case where $N = 1$.

10.3 Mathematical Preliminaries

Some of the mathematical tools needed to derive optimal control of a stochastic system are introduced in this section. In Section 10.2 it was shown that it is common for a conditional expectation to be minimized.

It turns out that it is crucial to specify the information that is needed at time t to determine the input signal $u(t)$.

In order to develop a more formal treatment, let x and y be two stochastic variables and let u be a decision variable. Also define a criterion function $\ell(x, y, u)$. The expected loss is

$$V = \mathbf{E}\, \ell(x, y, u) \,, \tag{10.39}$$

where expectation is with respect to x and y.

Now examine the optimization of V with respect to u. It is important to distinguish between the case when u is a function of both x and y ("complete state information") and the case when u is a function of y only ("incomplete state information").

Lemma 10.1 *Assume that the function $\ell(x, y, u)$ has a unique minimum with respect to u for all x and y, and denote it by $u^\circ(x, y)$. Then*

$$\min_{u(x,y)} \mathbf{E}\, \ell(x, y, u) = \mathbf{E}\, \ell(x, y, u^\circ(x, y)) = \mathbf{E}\, \min_u \ell(x, y, u) \,. \tag{10.40}$$

Remark. The minimization in the left-hand side is over all *functions* of x and y. The minimization in the right-hand side is over all *variables* u. The lemma means that the minimization and expectation operators commute. □

Proof Apparently, for any u:

$$\ell(x, y, u) \geq \ell(x, y, u^\circ(x, y)) = \min_u \ell(x, y, u) \,,$$

and hence

$$\mathbf{E}\, \ell(x, y, u) \geq \mathbf{E}\, \min_u \ell(x, y, u) \,.$$

Here, the right-hand side is a constant. The left-hand side depends on u. If the left-hand side is minimized over all functions $u(x, y)$, this gives

$$\min_{u(x,y)} \mathbf{E}\,\ell(x,y,u) \geq \mathbf{E}\,\min_{u}\ell(x,y,u)$$

$$= \mathbf{E}\,\ell(x,y,u^\circ(x,y))\,. \tag{10.41}$$

However, as $u^\circ(x,y)$ is a permitted control function, it also holds that

$$\min_{u(x,y)} \mathbf{E}\,\ell(x,y,u) \leq \mathbf{E}\,\ell(x,y,u)_{|u=u^\circ(x,y)}$$

$$= \mathbf{E}\,\ell(x,y,u^\circ(x,y))\,. \tag{10.42}$$

Combining (10.41) and (10.42) gives exactly (10.40). ∎

Next, consider the case when V is to be minimized by functions of y only.

Lemma 10.2 *Assume that the function*

$$f(y,u) \stackrel{\triangle}{=} \mathbf{E}\,[\ell(x,y,u)|y] \tag{10.43}$$

has a unique minimum, say $u^\circ(y)$, for all y. Then

$$\min_{u(y)} \mathbf{E}\,\ell(x,y,u) = \mathbf{E}_y\{\min_u \mathbf{E}\,[\ell(x,y,u)|y]\}\,, \tag{10.44}$$

where $\mathbf{E}_y$ denotes expectation with respect to y.

Proof Apparently:

$$f(y,u) \geq f(y,u^\circ(y)) = \min_u f(y,u)\,,$$

and hence

$$\mathbf{E}\,\ell(x,y,u) = \mathbf{E}_y f(y,u) \geq \mathbf{E}_y \min_u f(y,u) = \mathbf{E}_y f(y,u^\circ(y))$$

for any u. As the right-hand side is a constant, the inequality is maintained if the left-hand side is minimized:

$$\min_{u(y)} \mathbf{E}\,\ell(x,y,u) \geq \mathbf{E}_y\{\min_u f(y,u)\}\,. \tag{10.45}$$

As $u^\circ(y)$ is a possible decision law, it is also true that

$$\min_{u(y)} \mathbf{E}\,\ell(x,y,u) \leq \mathbf{E}\,(\ell(x,y,u^\circ(y)) = \mathbf{E}_y\{\min_u f(y,u)\}\,. \tag{10.46}$$

Combining (10.45) and (10.46) now gives (10.44). ∎

10.4 Dynamic Programming

In this section, the optimal control of a stochastic dynamic system is studied. The treatment will be based on the principle of dynamic programming. For didactic reasons, the simpler case of optimal control of a deterministic system is treated first.

10.4.1 Deterministic Systems

Consider a nonlinear system in state space form:

$$x(t+1) = f(x(t), u(t)) ,$$

(10.47)

and assume that its performance can be measured with the loss function, or performance index

$$J = \sum_{t=0}^{N-1} \ell(x(t), u(t)) .$$

(10.48)

The optimal control problem is to determine a feedback $u(t)$ as a *function* of the current state $x(t)$, so that J becomes as small as possible. This, of course, requires that noise-free measurements of the full state vector are available.

In the dynamic programming approach for solving this optimal control problem, the optimization is done recursively backwards in time.

Lemma 10.3 *Consider the system of (10.47) and the performance index J (10.48), which is to be optimized over all functions $u(t) = u(x(t))$. The optimal control is given by the functional equation*

$$V(x(t), t) = \min_{u(t)} \left[\ell(x(t), u(t)) + V(f(x(t), u(t)), t+1) \right] ,$$

(10.49)

where the minimization is over all functions $u(t) = u(x(t))$. The function $V(x(t), t)$ is the minimal loss when the system starts at time t in the state $x(t)$.

Proof Consider first the situation at time $t = N - 1$. Then, obviously, the only influence of the control action $u(N-1)$ is on the term $\ell(x(N-1), u(N-1))$ in the performance index, and hence

$$u(N-1) = \arg \min_{u(N-1)} \ell(x(N-1), u(N-1)) .$$

Note that this will give $u(N-1)$ as a function of $x(N-1)$. Assume for a moment that all optimal control actions $u(t+1), \ldots, u(N-1)$ have been found, and seek $u(t)$. Introduce

$$V(z, t) = \min_{u(t), \ldots, u(N-1)} \sum_{k=t}^{N-1} \ell(x(k), u(k))|_{x(t)=z}$$

(10.50)

as the minimal loss from time t onwards, starting with $x(t) = z$. One then finds that

$$V(z,t) = \min_{u(t)} \left\{ \min_{u(t+1),\dots,u(N-1)} \left[\ell(x(t),u(t)) \right. \right.$$

$$\left. \left. + \sum_{k=t+1}^{N-1} \ell(x(k),u(k)) \right]_{|x(t)=z} \right\} .$$

Of the control variables, only $u(t)$ will influence the term $\ell(x(t),u(t))$. Hence,

$$V(z,t) = \min_{u(t)} \left[\ell(x(t),u(t)) \right.$$

$$+ \min_{u(t+1),\dots,u(N-1)} \left. \sum_{k=t+1}^{N-1} \ell(x(k),u(k)) \right]_{|x(t)=z}$$

$$= \min_{u(t)} \left[\ell(x(t),u(t)) + V(x(t+1),t+1) \right]_{|x(t)=z}$$

$$= \min_{u(t)} \left[\ell(x(t),u(t)) + V(f(x(t),u(t)),t+1) \right]_{|x(t)=z} . \quad \blacksquare$$

Remark Note that (10.49) is a recursion for finding the optimal control. Its continuous-time counterpart is often called the Hamilton–Jacobi(–Bellman) equation, whereas in stochastic contexts it is referred to as the *Bellman equation*. It looks misleadingly simple. In most cases, it cannot be solved analytically. For numerical solutions, one has to discretize the state space and the control variable, and also the size of the problems becomes very large for very simple problems.

In an implicit form, (10.49) also expresses the *principle of optimality*. Assume that the optimal way of controlling the system starting at $x(t+1) = \bar{x}$ has been found. When optimizing $u(t)$ in (10.49), the control action $u(t)$ will have an influence on the whole future behaviour of the system. However, for any alternative where $x(t+1) = \bar{x}$, we already know how to proceed with future control actions from time $t+1$ onwards in an optimal way. $\square$

10.4.2 Stochastic Systems

For an optimal control problem in the stochastic case, it becomes crucial to specify the available information; that is, what variables are accessible when the control action $u(t)$ is to be determined.

Consider first the case of *complete state information*. Assume that the system dynamics is given by

$$x(t+1) = f(x(t),u(t)) + v(t) , \tag{10.51}$$

where $v(t)$ is a white noise sequence, and $f(x,u)$ is a nonlinear function. It is assumed that the full state vector $x(t)$ is measurable. The performance index of (10.48) has to be modified. In its previous form it is a random variable, taking a value that depends on the noise realization. A natural extension of (10.48) to the case of stochastic systems is to minimize the *expected* loss, that is:

$$J = \mathbf{E} \sum_{t=0}^{N-1} \ell(x(t), u(t)) \,. \tag{10.52}$$

In this case the following lemma holds.

Lemma 10.4 *Consider the system of (10.51), where $v(t)$ is white noise, and the performance index J (10.52) is to be optimized over all functions $u(t) = u(x(t))$. The optimal control is given by the functional equation*

$$\begin{aligned} V(x(t), t) = \min_{u(t)} \mathbf{E} \, [\ell(x(t), u(t)) \\ + V(x(t+1), t+1)|x(t)] \,. \end{aligned} \tag{10.53}$$

In (10.53) the minimization is over all such functions and $\mathbf{E}\,[\cdot|x(t)]$ denotes conditional expectation.

Proof When minimizing this J, Lemma 10.1 will be useful. The minimization is rewritten as

$$\min_{u(t)} J = \min_{u(t)} \mathbf{E} \sum_{k=0}^{N-1} \ell(x(k), u(k))$$

$$= \min_{u(t)} \mathbf{E}_{x(t)} \left(\mathbf{E} \left[\sum_{k=0}^{N-1} \ell(x(k), u(k)) \, \bigg| \, x(t) \, \right] \right),$$

where the inner expectation is conditioned on $x(t)$ and the outer expectation is taken with respect to $x(t)$ (see Lemma 2.2). Applying Lemma 10.1 next gives

$$\min_{u(t)} J = \mathbf{E}_{x(t)} \left(\min_{u(t)} \mathbf{E} \left[\sum_{k=0}^{N-1} \ell(x(k), u(k)) \, |x(t) \right] \right).$$

Obviously, $u(t)$ will only influence the future behaviour of the system. It thus follows that $u(t)$ at time t can be determined as

$$u(t) = \arg\min_{u(t)} \mathbf{E} \left[\sum_{k=t}^{N-1} \ell(x(k), u(k)) \, |x(t) \right].$$

To proceed, introduce the "loss-to-go":

$$V(z, t) = \min_{u(t), \dots, u(N-1)} \mathbf{E} \left[\sum_{k=t}^{N-1} \ell(x(k), u(k)) \, |x(t) = z \right]. \tag{10.54}$$

As with the development for deterministic systems, one now has

$$
\begin{aligned}
V(z,t) &= \min_{u(t)} \min_{u(t+1),\dots,u(N-1)} \mathbf{E}\left[\sum_{k=t}^{N-1} \ell(x(k),u(k))\,|x(t)\right] \\
&= \min_{u(t)} \left\{ \min_{u(t+1),\dots,u(N-1)} \mathbf{E}\left[\ell(x(t),u(t))\right.\right. \\
&\qquad \left.\left. + \sum_{k=t+1}^{N-1} \ell(x(k),u(k))|x(t)\right]\right\} \\
&= \min_{u(t)} \left\{ \mathbf{E}\left[\ell(x(t),u(t))\,|x(t)\right]\right. \\
&\qquad \left. + \min_{u(t+1),\dots,u(N-1)} \mathbf{E}\left[\sum_{k=t+1}^{N-1} \ell(x(k),u(k))\,|x(t)\right]\right\} \\
&= \min_{u(t)} \left\{ \mathbf{E}\left[\ell(x(t),u(t))|x(t)\right] + \mathbf{E}\left[V(x(t+1),t+1)|x(t)\right]\right\} \\
&= \min_{u(t)} \mathbf{E}\left[\ell(x(t),u(t)) + V(x(t+1),t+1)|x(t)\right]. \qquad \blacksquare
\end{aligned}
$$

Equation (10.53) is the Bellman equation, which defines the optimal control for a stochastic system. Note that the computational complexity has increased even further compared with the deterministic case. Now, the conditional pdf $p(x(t+1)|x(t))$ also has to be computed in order to evaluate the conditional expectation in (10.53). How this can be done recursively was explained in Section 5.4.

The case of *incomplete state information* will be briefly discussed next. Assume that the dynamics are

$$
\begin{aligned}
x(t+1) &= f(x(t),u(t)) + v(t)\,, \\
y(t) &= h(x(t)) + e(t)\,,
\end{aligned}
\qquad (10.55)
$$

and that the control action must be a function of available output measurements. The noise sequences $\{v(t)\}$ and $\{e(t)\}$ are assumed to be white and mutually independent. Compared with the previous case of complete state information, the present case becomes more difficult in that $y(t)$ is normally of smaller dimension than the full state vector $x(t)$ and in addition is perturbed by noise. Assume that the control action is constrained to be a function of previous outputs, so that $u(t) = F(Y^{t-1})$.

In a similar way to the treatment of complete state information, one can find that in this case (see also Lemma 10.2)

$$
u(t) = \arg\min_{u(t)} \mathbf{E}\left[\sum_{k=t}^{N-1} \ell(x(k),u(k))\,\Big|Y^{t-1}\right]. \qquad (10.56)
$$

Introduce

$$
V(\xi(t),t) = \min_{u(t),\dots,u(N-1)} \mathbf{E}\left[\sum_{k=t}^{N-1} \ell(x(k),u(k))\,\Big|Y^{t-1}\right], \qquad (10.57)
$$

where $\xi(t)$ is the *hyperstate* of the system. In the most general case, the hyperstate is the full conditional pdf $p(x(t)|Y^{t-1})$. As is known from Chapter 6, in the linear and Gaussian cases, this conditional pdf will be Gaussian, and it will then be sufficient to include its mean and covariance in the hyperstate $\xi(t)$.

The generalization of the Bellman equation (10.53) to the case of incomplete state information can be shown to give

$$V(\xi(t), t) = \min_{u(t)} \mathbf{E}\left[\ell(x(t), u(t)) + V(\xi(t+1), t+1)|Y^{t-1}\right]. \quad (10.58)$$

The following example illustrates the computational complexity of the Bellman equation (10.53). In order to describe the problem in some detail, a discretized example is chosen.

Example 10.1 Consider the system

$$x(t+1) = x(t) + u(t) + v(t) ,$$

where the possible control actions are $u(t) = 0, \pm 1$. Further, $v(t)$ is white noise with a discrete pdf:

$$P(v(t) = -1) = 0.3 ,$$
$$P(v(t) = 1) = 0.7 .$$

The initial value is a random variable that takes one of the values $-1, 0$ or 1. The state $x(t)$ is assumed to be available at time t; that is, this is a problem with complete state information. The performance index is chosen as

$$J = \mathbf{E}\left[x^2(2) + u^2(1) + u^2(0)\right] .$$

In the first phase, the control action $u(1)$ must be determined as a function of $x(1)$. In addition, the minimal loss-to-go, $V(x(1), 1)$, will be computed.

Owing to the specifications on possible values of $x(0)$, $u(1)$ and $v(0)$, it is found that the state $x(1)$ can take any of the values $-3, -2, -1, 0, 1, 2$ and 3. All these cases have to be evaluated separately. Introduce, further, the notation

$$J_1 = \left[x^2(2) + u^2(1)\right] ,$$
$$\bar{J}_1 = \mathbf{E}\left[x^2(2) + u^2(1)\right] ,$$

where expectation is with respect to $v(1)$.

Consider first the case when $x(1) = -3$. The results of different outcomes are summarized in Table 10.1.

The optimal control actions $u(1)$, assuming $x(1) = -3$, is the one giving the smallest value of $\bar{J}_1$ (see (10.56)). In this case, it is apparent that $u(1) = 1$ is optimal if $x(1) = -3$, and $V(-3, 1) = \min([14.8, 7.6, 4.4]) = 4.4$.

The above procedure must be repeated for all possible values of $x(1)$. The calculations are presented in Table 10.2.

For each value of $x(1)$, it is now straightforward to find the value of $u(1)$ that gives the smallest value of the cost-to-go, $\bar{J}_1$, as well as this minimum

Table 10.1. Evaluation of control actions at $t = 1$, for $x(1) = -3$

$u(1)$	$v(1)$	$x(2)$	$p(x(2))$	J_1	$\overline{J}_1$
-1	-1	-5	0.3	$(-5)^2 + (-1)^2 = 26$	$0.3 \times 26 + 0.7 \times 10 = 14.8$
	1	-3	0.7	$(-3)^2 + (-1)^2 = 10$	
0	-1	-4	0.3	$(-4)^2 + (0)^2 = 16$	$0.3 \times 16 + 0.7 \times 4 = 7.6$
	1	-2	0.7	$(-2)^2 + (0)^2 = 4$	
1	-1	-3	0.3	$(-3)^2 + 1^2 = 10$	$0.3 \times 10 + 0.7 \times 2 = 4.4$
	1	-1	0.7	$(-1)^2 + 1^2 = 2$	

value of $\overline{J}_1$. The findings are summarized in Table 10.3, which also describes the feedback of $u(1)$ as a function of $x(1)$.

One single time step of the Bellman equation has now been completed!

In this example, there are only two time steps (*i.e.* $N = 2$), and the remaining one must be considered as well. In that stage, the control action $u(0)$ is to be found as a function of $x(0)$. One can then proceed analogously to the previous calculations. There will be fewer cases to consider as the constraint $x(0) = -1, 0$ or 1 is imposed. When evaluating the minimal loss $\overline{J}_0^*$, the contribution $\overline{J}_1^*$ must be included. The calculations are summarized in Table 10.4, where

$$J_0 = u^2(0) + \overline{J}_1^*,$$
$$\overline{J}_0^* = \mathbf{E}\left[u^2(0) + \overline{J}_1^*\right].$$

The optimal control actions at time $t = 0$ are then easily found, as well as the minimal loss (see Table 10.5).

To summarize so far, the optimal control law is given, in tabulated form, in Tables 10.3 and 10.5. These tables describe how $u(t)$ is a function of the current state, $x(t)$.   □

It should be clear from the above example that the use of dynamic programming requires a huge amount of computation. In more realistic examples, it is necessary to use much finer grids of possible u and x values. If N is chosen large, the computational burden will increase still further. It is exceptional for the optimal regulator to be found in an analytical form. One important such case is linear systems with Gaussian distributed disturbances and quadratic criteria. This will be treated in Chapter 11 (although a somewhat different route will be taken in order to derive the optimal controller; also see Exercise 10.4).

Table 10.2. Evaluation of control actions at $t = 1$

$x(1)$	$u(1)$	$v(1)$	$x(2)$	$p(x(2))$	J_1	$\bar{J}_1$
-2	-1	-1	-4	0.3	$(-4)^2 + (-1)^2 = 17$	$0.3 \times 17 + 0.7 \times 5$
		1	-2	0.7	$(-2)^2 + (-1)^2 = 5$	$= 8.6$
	0	-1	-3	0.3	$(-3)^2 + 0^2 = 9$	$0.3 \times 9 + 0.7 \times 1$
		1	-1	0.7	$(-1)^2 + 0^2 = 1$	$= 3.4$
	1	-1	-2	0.3	$(-2)^2 + 1^2 = 5$	$0.3 \times 5 + 0.7 \times 1$
		1	0	0.7	$0^2 + 1^2 = 1$	$= 2.2$
-1	-1	-1	-3	0.3	$(-3)^2 + (-1)^2 = 10$	$0.3 \times 10 + 0.7 \times 2$
		1	-1	0.7	$(-1)^2 + (-1)^2 = 2$	$= 4.4$
	0	-1	-2	0.3	$(-2)^2 + 0^2 = 4$	$0.3 \times 4 + 0.7 \times 0$
		1	0	0.7	$0^2 + 0^2 = 0$	$= 1.2$
	1	-1	-1	0.3	$(-1)^2 + 1^2 = 2$	$0.3 \times 2 + 0.7 \times 2$
		1	1	0.7	$1^2 + 1^2 = 2$	$= 2.0$
0	-1	-1	-2	0.3	$(-2)^2 + (-1)^2 = 5$	$0.3 \times 5 + 0.7 \times 1$
		1	0	0.7	$0^2 + (-1)^2 = 1$	$= 2.2$
	0	-1	-1	0.3	$(-1)^2 + 0^2 = 1$	$0.3 \times 1 + 0.7 \times 1$
		1	1	0.7	$1^2 + 0^2 = 1$	$= 1.0$
	1	-1	0	0.3	$0^2 + 1^2 = 1$	$0.3 \times 1 + 0.7 \times 5$
		1	2	0.7	$2^2 + 1^2 = 5$	$= 3.8$
1	-1	-1	-1	0.3	$(-1)^2 + (-1)^2 = 2$	$0.3 \times 2 + 0.7 \times 2$
		1	1	0.7	$1^2 + (-1)^2 = 2$	$= 2.0$
	0	-1	0	0.3	$0^2 + 0^2 = 0$	$0.3 \times 0 + 0.7 \times 4$
		1	2	0.7	$2^2 + 0^2 = 4$	$= 2.8$
	1	-1	1	0.3	$1^2 + 1^2 = 2$	$0.3 \times 2 + 0.7 \times 10$
		1	3	0.7	$3^2 + 1^2 = 10$	$= 7.6$
2	-1	-1	0	0.3	$0^2 + (-1)^2 = 1$	$0.3 \times 1 + 0.7 \times 5$
		1	2	0.7	$2^2 + (-1)^2 = 5$	$= 3.8$
	0	-1	1	0.3	$1^2 + 0^2 = 1$	$0.3 \times 1 + 0.7 \times 9$
		1	3	0.7	$3^2 + 0^2 = 9$	$= 6.6$
	1	-1	2	0.3	$2^2 + 1^2 = 5$	$0.3 \times 5 + 0.7 \times 17$
		1	4	0.7	$4^2 + 1^2 = 17$	$= 13.4$
3	-1	-1	1	0.3	$1^2 + (-1)^2 = 2$	$0.3 \times 2 + 0.7 \times 10$
		1	3	0.7	$3^2 + (-1)^2 = 10$	$= 7.6$
	0	-1	2	0.3	$2^2 + 0^2 = 4$	$0.3 \times 4 + 0.7 \times 16$
		1	4	0.7	$4^2 + 0^2 = 16$	$= 12.4$
	1	-1	3	0.3	$3^2 + 1^2 = 10$	$0.3 \times 10 + 0.7 \times 26$
		1	5	0.7	$5^2 + 1^2 = 26$	$= 21.2$

10.5 Some Stochastic Controllers

By using the principle of optimality, it was shown in Section 10.4.2 how an optimal control law can, in principle, be found using the Bellman equation.

Table 10.3. Optimal control actions at $t = 1$

$x(1)$	$u(1)$	$\bar{J}_1^*$
-3	1	4.4
-2	1	2.2
-1	0	1.2
0	0	1.0
1	-1	2.0
2	-1	3.8
3	-1	7.6

Table 10.4. Evaluation of control actions at $t = 0$

$x(0)$	$u(0)$	$v(0)$	$x(1)$	$p(x(1))$	$\bar{J}_1^*$	J_0	$\bar{J}_0$
-1	-1	-1	-3	0.3	4.4	5.4	$0.3 \times 5.4 + 0.7 \times 2.2 = 3.16$
		1	-1	0.7	1.2	2.2	
	0	-1	-2	0.3	2.2	2.2	$0.3 \times 2.2 + 0.7 \times 1.0 = 1.36$
		1	0	0.7	1.0	1.0	
	1	-1	-1	0.3	1.2	2.2	$0.3 \times 2.2 + 0.7 \times 3.0 = 2.76$
		1	1	0.7	2.0	3.0	
0	-1	-1	-2	0.3	2.2	3.2	$0.3 \times 3.2 + 0.7 \times 2.0 = 2.36$
		1	0	0.7	1.0	2.0	
	0	-1	-1	0.3	1.2	1.2	$0.3 \times 1.2 + 0.7 \times 2.0 = 1.76$
		1	1	0.7	2.0	2.0	
	1	-1	0	0.3	1.0	2.0	$0.3 \times 2.0 + 0.7 \times 4.8 = 3.96$
		1	2	0.7	3.8	4.8	
1	-1	-1	-1	0.3	1.2	2.2	$0.3 \times 2.2 + 0.7 \times 3.0 = 2.76$
		1	1	0.7	2.0	3.0	
	0	-1	0	0.3	1.0	1.0	$0.3 \times 1.0 + 0.7 \times 3.8 = 2.96$
		1	2	0.7	3.8	3.8	
	1	-1	1	0.3	2.0	3.0	$0.3 \times 3.0 + 0.7 \times 8.6 = 6.92$
		1	3	0.7	7.6	8.6	

Table 10.5. Optimal control actions at $t = 0$

$x(0)$	$u(0)$	$\bar{J}_0^*$
-1	0	1.36
0	0	1.76
1	-1	2.76

As this approach requires a tremendous amount of computation for realistic problems, it is also of interest to consider some suboptimal schemes. The purpose of this section is to indicate some approaches and aspects of this kind.

It is generally assumed here that there is incomplete state information. The system dynamics are

$$x(t+1) = f(x(t), u(t)) + v(t) ,$$
$$y(t) = h(x(t)) + e(t) ,$$

(10.59)

while the criterion is

$$J = \mathbf{E} \sum_{t=0}^{N-1} \ell(x(t), u(t)) .$$

(10.60)

10.5.1 Dual Control

In the case of incomplete state information, the control action has two effects, with contradictory aims.

- On the one hand, $u(t)$ should be chosen to render $\mathbf{E}\,\ell(x(t), u(t))$ small, as well as the future terms $\mathbf{E}\,\ell(x(t+k), u(t+k))$, $k > 0$, in the performance index. This may often imply that $u(t)$ should be "small".
- On the other hand, there is also a reason for choosing $u(t)$ "large", since then it may dominate over the disturbances and it may be possible to obtain more precise information about the states. Such information is, in principle, collected in the conditional pdf $p(x(t)|Y^t)$ and may be used in the form of the conditional mean $\mathbf{E}\,[x(t)|Y^t]$. This effect of the input is sometimes called probing.

The concept of "dual control" concerns this double role of the input. It is interesting and useful to regard this double objective of the control law. It is worth mentioning that only in extreme cases can the explicit optimal control law be found by reasonable effort.

10.5.2 Certainty Equivalence Control

The concept of certainty equivalence was mentioned in Section 10.2.3. The basic idea is that the estimation and control problems are decoupled. The certainty equivalence principle can be formulated in alternative ways. The following is chosen here. The unknown states are estimated, for example, with an EKF, giving $\hat{x}(t+1|t)$, $t = 0, 1, \ldots$. Then, a deterministic optimal control problem is formulated, replacing true state vectors by state estimates; that is, the problem controlling the system

$$\hat{x}(t+1|t) = f(\hat{x}(t|t-1), u(t)) ,$$

(10.61)

such that the criterion

$$\hat{J} = \sum_{t=o}^{N-1} \ell(\hat{x}(t|t-1), u(t))$$

(10.62)

is minimized, is sought.

As was shown in Section 10.2, there are cases where the certainty equivalence principle gives optimal control and other cases where it does not. The normal situation is that it gives only a suboptimal performance. An important exception, when the principle leads to truly optimal control, is the case of a linear system, a quadratic performance index and Gaussian distributed disturbances (see Chapter 11).

10.5.3 Cautious Control

A simple example of a cautious controller was examined in Section 10.2.7. Here, a somewhat more general case is treated. The system to be controlled is assumed to have some unknown parameters. These are modelled as random variables with known means and covariances. Replacing the unknown variables by the mean values would lead to a certainty equivalence control, whereas also taking the covariance (*i.e.* the uncertainty of the mean values as parameter estimates) into account will give a cautious controller.

Consider a system with random and time-varying parameters, modelled as

$$y(t) + a_1(t-1)y(t-1) + \ldots + a_n(t-1)y(t-n)$$
$$= b_1(t-1)u(t-1) + \ldots + b_n(t-1)u(t-n) + e(t) , \qquad (10.63)$$

where $e(t)$ is white Gaussian noise of zero mean and variance λ^2. The model can also be written as a time-varying linear regression

$$y(t) = \varphi^T(t)\theta(t-1) + e(t) , \qquad (10.64)$$

where

$$\varphi^T(t) = [-y(t-1)\ldots - y(t-n)\ \ u(t-1)\ldots u(t-n)] ,$$
$$\theta^T(t-1) = [a_1(t-1)\ldots a_n(t-1)\ \ b_1(t-1)\ldots b_n(t-1)] . \qquad (10.65)$$

The parameter vector is assumed to vary according to the Markov model

$$\theta(t+1) = F\theta(t) + v(t) , \qquad (10.66)$$

where $v(t)$ is Gaussian, white, of zero mean and covariance R_1. As a control objective, it is assumed that a known reference signal should be followed in the sense that

$$J = \mathbf{E}\left[\{y(t+1) - r(t)\}^2 | Y^t\right] \qquad (10.67)$$

is to be minimized by the control actions at time t. For known parameters, this is a so-called minimal variance regulator problem, which will be dealt with in more detail in Section 11.5.2.

It turns out that the mean square optimal estimate of the unknown parameters is given by the standard Kalman filter (with (10.66) representing the system dynamics and (10.64) the measurement model), and in fact

$$p(\theta(t)|Y^t) = \gamma(\theta; \hat{\theta}(t), P(t)) , \tag{10.68}$$

where $\gamma(x; m, P)$ is the pdf of a Gaussian distributed random vector x with mean m and covariance matrix P, and $\hat{\theta}(t)$ is the usual state estimator.

The criterion J (10.67) can now be rewritten as follows, using the fact that $e(t + 1)$ is white, Lemma 2.1, and that $\varphi(t + 1)$ is known once Y^t is available:

$$\begin{aligned} J &= \mathbf{E}\left[\{\varphi^T(t+1)\theta(t) + e(t+1) - r(t)\}^2 | Y^t\right] \\ &= \lambda^2 + \mathbf{E}\left[\{\varphi^T(t+1)\theta(t) - r(t)\}^2 | Y^t\right] \\ &= \lambda^2 + [\varphi^T(t+1)\hat{\theta}(t) - r(t)]^2 + \varphi^T(t+1)P(t)\varphi(t+1) . \end{aligned} \tag{10.69}$$

This expression is now to be minimized with respect to $u(t)$, which appears as component $n + 1$ of $\varphi(t + 1)$. The criterion J depends in an easy way on $u(t)$ and the minimization can be done analytically.

In the simplified case, when the parameters are known (or treated as known – the certainty equivalence principle!), the last term of (10.69) is neglected and the control law is in an implicit form given by

$$\varphi^T(t+1)\hat{\theta}(t) = r(t) . \tag{10.70}$$

Assuming that $\hat{b}_1(t) \neq 0$, spelling out this relation yields

$$\begin{aligned} u(t) &= \frac{r(t) + \hat{a}_1(t)y(t) + \ldots + \hat{a}_n(t)y(t - n + 1)}{\hat{b}_1(t)} \\ &\quad - \frac{\hat{b}_2(t)u(t - 1) + \ldots + \hat{b}_n(t)u(t - n + 1)}{\hat{b}_1(t)} . \end{aligned} \tag{10.71}$$

In order to treat the general case of minimizing J in (10.69), it is convenient to rewrite the $\varphi(t + 1)$ vector as

$$\varphi(t+1) = \tilde{\varphi}(t+1) + \varepsilon u(t) . \tag{10.72}$$

Here, $\tilde{\varphi}(t + 1)$ is obtained from $\varphi(t + 1)$ by replacing $u(t)$ with zero, while ε is a unit vector with a one at the position $n + 1$. Using (10.69), the criterion is rewritten as

$$\begin{aligned} J &= \lambda^2 + [\tilde{\varphi}^T(t+1)\hat{\theta}(t) + u(t)\varepsilon^T\hat{\theta}(t) - r(t)]^2 \\ &\quad + [\tilde{\varphi}^T(t+1)P(t)\tilde{\varphi}(t+1) + 2u(t)\varepsilon^T P(t)\tilde{\varphi}(t+1) + u^2(t)\varepsilon^T P(t)\varepsilon] . \end{aligned} \tag{10.73}$$

Minimization of J with respect to $u(t)$ gives

$$u(t) = -\frac{\varepsilon^T\hat{\theta}(t)[\tilde{\varphi}^T(t+1)\hat{\theta}(t) - r(t)] + \varepsilon^T P(t)\tilde{\varphi}(t+1)}{[\varepsilon^T\hat{\theta}(t)]^2 + \varepsilon^T P(t)\varepsilon} . \tag{10.74}$$

Note that $\varepsilon^T\hat{\theta}(t) = \hat{b}_1(t)$, $\varepsilon^T P(t)\varepsilon = \text{var}(\hat{b}_1(t))$. It is worth adding that there is no guarantee that the regulator (10.74) always yields a stable closed loop system.

Exercises

Exercise 10.1 As an illustration of (10.30), consider the following case
($t = 1$):

$$x = ax_0 , \qquad x_0 = 1 ,$$
$$y = x + e .$$

Let a and e be jointly Gaussian

$$\begin{pmatrix} a \\ e \end{pmatrix} \sim N \left(\begin{pmatrix} m \\ 0 \end{pmatrix} , \begin{pmatrix} \sigma_a^2 & 0 \\ 0 & \sigma_e^2 \end{pmatrix} \right) .$$

Find $E[a|y] = E[x|y]$ and $E[ax|y]$.

Exercise 10.2 Consider (10.74) for the cautious controller. Verify that it
simplifies to (10.37) for the case treated in Section 10.2.7.

Exercise 10.3 Consider the cautious controller in (10.74). Assume that the
coefficient $\hat{b}_1(t)$ of $\hat{\theta}(t)$ is completely known. Show that the cautious controller
coincides with the certainty equivalence regulator in this case.

Exercise 10.4 Consider the one-dimensional system

$$x(t + 1) = x(t) + u(t) + v(t) ,$$

where $v(t)$ is white noise of zero mean and unit variance. Assume that the
system is to be controlled such that the criterion

$$J = E \sum_{t=0}^{3} [x^2(t) + u^2(t)]$$

is minimized. Use the dynamic programming approach to derive the optimal
feedback law.

Exercise 10.5 Consider the system

$$x(t + 1) = ax(t) + bu(t) + v(t) ,$$
$$t = 0, 1, \ldots$$
$$y(t) = x(t) .$$

Assume that

- a is a known constant,
- b is an unknown random variable with nonzero mean,
- $v(t)$ is white noise, of zero mean and variance r_2, independent of b.

Consider the criterion

$$J = \mathbf{E}\, x^2(N)$$

to be minimized over all regulators of the form

$$u(t) = f(x(t)) \ .$$

Find the optimal regulator.

Bibliography

Section 10.2 is based on the book

Aoki. M., 1967. *Optimization of Stochastic Systems.* Academic Press, New York.

which also gives a detailed treatment of these problems.
Section 10.3 is based on

Åström. K.J., 1970. *Introduction to Stochastic Control.* Academic Press, New York.

which is strongly recommended for further reading.
For a deeper treatment of dynamic programming, see

Bellman. R., 1957. *Dynamic Programming.* Princeton University Press, Princeton, NJ.

Bellman, R., 1961. *Adaptive Control Processes – A Guided Tour.* Princeton University Press, Princeton, NJ.

Fleming, W.H., Rishel, R.W., 1975. *Deterministic and Stochastic Optimal Control.* Springer-Verlag, New York.

Whittle, P.W., 1982. *Optimization over Time.* Wiley, Chichester.

For some further texts and deeper treatments on optimal control of stochastic systems, see also

Åström. K.J., Wittenmark, B., 1995. *Adaptive Control,* second ed. Addison Wesley, Reading, MA.

Kumar. P.R., Varaiya, P., 1986. *Stochastic Systems: Estimation, Identification and Adaptive Control.* Prentice Hall, Englewood Cliffs, NJ.

11. Linear Quadratic Gaussian Control

11.1 Introduction

The problem to be coped with in this chapter will lead to the celebrated separation theorem. The basic setup has three essential ingredients:

- the system is linear
- the criterion is quadratic
- the disturbances are Gaussian.

The problem is often referred to as linear quadratic Gaussian control or LQG control.

To be specific, let the system be given by

$$x(t+1) = Fx(t) + Gu(t) + v(t) ,$$
$$y(t) = Hx(t) + e(t) ,$$

(11.1)

where $v(t)$ and $e(t)$ are jointly Gaussian, white noise sequences of zero mean and

$$\mathbf{E} \begin{pmatrix} v(t) \\ e(t) \end{pmatrix} (v^T(t) \ \ e^T(t)) = \begin{pmatrix} R_1 & R_{12} \\ R_{21} & R_2 \end{pmatrix} ,$$

(11.2)

The covariance matrix in (11.2) is symmetric, so it must hold that $R_{21} = R_{12}^T$. Further, assume that $R_2 > 0$.

The criterion function, or performance index, is

$$\ell = x^T(N)Q_0x(N) + \sum_{t=t_0}^{N-1} (x^T(t) \ \ u^T(t)) \begin{pmatrix} Q_1 & Q_{12} \\ Q_{21} & Q_2 \end{pmatrix} \begin{pmatrix} x(t) \\ u(t) \end{pmatrix} ,$$

(11.3)

where the weighting matrices satisfy

$$Q_0 \geq 0 , \quad Q_2 > 0 , \quad \begin{pmatrix} Q_1 & Q_{12} \\ Q_{21} & Q_2 \end{pmatrix} \geq 0 .$$

(11.4)

Without restrictions we let the three matrices in (11.4) be symmetric, which in particular implies $Q_{21} = Q_{12}^T$.

The initial value $x(t_0)$ is assumed to be Gaussian distributed

$$x(t_0) \sim \mathbf{N}(m, R_0) \,, \tag{11.5}$$

and independent of the noise sequences $\{v(t)\}$ and $\{e(t)\}$.

The regulator that minimizes the expected loss, $\mathbf{E}\,\ell$, will be derived in the following three cases:

- Deterministic systems. (No noise sources at all are present.)
- Complete state information. (At time t, $x(t)$ is fully available. Phrased differently, $e(t) \equiv 0$, $H = I$.)
- Incomplete state information. (The problem as stated above.)

These three situations will be treated in the following section.

11.2 The Optimal Controllers

11.2.1 Optimal Control of Deterministic Systems

Consider now the situation of no disturbances present in the system. The problem is then often referred to as linear quadratic (LQ) control. A full *derivation* of the optimal regulator will not be given here, but rather it will be postulated that it can be found through the solution of a certain Riccati equation, and this statement will be proved in detail. In Section 11.A, however, we do provide a derivation of the optimal regulator and the Riccati equation for the case of complete state information (no measurement noise), using the Hamilton–Jacobi–Bellman equation (10.53) of Lemma 10.4.

Lemma 11.1 *Consider the system of (11.1). Assume that the difference equation*

$$S(t) = F^T S(t+1)F + Q_1 - [F^T S(t+1)G + Q_{12}]$$
$$\times [G^T S(t+1)G + Q_2]^{-1}[G^T S(t+1)F + Q_{21}]\,, \tag{11.6}$$
$$S(N) = Q_0 \,.$$

has a nonnegative definite solution for $t_0 \leq t \leq N$. Introduce

$$L(t) = [G^T S(t+1)G + Q_2]^{-1}[G^T S(t+1)F + Q_{21}] \,. \tag{11.7}$$

Then

$$\ell = x^T(t_0)S(t_0)x(t_0)$$
$$+ \sum_{t=t_0}^{N-1} [u(t) + L(t)x(t)]^T [G^T S(t+1)G + Q_2][u(t) + L(t)x(t)]$$
$$+ \sum_{t=t_0}^{N-1} \left(2v^T(t)S(t+1)[Fx(t) + Gu(t)] + v^T(t)S(t+1)v(t)\right) \,. \tag{11.8}$$

Proof Start with the identity

$$x^T(N)Q_0 x(N) = x^T(N)S(N)x(N) = x^T(t_0)S(t_0)x(t_0)$$
$$+ \sum_{t=t_0}^{N-1} [x^T(t+1)S(t+1)x(t+1) - x^T(t)S(t)x(t)] .$$

(11.9)

Now note that

$$x^T(t+1)S(t+1)x(t+1) = [Fx(t) + Gu(t) + v(t)]^T S(t+1)$$
$$\times [Fx(t) + Gu(t) + v(t)] ,$$
$$x^T(t)S(t)x(t) = x^T(t)[F^T S(t+1)F + Q_1$$
$$- L^T(t)(Q_2 + G^T S(t+1)G)L(t)]x(t) .$$

Using these expressions in (11.9) gives

$$x^T(N)Q_0 x(N) = x^T(t_0)S(t_0)x(t_0)$$
$$+ \sum_{t=t_0}^{N-1} (2v^T(t)S(t+1)[Fx(t) + Gu(t)]$$
$$+ v^T(t)S(t+1)v(t)$$
$$+ x^T(t)[-Q_1 + L^T(t)\{Q_2 + G^T S(t+1)G\}L(t)]x(t)$$
$$+ 2x^T(t)F^T S(t+1)Gu(t) + u^T(t)G^T S(t+1)Gu(t)) .$$

Hence set

$$\ell = x^T(t_0)S(t_0)x(t_0) + \sum_{t=t_0}^{N-1} (2v^T(t)S(t+1)[Fx(t) + Gu(t)]$$
$$+ v^T(t)S(t+1)v(t))$$
$$+ \sum_{t=t_0}^{N-1} (x^T(t)L^T(t)[Q_2 + G^T S(t+1)G]L(t)x(t)$$
$$+ 2x^T(t)[Q_{12} + F^T S(t+1)G]u(t) + u^T(t)[Q_2 + G^T S(t+1)G]u(t))$$
$$= x^T(t_0)S(t_0)x(t_0) + \sum_{t=t_0}^{N-1} (2v^T(t)S(t+1)[Fx(t) + Gu(t)]$$
$$+ v^T(t)S(t+1)v(t))$$
$$+ \sum_{t=t_0}^{N-1} [u(t) + L(t)x(t)]^T [Q_2 + G^T S(t+1)G][u(t) + L(t)x(t)] ,$$

which is (11.8). ∎

When dealing with stochastic systems, it is useful to evaluate *the expected value* of the criterion function. This is dealt with in the following lemma.

Lemma 11.2 *Consider the system in (11.1) and the criterion function ℓ, (11.3). Assume that either*

(i) $u(t)$ is a function of the current state $x(t)$

or

(ii) $u(t)$ is a function of Y^{t-k}, $k \geq 0$, all output values up to including time $t - k$. Further, in case $k = 0$, assume that $R_{12} = 0$.

Then the expected value of the criterion function fulfils

$$\mathbf{E}\,\ell = \mathbf{E}\left\{x^T(t_0)S(t_0)x(t_0)\right\} + \mathbf{E}\sum_{t=t_0}^{N-1}\left\{v^T(t)S(t+1)v(t)\right\}$$

$$+\mathbf{E}\left(\sum_{t=t_0}^{N-1}[u(t)+L(t)x(t)]^T[Q_2+G^TS(t+1)G][u(t)+L(t)x(t)]\right)$$

$$= m^T S(t_0)m + \mathrm{tr}\,S(t_0)R_0 + \sum_{t=t_0}^{N-1}\mathrm{tr}\,S(t+1)R_1$$

$$+\mathbf{E}\left(\sum_{t=t_0}^{N-1}[u(t)+L(t)x(t)]^T[Q_2+G^TS(t+1)G][u(t)+L(t)x(t)]\right).$$

$$(11.10)$$

Proof The expected value of $x^T(t_0)S(t_0)x(t_0)$ is evaluated using Lemma 2.1. It remains to show that $v(t)$ is uncorrelated with $z(t) \triangleq Fx(t) + Gu(t)$.

In case (i), $u(t)$ is apparently a function of $x(t)$, and hence by the system equation a function of $v(s)$ for $s \leq t - 1$, which is uncorrelated with $v(t)$.

In case (ii) it follows as above that $x(t)$ is uncorrelated with $v(t)$. If $k > 0$, then $u(t)$ is a function of Y^{t-k}, and hence a function of X^{t-k} and V^{t-k}, both of which are uncorrelated with $v(t)$. If $k = 0$ and $R_{12} = 0$, then $u(t)$ is a function of X^t and E^t. Noting that $e(t)$ in this case is uncorrelated with $v(t)$, the desired statement follows. ∎

In what follows in this section, assume that the assumptions of Lemma 11.1 are satisfied. Next, consider optimal control of deterministic systems.

Theorem 11.1 *Consider the system of (11.1) with no process noise $(v(t) \equiv 0)$ and complete state information $(y(t) \equiv x(t))$. Let possible controllers be of the form $u(t) = f(x(t))$. Then the optimal controller is a linear time-varying feedback*

$$u(t) = -L(t)x(t)\,,$$

$$(11.11)$$

where $L(t)$ is given by (11.7). The minimum value of the criterion is given by

$$\min V = m^T S(t_0)m + \operatorname{tr} S(t_0)R_0 \,, \tag{11.12}$$

with $S(t)$ being given by (11.6).

Proof Using Lemma 11.1:

$$V = \mathbf{E}\,\ell = \mathbf{E}\left[x^T(t_0)S(t_0)x(t_0)\right]$$
$$+ \sum_{t=t_0}^{N-1} \mathbf{E}\left[u(t) + L(t)x(t)\right]^T [G^T S(t+1)G + Q_2][u(t) + L(t)x(t)] \,.$$

Owing to the assumptions:

$$G^T S(t+1)G + Q_2 > 0$$

holds. Hence all terms in the sum are positive. It follows from Lemma 10.1 that the optimal control is given by (11.11). The expression in (11.12) for $\min V$ next follows easily from Lemma 2.1.   ∎

Remark It is no restriction to consider controllers of the form $u(t) = f(x(t))$. As $x(t)$ is the *state* variable, it includes all information about the past trajectories that are of any importance for the future. The most general controller is therefore a function of the current state vector.   □

Under the given assumptions, in particular $Q_2 > 0$, the Riccati equation of (11.6) has a unique solution. It is shown in Section 11.3 that it is dual to the Riccati equation associated with the optimal state estimation. Hence, results such as those of Section 6.7 can be derived for the stationary case (now corresponding to $N \to \infty$). In particular, $S(t)$ will converge, as $N - t \to \infty$, to a constant, positive definite matrix, if

- $Q_2 > 0$
- (F, G) is stabilizable
- (F, C) is detectable, where Q_1 is factorized as $Q_1 = C^T C$.

11.2.2 Optimal Control with Complete State Information

In this case, allow process noise but maintain the assumption that all state variables are measurable.

Theorem 11.2 *Consider the system of (11.1) and assume that $y(t) \equiv x(t)$. Let possible controllers be of the form $u(t) = f(x(t))$. The optimal regulator is then*

$$u(t) = -L(t)x(t) \ . \tag{11.13}$$

The minimum value of the criterion is given by

$$\min \mathbf{E}\,\ell = m^T S(t_0)m + \operatorname{tr} S(t_0)R_0 + \sum_{t=t_0}^{N-1} \operatorname{tr} S(t+1)R_1 \ . \tag{11.14}$$

The matrices $L(t)$ and $S(t)$ are given by Lemma 11.1.

Proof Using Lemma 11.2 it follows directly that

$$\mathbf{E}\,\ell = m^T S(t_0)m + \operatorname{tr} S(t_0)R_0 + \sum_{t=t_0}^{N-1} \operatorname{tr} S(t+1)R_1$$

$$+ \mathbf{E}\left(\sum_{t=t_0}^{N-1} [u(t) + L(t)x(t)]^T [Q_2 + G^T S(t+1)G][u(t) + L(t)x(t)]\right) \ .$$

The first three terms are independent of the control. Each term in the final sum is positive. It thus follows from Lemma 10.1 that the optimal regulator is given by (11.13). The expression (11.14) for the minimal loss is then trivial. ∎

Remark In Section 11.A we *derive* the results (11.13) and (11.14), rather than proving them. □

11.2.3 Optimal Control with Incomplete State Information

Now the scene is set to cope with the general case when the input is to be determined from noisy output measurements only.

Theorem 11.3 (The separation theorem) *Consider the system in (11.1). Let possible controllers be of the form $u(t) = f(Y^{t-k})$ where $k \geq 0$, $Y^t = [y(t_0)^T, \ldots , y(t)^T]^T$. If $k = 0$ assume further that $R_{12} = 0$. The optimal regulator is*

$$u(t) = -L(t)\hat{x}(t|t-k) \ , \tag{11.15}$$

and the minimal value of the criterion is

$$\min \mathbf{E}\,\ell = m^T S(t_0)m + \operatorname{tr}\ S(t_0)R_0$$
$$+ \sum_{t=t_0}^{N-1} \operatorname{tr}\ S(t+1)R_1$$
$$+ \sum_{t=t_0}^{N-1} \operatorname{tr}\ \left\{ P(t|t-k)L^T(t) \right.$$
$$\left. \times [G^T S(t+1)G + Q_2]L(t) \right\}, \tag{11.16}$$

where

$$P(t|t-k) = \mathbf{E}\,[x(t) - \hat{x}(t|t-k)][x(t) - \hat{x}(t|t-k)]^T. \tag{11.17}$$

The matrices $L(t)$ and $S(t)$ are given by (11.6) and (11.7). In (11.15), $\hat{x}(t|t-k)$ is the optimal k-step predictor, as given by (6.39).

Proof The expected value of ℓ can be rewritten as (11.10). However, the last term can no longer be brought to zero. Use of Lemma 10.2 will give

$$\min_{u(t)=f(Y^{t-k})} \mathbf{E} \sum_{t=t_0}^{N-1} [u(t) + L(t)x(t)]^T [Q_2 + G^T S(t+1)G][u(t) + L(t)x(t)]$$

$$= \mathbf{E}_y \left(\min_{u(t)=f(Y^{t-k})} \mathbf{E} \left\{ \sum_{t=t_0}^{N-1} [u(t) + L(t)x(t)]^T \right. \right.$$
$$\times [Q_2 + G^T S(t+1)G][u(t) + L(t)x(t)]|Y^{t-k} \right\} \bigg)$$

$$= \mathbf{E}_y \left(\min_{u(t)=f(Y^{t-k})} \mathbf{E} \left\{ \sum_{t=t_0}^{N-1} [u(t) + L(t)\hat{x}(t|t-k)]^T \right. \right.$$
$$\times [Q_2 + G^T S(t+1)G][u(t) + L(t)\hat{x}(t|t-k)] \right\}$$

$$+ \sum_{t=t_0}^{N-1} \operatorname{tr}\ L^T(t)[Q_2 + G^T S(t+1)G]L(t)P(t|t-k) \bigg). \tag{11.18}$$

(Recall that $\hat{x}(t|t-k)$ and $x(t) - \hat{x}(t|t-k)$ are uncorrelated.) The second sum in (11.18) does not depend on the control. The first sum in (11.18) is brought to zero by the control (11.15), which is thus proved to be optimal. The minimal value (11.16) of the criterion then follows directly from (11.18). ∎

Remark 1 The theorem is called the separation theorem because it is allowed to separate control and estimation. Note that the same feedback vector $L(t)$ is used as when $x(t)$ is known; see (11.11). In the case of incomplete state information, *optimal* performance is obtained by substituting the exact state vector by the optimal state estimate.   □

Remark 2 The regulator (11.15) has been shown to be the optimal one in the linear quadratic case for Gaussian distributed disturbances. It has another optimality property as well. Let the assumption on Gaussian distributions be relaxed, so that arbitrarily distributed disturbances are considered. Assume that the regulator is constrained to be linear. Under these assumptions, the regulator (11.15) is still optimal. Conceptually, this statement is closely related to the fact that the Kalman filter (*i.e.* the conditional mean) and the LLMS filter coincide.   □

11.3 Duality Between Estimation and Control

It has been shown that the solutions to both optimal linear control and optimal state estimation are given by the solutions to the Riccati equations. See Sections 11.2.1 and 6.4. The connection between these two problems and their solutions is now established.

It is appropriate first to repeat the estimation problem and its solution. Consider the system

$$x(t+1) = Fx(t) + v(t) ,$$
$$y(t) = Hx(t) + e(t) ,$$
                    (11.19)

where $x(t_0) \in \mathbf{N}(m, R_0)$ is independent of the white noise sequence

$$\begin{pmatrix} v(t) \\ e(t) \end{pmatrix} \in \mathbf{N} \left(\begin{pmatrix} 0 \\ 0 \end{pmatrix} , \begin{pmatrix} R_1 & R_{12} \\ R_{21} & R_2 \end{pmatrix} \right) .$$
                    (11.20)

The optimal state estimator $\hat{x}(t|t-1)$ is given by

$$\hat{x}(t+1|t) = F\hat{x}(t|t-1) + K(t)[y(t) - H\hat{x}(t|t-1)] ,$$
$$K(t) = [FP(t)H^T + R_{12}][HP(t)H^T + R_2]^{-1} ,$$
$$P(t+1) = FP(t)F^T + R_1 - K(t)[HP(t)F^T + R_{21}] ,$$
$$\hat{x}(t_0|t_0 - 1) = m ,$$
$$P(t_0) = R_0 .$$
                    (11.21)

Now introduce the system

$$\overline{x}(\overline{t}+1) = \overline{F}\overline{x}(\overline{t}) + \overline{G}\overline{u}(\overline{t}) ,$$
                    (11.22)

where

$$\bar{t} = t_0 + N - t \,,$$
$$\overline{F} = F^T \,,$$
$$\overline{G} = H^T \,. \tag{11.23}$$

Introduce, further, the performance index

$$\ell = \overline{x}^T(N)\overline{Q}_0\overline{x}(N) + \sum_{\bar{t}=t_0}^{N-1} (\overline{x}^T(\bar{t}) \ \overline{u}^T(\bar{t})) \begin{pmatrix} \overline{Q}_1 & \overline{Q}_{12} \\ \overline{Q}_{21} & \overline{Q}_2 \end{pmatrix} \begin{pmatrix} \overline{x}(\bar{t}) \\ \overline{u}(\bar{t}) \end{pmatrix} \,, \tag{11.24}$$

with

$$\overline{Q}_0 = R_0 \,,$$
$$\overline{Q}_1 = R_1 \,,$$
$$\overline{Q}_{12} = R_{12} \,,$$
$$\overline{Q}_2 = R_2 \,. \tag{11.25}$$

The control law that minimizes $\mathbf{E}\,\ell$ is given by (see Theorem 11.1)

$$\overline{u}(\bar{t}) = -\overline{L}(\bar{t})\overline{x}(\bar{t}) \,, \tag{11.26}$$

with

$$\overline{L}(\bar{t}) = [\overline{Q}_2 + \overline{G}^T\overline{S}(\bar{t}+1)\overline{G}]^{-1}[\overline{G}^T\overline{S}(\bar{t}+1)\overline{F} + \overline{Q}_{21}] \,,$$
$$\overline{S}(\bar{t}) = \overline{F}^T\overline{S}(\bar{t}+1)\overline{F} + \overline{Q}_1 - \overline{L}(\bar{t})^T[\overline{G}^T\overline{S}(\bar{t}+1)\overline{F} + \overline{Q}_{21}] \,, \tag{11.27}$$
$$\overline{S}(N) = \overline{Q}_0 \,.$$

Now introduce

$$P^*(t) = \overline{S}(\bar{t}) \,,$$
$$K^*(t) = \overline{L}^T(\bar{t}-1) \,. \tag{11.28}$$

Then from (11.23), (11.25) and (11.27):

$$P^*(t_0) = \overline{S}(N) = \overline{Q}_0 = R_0 = P(t_0) \,, \tag{11.29}$$

$$P^*(t+1) = \overline{S}(\bar{t}-1) = \overline{F}^T\overline{S}(\bar{t})\overline{F} + \overline{Q}_1$$
$$\qquad -[\overline{Q}_{21} + \overline{G}^T\overline{S}(\bar{t})\overline{F}]^T[\overline{G}^T\overline{S}(\bar{t})\overline{G} + \overline{Q}_2]^{-1}[\overline{G}^T\overline{S}(\bar{t})\overline{F} + \overline{Q}_{21}]$$
$$\qquad = FP^*(t)F^T + R_1 - [R_{21} + HP^*(t)F^T]^T$$
$$\qquad \times[HP^*(t)H^T + R_2]^{-1}[R_{21} + HP^*(t)F^T] \,, \tag{11.30}$$

$$K^*(t) = [\overline{F}^T\overline{S}(\bar{t})\overline{G} + \overline{Q}_{12}][\overline{G}^T\overline{S}(\bar{t})\overline{G} + \overline{Q}_2]^{-1}$$
$$\qquad = [FP^*(t)H^T + R_{12}][HP^*(t)H^T + R_2]^{-1} \,. \tag{11.31}$$

Hence it is found that $P^*(t)$, $K^*(t)$ coincide fully with $P(t)$, $K(t)$ (see (11.21)). This shows that the estimation and control problems are *dual*. In loose terms, one can go from one of the problems to the other by transposing the matrices and reversing the time direction (see (11.23)). In particular, it is possible to apply the analysis of Section 6.7 to determine the time-invariant controllers corresponding to the limiting case $N \rightarrow \infty$.

11.4 Closed Loop System Properties

Some fundamental properties of the closed loop system under LQG control are examined in this section. The time-invariant case only will be considered, implying that $t_0 \to -\infty$, $N \to \infty$.

11.4.1 Representations of the Regulator

It was found in Section 11.2.3 that the optimal regulator is

$$u(t) = -L\hat{x}(t|t - k) , \tag{11.32}$$

where the state estimate $\hat{x}(t|t - k)$ is given in Section 6.4. In practice, one would choose $k = 0$, or possibly $k = 1$, if the computation of the new control value $u(t)$ takes a noticeable part of a sampling interval. The state estimate in (11.32) is implicitly a function of old input and output values. Hence, it must be possible to find a representation for the regulator in "usual" forms also, such as a state space representation or a transfer function.

Lemma 11.3 *Consider the regulator*

$$u(t) = -L\hat{x}(t|t - k) ,$$
$$\hat{x}(t + 1|t) = F\hat{x}(t|t - 1) + Gu(t) + K_p[y(t) - H\hat{x}(t|t - 1)] , \tag{11.33}$$
$$\hat{x}(t|t) = \hat{x}(t|t - 1) + K_f[y(t) - H\hat{x}(t|t - 1)] .$$

For $k = 1$, the regulator can be represented as

$$\hat{x}(t + 1|t) = (F - K_pH - GL)\hat{x}(t|t - 1) + K_py(t) ,$$
$$u(t) = -L\hat{x}(t|t - 1) . \tag{11.34}$$

The corresponding transfer function becomes

$$G(z) = -L[zI - F + K_pH + GL]^{-1}K_p . \tag{11.35}$$

For $k = 0$, the regulator can be written as

$$\hat{x}(t + 1|t) = [F - GL - (K_p - GLK_f)H]\hat{x}(t|t - 1)$$
$$+ (K_p - GLK_f)y(t) , \tag{11.36}$$
$$u(t) = -L(I - K_fH)\hat{x}(t|t - 1) - LK_fy(t) ,$$

and the associated transfer function is

$$G(z) = -LK_f - L(I - K_fH)[zI - F + GL + (K_p - GLK_f)H]^{-1}$$
$$\times [K_p - GLK_f] . \tag{11.37}$$

Proof Equations (11.32) and (11.33) give (11.34) directly, and (11.35) is then trivial. For $k = 0$, it is possible to write

$$\hat{x}(t+1|t) = (F - K_p H)\hat{x}(t|t-1) + K_p y(t)$$
$$-GL[(I - K_f H)\hat{x}(t|t-1) + K_f y(t)]$$
$$= [F - GL - (K_p - GLK_f)H]\hat{x}(t|t-1) + (K_p - GLK_f)y(t) ,$$
$$u(t) = -L\hat{x}(t|t) = -L[(I - K_f H)\hat{x}(t|t-1) + K_f y(t)] ,$$

which is (11.36). Then (11.37) follows directly. ∎

Remark 1 Note that no condition has been imposed on the true dynamics of the system. Only the structure in (11.33) of the regulator has been postulated. It is not required that $\hat{x}(t|t-1)$ is the optimal state estimate. □

Remark 2 Assume that $K_p = FK_f$ holds. This is a relatively mild assumption. (For example, assume that the regulator is obtained by applying the LQG problem to a nominal model with uncorrelated process and measurement noise, so that $R_{12} = 0$.) Under this assumption, (11.36) can be simplified to

$$\hat{x}(t+1|t) = (F - GL)(I - K_f H)\hat{x}(t|t-1) + (F - GL)K_f y(t) ,$$
$$u(t) = -L(I - K_f H)\hat{x}(t|t-1) - LK_f y(t) . \tag{11.38}$$

□

11.4.2 Representations of the Closed Loop System

The following lemma characterizes the closed loop system.

Lemma 11.4 *Consider the system*

$$x(t+1) = Fx(t) + Gu(t) + v(t) ,$$
$$y(t) = Hx(t) + e(t) . \tag{11.39}$$

Assume that the system is controlled as

$$u(t) = -L\hat{x}(t|t-k) + u_r(t) , \qquad \text{with } k = 0 \text{ or } 1 ,$$
$$\hat{x}(t+1|t) = F\hat{x}(t|t-1) + Gu(t) + K_p[y(t) - H\hat{x}(t|t-1)] , \tag{11.40}$$
$$\hat{x}(t|t) = \hat{x}(t|t-1) + K_f[y(t) - H\hat{x}(t|t-1)] .$$

Introduce the one-step prediction error

$$\tilde{x}(t|t-1) = x(t) - \hat{x}(t|t-1) . \tag{11.41}$$

For $k = 1$, the closed loop system can be written as

$$\begin{pmatrix} x(t+1) \\ \tilde{x}(t+1|t) \end{pmatrix} = \begin{pmatrix} F - GL & GL \\ 0 & F - K_p H \end{pmatrix} \begin{pmatrix} x(t) \\ \tilde{x}(t|t-1) \end{pmatrix}$$
$$+ \begin{pmatrix} G \\ 0 \end{pmatrix} u_r(t) + \begin{pmatrix} I & 0 \\ I & -K_p \end{pmatrix} \begin{pmatrix} v(t) \\ e(t) \end{pmatrix} . \tag{11.42}$$

For k = 0:

$$\begin{pmatrix} x(t+1) \\ \tilde{x}(t+1|t) \end{pmatrix} = \begin{pmatrix} F - GL & GL(I - K_f H) \\ 0 & F - K_p H \end{pmatrix} \begin{pmatrix} x(t) \\ \tilde{x}(t|t-1) \end{pmatrix}$$
$$+ \begin{pmatrix} G \\ 0 \end{pmatrix} u_r(t) + \begin{pmatrix} I & -GLK_f \\ I & -K_p \end{pmatrix} \begin{pmatrix} v(t) \\ e(t) \end{pmatrix} \qquad (11.43)$$

holds.

Proof Independently of k, the prediction error satisfies

$$\begin{aligned} \tilde{x}(t+1|t) &= x(t+1) - \hat{x}(t+1|t) = Fx(t) + Gu(t) + v(t) \\ &\quad - F\hat{x}(t|t-1) - Gu(t) - K_p[Hx(t) + e(t) - H\hat{x}(t|t-1)] \\ &= (F - K_p H)\tilde{x}(t|t-1) + v(t) - K_p e(t) . \end{aligned}$$

This is the second block row of (11.42) and (11.43). For $k = 1$, it is also found that

$$\begin{aligned} x(t+1) &= Fx(t) - GL\hat{x}(t|t-1) + Gu_r(t) + v(t) \\ &= Fx(t) - GL[x(t) - \tilde{x}(t|t-1)] + Gu_r(t) + v(t) \\ &= (F - GL)x(t) + GL\tilde{x}(t|t-1) + Gu_r(t) + v(t) , \end{aligned}$$

and $k = 0$ gives

$$\begin{aligned} x(t+1) &= Fx(t) - GL\hat{x}(t|t) + Gu_r(t) + v(t) \\ &= Fx(t) - GL[(I - K_f H)\{x(t) - \tilde{x}(t|t-1)\} \\ &\quad + K_f\{Hx(t) + e(t)\}] + Gu_r(t) + v(t) \\ &= (F - GL)x(t) + GL(I - K_f H)\tilde{x}(t|t-1) \\ &\quad + Gu_r(t) + v(t) - GLK_f e(t) . \quad \blacksquare \end{aligned}$$

Remark 1 There is no assumption on the noise sources $v(t)$ and $e(t)$. Further, the result is purely algebraic, starting with the representations in (11.39) and (11.40). There is no requirement that the regulator should be the solution to an LQG problem. □

Remark 2 The transfer function from the reference signal $u_r(t)$ to the output $y(t)$ is readily found. For (11.42)

$$H(z) = (H \quad 0) \begin{pmatrix} zI - (F - GL) & -GL \\ 0 & zI - (F - K_p H) \end{pmatrix}^{-1} \begin{pmatrix} G \\ 0 \end{pmatrix}$$
$$= H[zI - (F - GL)]^{-1} G . \qquad (11.44)$$

holds. The same result is obtained for (11.43). Note that this is exactly the result that would be obtained if the full state could be measured, and the feedback $u(t) = -Lx(t)$ was applied. Replacing $x(t)$ in the feedback law by an estimate will thus not change the closed loop transfer function. It is not difficult to see that the states $\tilde{x}(t|t-1)$ are not controllable from $u_r(t)$.   □

Corollary Both for $k = 0$ and for $k = 1$, the closed loop poles are given by the eigenvalues of the matrices $F - GL$ and $F - K_pH$.

Proof Immediate from (11.42) and (11.43).   ■

Note that in case the regulator of (11.40) really is the solution to an LQG problem for the system of (11.39), then the closed loop system will be asymptotically stable. In Lemma 6.3 it was proved that $F - K_pH$ has all eigenvalues inside the unit circle under weak assumptions. The similar property of $F - GL$ follows from the duality results of Section 11.3.

11.4.3 The Closed Loop Poles

Introduce some notation for the characteristic polynomials of the open loop and the closed loop systems:

$$\begin{aligned}
\phi_{ol}(z) &\triangleq \det[zI - F], \\
\phi_{cl}(z) &\triangleq \det[zI - (F - GL)].
\end{aligned} \tag{11.45}$$

The two polynomials are connected as follows.

Lemma 11.5 *The relation*

$$\phi_{cl}(z) = \phi_{ol}(z) \det[I + L(zI - F)^{-1}G] \tag{11.46}$$

holds.

Proof Simple calculations give

$$\begin{aligned}
\phi_{cl}(z) &= \det[(zI - F)\{I + (zI - F)^{-1}GL\}] \\
&= \det[zI - F] \det[I + (zI - F)^{-1}GL] \\
&= \phi_{ol}(z) \det[I + L(zI - F)^{-1}G].
\end{aligned}$$

In the last line, Lemma 6.9 was used.   ■

Note that (11.46) does *not* imply that $\phi_{ol}(z)$ is a factor of $\phi_{cl}(z)$. In other words, the open loop poles are not a subset of the closed loop poles.

A more useful result can be derived by assuming that the feedback vector L originates from an LQ(G) problem.

Lemma 11.6 *Assume that*

$$\begin{aligned}
S &= F^TSF + Q_1 - (F^TSG + Q_{12})(G^TSG + Q_2)^{-1} \\
&\quad \times (Q_{21} + G^TSF), \tag{11.47} \\
L &= (G^TSG + Q_2)^{-1}(Q_{21} + G^TSF). \tag{11.48}
\end{aligned}$$

Then

$$[I + L(zI - F)^{-1}G]^T[G^TSG + Q_2][I + L(z^{-1}I - F)^{-1}G]$$
$$\equiv Q_2 + G^T(zI - F^T)^{-1}Q_1(z^{-1}I - F)^{-1}G$$
$$+ G^T(zI - F^T)^{-1}Q_{12} + Q_{21}(z^{-1}I - F)^{-1}G. \qquad (11.49)$$

Proof Straightforward, but somewhat tedious, calculations give

$$[I + L(zI - F)^{-1}G]^T[G^TSG + Q_2][I + L(z^{-1}I - F)^{-1}G]$$
$$\equiv G^TSG + Q_2 + G^T(zI - F^T)^{-1}(Q_{21} + G^TSF)^T$$
$$+ (Q_{21} + G^TSF)(z^{-1}I - F)^{-1}G$$
$$+ G^T(zI - F^T)^{-1}(Q_{21} + G^TSF)^T(G^TSG + Q_2)^{-1}$$
$$\times (Q_{21} + G^TSF)(z^{-1}I - F)^{-1}G$$
$$\equiv Q_2 + G^T(zI - F^T)^{-1}Q_{12} + Q_{21}(z^{-1}I - F)^{-1}G$$
$$+ G^T(zI - F^T)^{-1}\left[(zI - F^T)S(z^{-1}I - F) + F^TS(z^{-1}I - F)\right.$$
$$+ (zI - F^T)SF + Q_1 - S + F^TSF\left.\right](z^{-1}I - F)^{-1}G$$
$$= Q_2 + G^T(zI - F^T)^{-1}Q_{12} + Q_{21}(z^{-1}I - F)^{-1}G$$
$$+ G^T(zI - F^T)^{-1}Q_1(z^{-1}I - F)^{-1}G. \qquad \blacksquare$$

Remark Using the duality results of Section 11.3, it is possible to give an interpretation of (11.49) in the light of Theorem 4.2. The left-hand side is the spectrum of the output derived from the innovation form, whereas the right-hand side is the spectrum from an arbitrary (given) form. Compare with the last part of the proof of Theorem 4.2. □

Lemmas 11.5 and 11.6 can be combined to give insight into how the weighting matrices Q_1, Q_{12} and Q_2 influence the closed loop pole positions. In fact, (11.46) and (11.48) give

$$\det[Q_2 + G^T(zI - F^T)^{-1}Q_1(z^{-1}I - F)^{-1}G$$
$$+ G^T(zI - F^T)^{-1}Q_{12} + Q_{21}(z^{-1}I - F)^{-1}G]$$
$$= \det[G^TSG + Q_2]\frac{\phi_{cl}(z)\phi_{cl}(z^{-1})}{\phi_{ol}(z)\phi_{ol}(z^{-1})}. \qquad (11.50)$$

Note that the left-hand side is given by the system description and the weighting matrices only. Further, the determinant appearing in the right-hand side works simply as a scaling factor.

11.5 Linear Quadratic Gaussian Design by Polynomial Methods

11.5.1 Problem Formulation

Assuming that an LQG criterion as in (11.3) penalizes a weighted sum of output and input variances, it is possible to handle the LQG problem using polynomial methods. This is much in parallel with how polynomial methods were used in Chapter 7 for treating the problems of general state estimation in Chapter 6 when only a particular signal was to be estimated. Further paralleling Chapter 7, we will treat the stationary case. The system will be an ARMAX model

$$A(q)y(t) = B(q)u(t) + C(q)e(t) \tag{11.51}$$

where

$$A(q) = q^n + a_1 q^{n-1} + \ldots + a_n \, ,$$
$$B(q) = b_0 q^m + b_1 q^{m-1} + \ldots + b_m, \qquad b_0 \neq 0 \, , \tag{11.52}$$
$$C(q) = q^n + c_1 q^{n-1} + \ldots + c_n \, ,$$

and

$$k = n - m > 0 \tag{11.53}$$

is the *pole excess*. Assume that $e(t)$ is zero mean white noise of variance λ^2 and that $B(z)$ *and* $C(z)$ have all zeros strictly inside the unit circle. Apparently, there is a delay of k sampling intervals from the input to the output.

The general criterion to be treated is

$$V = \lim_{N \to \infty} \frac{1}{N} \mathbf{E} \sum_{t=0}^{N-1} [y^2(t) + \rho u^2(t)] \, . \tag{11.54}$$

In terms of Section 11.1, this means that

$$\begin{aligned}
Q_1 &= H^T H \, , \\
Q_{12} &= 0 \, , \\
Q_2 &= \rho \, , \\
N &\to \infty \, ,
\end{aligned} \tag{11.55}$$

and that the case of incomplete state information is treated.

The specific case of $\rho = 0$ is often referred to as minimum variance control, and is a bit easier to handle than the general case. We will start by treating this special case, for minimal, finite, and infinite N.

11.5.2 Minimum Variance Control

In treating the minimal variance control, let us first consider the case of only one term in the criterion. Recall that the system (11.51) is assumed to have a delay of k samples. Hence it makes sense to determine the input $u(t)$ as time t so that the output variance at time $t + k$, that is

$$V = \mathbf{E}\, y^2(t + k) \tag{11.56}$$

is minimized.

When dealing with this problem we will, for the time being, assume that the system is minimum phase, that is the polynomial $B(z)$ has all zeros strictly inside the unit circle.

The optimal regulator will be strongly tied to the optimal prediction. In fact, it is known from Section 5.3 that the predictor $\hat{y}(t+k|t)$ and the prediction error $\tilde{y}(t+k) = y(t+k) - \hat{y}(t+k|t)$ are uncorrelated. Hence

$$
\begin{aligned}
V &= \mathbf{E}\left[\hat{y}(t+k|t) + \tilde{y}(t+k)\right]^2 \\
&= \mathbf{E}\left[\hat{y}(t+k|t)\right]^2 + \mathbf{E}\left[\tilde{y}(t+k)\right]^2 \\
&\geq \mathbf{E}\left[\tilde{y}(t+k)\right]^2 .
\end{aligned}
\tag{11.57}
$$

Now, recall Section 7.2.2, where the optimal predictor of an ARMA process was derived. In the present case, it is sufficient to modify the derivation to take care of the (known) contribution from the input signal. Similar to Section 7.2.2, introduce the polynomials

$$
\begin{aligned}
F(z) &= z^{k-1} + f_1 z^{k-2} + \ldots + f_{k-1} , \\
L(z) &= \ell_0 z^{n-1} + \ldots + \ell_{n-1} ,
\end{aligned}
\tag{11.58}
$$

from the predictor identity

$$
z^{k-1}C(z) \equiv A(z)F(z) + L(z) .
\tag{11.59}
$$

Then

$$
\begin{aligned}
y(t+k) &= \frac{B(q)}{A(q)}u(t+k) + \frac{C(q)}{A(q)}e(t+k) \\
&= \frac{B(q)}{A(q)}u(t+k) + \frac{A(q)F(q) + L(q)}{q^{k-1}A(q)}e(t+k) \\
&= \frac{B(q)}{A(q)}u(t+k) + F(q)e(t+1) + \frac{qL(q)}{A(q)}e(t) \\
&= F(q)e(t+1) + \frac{B(q)}{A(q)}u(t+k) \\
&\quad + \frac{qL(q)}{A(q)}\frac{1}{C(q)}\{A(q)y(t) - B(q)u(t)\} \\
&= F(q)e(t+1) + \frac{qL(q)}{C(q)}y(t) + \frac{B(q)}{A(q)C(q)}[q^k C(q) - qL(q)]u(t) \\
&= F(q)e(t+1) + \frac{qL(q)}{C(q)}y(t) + \frac{qB(q)F(q)}{C(q)}u(t) .
\end{aligned}
\tag{11.60}
$$

Noting that

$$
\begin{aligned}
\deg[qL] &= 1 + (n-1) = \deg C , \\
\deg[qBF] &= 1 + m + (k-1) = n = \deg C ,
\end{aligned}
$$

it can be seen that

$$
\begin{aligned}
\hat{y}(t+k|t) &= [qL(q)]/[C(q)]y(t) + [qB(q)F(q)]/[C(q)]u(t) , \\
\tilde{y}(t+k) &= F(q)e(t+1) .
\end{aligned}
\tag{11.61}
$$

It is obvious from (11.57) that the best one can hope for is to obtain the prediction $\hat{y}(t + k|t)$ equal to zero. This idea would give the regulator

$$u(t) = -\frac{L(q)}{B(q)F(q)}y(t) , \qquad (11.62)$$

which is the *minimum variance regulator*.

It is also instructive to compute the closed loop system. In order to facilitate a sensitivity analysis, assume that the regulator design is based on a nominal model (11.51) but that the true open loop system is given by

$$A_0(q)y(t) = B_0(q)u(t) + C_0(q)e(t) , \qquad (11.63)$$

and is subject to the degree conditions (11.52) and (11.53). The closed loop system is readily found from straightforward calculations. By successive simplification,

$$A_0(q)y(t) + \frac{B_0(q)L(q)}{B(q)F(q)}y(t) = C_0(q)e(t) ,$$

$$[A_0(q)B(q)F(q) + B_0(q)L(q)]y(t) = B(q)F(q)C_0(q)e(t) ,$$

$$[\{A_0(q)B(q) - A(q)B_0(q)\}F(q) + q^{k-1}B_0(q)C(q)]y(t)$$
$$= B(q)F(q)C_0(q)e(t) . \qquad (11.64)$$

Hence it can be seen that, when the nominal model coincides with the true system, this gives

$$B(q)C(q)y(t) = B(q)C(q)F(q)e(t - k + 1) . \qquad (11.65)$$

As it was assumed that both $B(z)$ and $C(z)$ have all zeros inside the unit circle, these poles and zeros can be cancelled to give finally

$$y(t) = F(q)e(t - k + 1)$$
$$= e(t) + f_1 e(t - 1) + \ldots + f_{k-1}e(t - k + 1) , \qquad (11.66)$$
$$\mathbf{E}\, y^2(t) = \lambda^2(1 + f_1^2 + \ldots + f_{k-1}^2) .$$

It is worth stressing that the assumption of a minimum phase system is crucial for the derived regulator to be meaningful. Otherwise, if $B(z)$ has any zero outside the unit circle, an unstable closed loop system would result.

Remark 1 The derivation can be generalized to a multivariable system

$$y(t) = G(q)u(t) + H(q)e(t)$$

in innovations form, with a $G(q)$ having a delay of k steps, and being minimum phase, in the sense that $[q^{-k}G(q)]^{-1}$ is asymptotically stable. The key point is still to require that the k-step predictor $\hat{y}(t + k|t)$ vanishes. $\square$

Remark 2 If the system is nonminimum phase, the optimal regulator becomes more complicated. Here, only the result is given, and a formal derivation is delayed until Section 11.5.3, where we treat the more general situation with minimizing a weighted sum of input and output variances. First factorize $B(z)$ as

$$B(z) = B_+(z)B_-(z) , \tag{11.67}$$

where B_+ has all zeros inside and B_- has all zeros outside the unit circle. Let B_- have degree r. Next solve the Diophantine equation

$$A(z)R(z) + B_-(z)P(z) \equiv C(z)B_-(z^{-1})z^{k+r-1} \tag{11.68}$$

with respect to $R(z)$ and $P(z)$, where $\deg R = k + r - 1$, $\deg P = n - 1$. The optimal regulator is then given by

$$u(t) = -\frac{P(q)}{R(q)B_+(q)}y(t) . \tag{11.69}$$

$\square$

The variance criterion that has been minimized is $\mathbf{E}\,y^2(t+k)$ at time t. It gives a "one-step optimal control". It is a remarkable fact that the same regulator also minimizes future output variances, as is now shown.

Lemma 11.7 *Consider the system*

$$A(q)y(t) = B(q)u(t) + C(q)e(t) \tag{11.70}$$

under the previous assumptions. The criteria

$$V_1 = \mathbf{E}\,y^2(t+k) \tag{11.71}$$

$$V_2^{(M)} = \mathbf{E}\,\frac{1}{M}\sum_{s=0}^{M-1} y^2(t+k+s) \tag{11.72}$$

(M a positive integer) are minimized by the same regulator.

Proof The optimal regulator for V_1 was derived previously in this section, with the optimal regulator being given by (11.62). Set

$$Q = \mathbf{E}\,y^2(t+k) = \lambda^2(1 + f_1^2 + \ldots + f_{k-1}^2) .$$

Using the optimality principle of dynamic programming gives

$$\min_{u(t),\ldots,u(t+M-1)} V_2^{(M)} = \min_{u(t),\ldots,u(t+M-2)} \left[\mathbf{E}\,\frac{1}{M}\sum_{s=0}^{M-2} y^2(t+k+s) \right.$$

$$\left. + \min_{u(t+M-1)} \mathbf{E}\,\frac{1}{M}y^2(t+k+M-1) \right]$$

$$= \min_{u(t),\ldots,u(t+M-2)} \left[\frac{M-1}{M}V_2^{(M-1)} + \frac{1}{M}Q \right]$$

$$= \frac{M-1}{M}\min_{u(t),\ldots,u(t+M-2)} V_2^{(M-1)} + \frac{1}{M}Q . \tag{11.73}$$

Note that Q is *completely independent* of the control actions, $u(t) \ldots u(t + M - 2)$. By iterating (11.73), it is found that the minimum variance regulator minimizes both V_1 and $V_2^{(M)}$.   ∎

Remark Note that it is crucial in the above lemma that the output variance *only* is penalized. For a general linear quadratic criterion, the corresponding one-step minimal loss Q *will depend* on previous control actions. Hence, the lemma can no longer hold under such conditions.   □

11.5.3 The General Case

Recall that the system considered is

$$A(q)y(t) = B(q)u(t) + C(q)e(t) \tag{11.74}$$

where we make the assumptions

$\deg A = \deg C = n$, $\deg B = m$, $k = n - m > 0$
$C(z)$ has all zeros strictly inside the unit circle
$A(z)$ and $B(z)$ are coprime (have no common factors)

and the criterion is

$$V = \lim_{N \to \infty} \frac{1}{N} \mathbf{E} \sum_{t=0}^{N-1} [y^2(t) + \rho u^2(t)] . \tag{11.75}$$

We will derive the optimal regulator relying on the state space methodology of Section 11.2.

We first note that the optimal regulator is given by

$$u(t) = -L\hat{x}(t|t) \tag{11.76}$$

according to Theorem 11.3, and due to Lemma 11.3 it is an nth-order linear system from $y(t)$ to $u(t)$. We can hence postulate that the optimal regulator has the form

$$u(t) = -\frac{S(q)}{R(q)}y(t) \tag{11.77}$$

where

$$R(q) = q^n + r_1 q^{n-1} + \ldots + r_n$$
$$S(q) = s_0 q^n + s_1 q^{n-1} + \ldots + s_n \tag{11.78}$$

To find the polynomials $R(q)$ and $S(q)$, we proceed in an indirect fashion. We first utilize the results of Section 11.4.2 to get the closed loop characteristics polynomial. Next, we compare with the polynomial implied by the open loop system (11.51) and the regulator (11.77). Equating the two expressions will eventually lead to a full determination of the regulator polynomials $R(q)$ and $S(q)$.

The first step is hence to characterize the closed loop poles. According to Lemma 11.4 they are the eigenvalues of $F - GL$ and $F - K_\mathrm{p}H$. When the criterion (11.75) is minimized, the closed loop poles are, in fact, independent of the state space representation that would be selected for representating the system in (11.51). Let

$$P(z) = \det(zI - (F - GL)). \tag{11.79}$$

Then the result (11.50) tells us that $P(z)$ can in fact be determined from the following spectral factorization

$$\rho + \frac{B(z)}{A(z)} \frac{B(z^{-1})}{A(z^{-1})} \equiv r \frac{P(z)}{A(z)} \frac{P(z^{-1})}{A(z^{-1})} \tag{11.80}$$

which we rewrite as

$$rP(z)P(z^{-1}) \equiv \rho A(z)A(z^{-1}) + B(z)B(z^{-1}) . \tag{11.81}$$

We make the assumption that

$$a_n = A(0) \neq 0 \tag{11.82}$$

which guarantees that the degree of the polynomial $P(z)$ is indeed n, that is $p_n = P(0) \neq 0$.

The polynomial $P(z)$ gives "half" of the closed loop characteristic polynomial. The characteristic polynomial of $F - K_\mathrm{p}H$ is easy to find. It was derived in Example 7.2 that

$$\det(zI - (F - K_\mathrm{p}H)) = C(z) . \tag{11.83}$$

Next we derive the closed loop system when the system in (11.51) is combined with the regulator in (11.77). We easily get

$$A(q)y(t) = -B(q)\frac{S(q)}{R(q)}y(t) + C(q)e(t)$$

and

$$[A(q)R(q) + B(q)S(q)]y(t) = R(q)C(q)e(t) . \tag{11.84}$$

Summing up the expressions derived for the closed loop poles, we thus have

$$A(z)R(z) + B(z)S(z) \equiv C(z)P(z) . \tag{11.85}$$

Note that this is a Diophantine equation of a similar type to (7.65). When converting (11.85) to a linear system of equations, there will be $2n$ equations, as both sides are polynomials of degree $2n$. The number of unknowns is, however, $2n + 1$ (n coefficients for $R(z)$ and $n + 1$ coefficients for $S(z)$). We thus have to find one more condition for determining the regulator. To do so, we make use of the following result.

Lemma 11.8 *Assume that $a_n = A(0) \neq 0$. Then it holds that $s_n = S(0) = 0$.*

Proof We will have to show that the transfer function (11.77) of the regulator has a zero in $z = 0$.

According to Lemma 11.3, the regulator can be represented in state space form with $R_{12} = 0$, leading to $K_p = FK_f$, see (6.35), as

$$z(t + 1) = F_r z(t) + G_r y(t)$$
$$u(t) = -H_r z(t) - J_r y(t) ,$$

(11.86)

where

$$F_r = F - GL + GLK_f H + K_p H = (F - GL)(I - K_f H) ,$$
$$G_r = K_p - GLK_f = (F - GL)K_f ,$$
$$H_r = L - LK_f H = L(I - K_f H) ,$$
$$J_r = LK_f .$$

(11.87)

To prove the lemma it is enough to examine the transfer function for argument $z = 0$:

$$\begin{aligned}
G_r(0) &= [J_r + H_r(zI - F_r)^{-1}G_r]_{|z=0} \\
&= J_r - H_r F_r^{-1} G_r \\
&= LK_f - L(I - K_f H)[(F - GL)(I - K_f H)]^{-1} \\
&\quad \times (F - GL)K_f = 0 ,
\end{aligned}$$

(11.88)

which ends the proof.   ∎

To sum up, we have found that in the case $a_n \neq 0$, we can conclude that $s_n = 0$. In this case the Diophantine equation has a unique solution with the given degrees.

Remark If $a_n = 0$, the computation of the optimal regulator gets more complicated and requires the solution of two coupled Diophantine equations.
                                                                         □

To summarize so far, the optimal regulator is found by solving the Diophantine equation (11.85), in the case $a_n \neq 0$. The polynomial $P(z)$ is first to be determined by solving the spectral factorization (11.81).

As illustrations of the procedure for determining the optimal regulator, let us reconsider the minimal variance regulator of Section 11.5.2.

Example 11.1 Consider first the case of a minimum phase system. Then we can conclude from (11.81) that $P(z)$ coincides with $B(z)$, save for a normalizing factor, and a factor z^{n-m} that adjusts the degree. Hence

$$P(z) = B(z)z^{n-m}/b_0 .$$

The Diophantine equation (11.85) gives

$$A(z)R(z) + B(z)S(z) \equiv C(z)B(z)z^{n-m}/b_0 .$$

As the second term and the right-hand side contain the factor $B(z)$, and this factor has no common zero with $A(z)$, we conclude that $B(z)$ must be a factor of $R(z)$. Hence we can write

$$R(z) = B(z)R_1(z)/b_0 ,$$

where the division with b_0 is made to keep the coefficient of z^n equal to one, and the degree of $R_1(z)$ is $k = n - m$. Inserting this into the Diophantine equation and cancelling the factor $B(z)/b_0$ gives

$$A(z)R_1(z) + b_0 S(z) \equiv C(z)z^{n-m} .$$

As $s_n = 0$, we can write $S(z) = S_1(z)z$ with $\deg S_1(z) = n - 1$. Then it follows that a factor z can be cancelled, so that $R_1(z) = zR_2(z)$, with $\deg R_2(z) = k - 1$. After cancelling the factor z, the Diophantine equation becomes

$$AR_2(z) + b_0 S_1(z) \equiv C(z)z^{k-1} ,$$

which coincides with the predictor identity (11.59).   □

Example 11.2 Next we consider the case with an arbitrary $B(z)$ polynomial, which we split as

$$B(z) = B_+(z)B_-(z) ,$$

where B_+ has all zeros inside, and B_- all zeros outside the unit circle. Let B_- have degree ℓ. The spectral factorization (11.81) gives in this case

$$P(z) = B_+(z)z^\ell B_-(z^{-1})z^k/b_0 ,$$

and the Diophantine equation becomes

$$A(z)R(z) + B_+(z)B_-(z)S(z) \equiv B_+(z)B_-(z^{-1})z^{k+\ell}C(z)/b_0 .$$

As before, we have $S(z) = zS_1(z)$ with $\deg S_1(z) = n - 1$. In this case we conclude that $zB_+(z)$ must be a factor of $R(z)$. Hence

$$R(z) = zB_+(z)R_1(z)/b_0 ,$$

where $\deg R_1(z) = n - 1 - (m - \ell) = k + \ell - 1$. The adjusted Diophantine equation becomes

$$A(z)R_1(z) + B_-(z)S(z)/b_0 \equiv B_-(z^{-1})C(z)z^{k+\ell-1} ,$$

which is precisely what we achieved with somewhat other notations in (11.68).   □

Example 11.3 As a further illustration, we consider minimizing

$$J = \mathbf{E}\, y^2(t) + 2\mathbf{E}\, u^2(t)$$

for the first-order system

$$y(t) - y(t-1) = u(t-1) + e(t) + ce(t-1)\,, \qquad |c| < 1\,.$$

We first solve the problem of using the polynomial formalism. The spectral factorization (11.81) becomes

$$r(z+p)(z^{-1}+p) \equiv 2(z-1)(z^{-1}-1) + 1$$

leading to

$$z^0 : r(1+p^2) = 5$$
$$z^1 : \qquad rp = -2\,,$$

with the solution $r = 4$, $p = -0.5$. The Diophantine equation (11.85) will hence be

$$(z-1)(z+r_1) + 1 \times s_0 z \equiv (z-0.5)(z+c)\,,$$

leading to

$$z^0 : \qquad -r_1 = -0.5c$$
$$z^1 : -1 + r_1 + s_0 = -0.5 + c\,,$$

with the solution $r_1 = 0.5c$, $s_0 = 0.5(1+c)$. Hence the optimal regulator will be

$$u(t) = -\frac{0.5(1+c)q}{q+0.5c}y(t)\,.$$

For illustration, the problem is also treated using the state space formalism. As there will be no delay in the regulator, it is convenient to choose a state space model with $R_{12} = 0$; compare with Theorem 11.3. One such possibility is

$$x(t+1) = \begin{pmatrix} 1 & 1 \\ 0 & 0 \end{pmatrix} x(t) + \begin{pmatrix} 1 \\ 0 \end{pmatrix} u(t) + v(t)\,,$$
$$y(t) = \begin{pmatrix} 1 & 0 \end{pmatrix} x(t)\,,$$

with

$$R_1 = \begin{pmatrix} 1 & c \\ c & c^2 \end{pmatrix} \lambda^2\,, \qquad R_2 = 0\,, \qquad R_{12} = 0\,,$$

$$Q_1 = \begin{pmatrix} 1 & 0 \\ 0 & 0 \end{pmatrix}\,, \quad Q_2 = 2\,.$$

Set

$$P = \begin{pmatrix} p_{11} & p_{12} \\ p_{12} & p_{22} \end{pmatrix}\,.$$

The Riccati equation for P becomes

$$P = R_1 + F[P - PH^T(HPH^T)^{-1}HP]F^T .$$

Here,

$$P - PH^T(HPH^T)^{-1}HP]$$

$$= \begin{pmatrix} p_{11} & p_{12} \\ p_{12} & p_{22} \end{pmatrix} - \begin{pmatrix} p_{11} \\ p_{12} \end{pmatrix} \begin{pmatrix} p_{11} & p_{12} \end{pmatrix} /p_{11}$$

$$= \begin{pmatrix} 0 & 0 \\ 0 & 1 \end{pmatrix} \mu ,$$

where

$$\mu = p_{22} - \frac{p_{12}^2}{p_{11}} .$$

The Riccati equation gives

$$P = R_1 + \begin{pmatrix} 1 & 1 \\ 0 & 0 \end{pmatrix} \begin{pmatrix} 0 & 0 \\ 0 & 1 \end{pmatrix} \mu \begin{pmatrix} 1 & 0 \\ 1 & 0 \end{pmatrix} = R_1 + \begin{pmatrix} \mu & 0 \\ 0 & 0 \end{pmatrix} ,$$

and one finds that

$$\mu = \lambda^2 c^2 - \frac{(\lambda^2 c)^2}{\lambda^2 + \mu} ,$$

which has solutions $\mu_1 = 0$ and $\mu_2 = c^2 - 1$. As $\mu_2 < 0$, we conclude that the pertinent solution is $\mu = 0$, and

$$P = \lambda^2 \begin{pmatrix} 1 & c \\ c & c^2 \end{pmatrix} ,$$

$$K_f = \begin{pmatrix} p_{11} \\ p_{12} \end{pmatrix} \frac{1}{p_{11}} = \begin{pmatrix} 1 \\ c \end{pmatrix} ,$$

$$K_p = FK_f = \begin{pmatrix} c+1 \\ 0 \end{pmatrix} .$$

The Riccati equation for S is solved in a similar way:

$$S = Q_1 + F^T[S - SG(G^T SG + Q_2)^{-1}G^T S]F$$

$$= \begin{pmatrix} 1 & 0 \\ 0 & 0 \end{pmatrix} + \begin{pmatrix} 1 & 0 \\ 1 & 0 \end{pmatrix} \left[\begin{pmatrix} s_{11} & s_{12} \\ s_{12} & s_{22} \end{pmatrix} - \right.$$

$$\left. - \begin{pmatrix} s_{11} \\ s_{12} \end{pmatrix} \begin{pmatrix} s_{11} & s_{12} \end{pmatrix} \frac{1}{s_{11}+2} \right] \begin{pmatrix} 1 & 1 \\ 0 & 0 \end{pmatrix}$$

$$= \begin{pmatrix} 1 & 0 \\ 0 & 0 \end{pmatrix} + \begin{pmatrix} 1 & 1 \\ 1 & 1 \end{pmatrix} \beta ,$$

with

$$\beta = s_{11} - \frac{s_{11}^2}{s_{11}+2} = \frac{2s_{11}}{s_{11}+2} .$$

The scalar β is next to be found:

$$\beta = \frac{2(1+\beta)}{1+\beta+2},$$

with solutions $\beta_1 = 1$ and $\beta_2 = -2$. As $\beta_1 > 0$, it is the relevant solution, and one gets

$$S = \begin{pmatrix} 2 & 1 \\ 1 & 1 \end{pmatrix}, \qquad L = \frac{1}{2}\begin{pmatrix} 1 & 1 \end{pmatrix}.$$

We finally apply (11.37) of Lemma 11.3 to get the optimal feedback. In this case we have

$$LK_f = \frac{1}{2}(1+c),$$

$$L(I - K_f H) = \frac{1}{2}\begin{pmatrix} 1 & 1 \end{pmatrix} - \frac{1}{2}(1+c)\begin{pmatrix} 1 & 0 \end{pmatrix}$$
$$= \frac{1}{2}\begin{pmatrix} -c & 1 \end{pmatrix},$$

$$F - GL - (K_p + GLK_f)H = (F - GL)(I - K_f H)$$
$$= \begin{pmatrix} 1/2 & 1/2 \\ 0 & 0 \end{pmatrix}\begin{pmatrix} 0 & 0 \\ -c & 1 \end{pmatrix}$$
$$= \frac{1}{2}\begin{pmatrix} -c & 1 \\ 0 & 0 \end{pmatrix},$$

$$K_p - GLK_f = (F - GL)K_f$$
$$= \begin{pmatrix} 1/2 & 1/2 \\ 0 & 0 \end{pmatrix}\begin{pmatrix} 1 \\ c \end{pmatrix}$$
$$= \frac{1}{2}\begin{pmatrix} 1+c \\ 0 \end{pmatrix}.$$

From the above expressions we can form the optimal regulator as

$$G(z) = \frac{1}{2}(1+c) + \frac{1}{2}\begin{pmatrix} -c & 1 \end{pmatrix}\begin{pmatrix} z + c/2 & -1/2 \\ 0 & z \end{pmatrix}^{-1}\begin{pmatrix} 1/2(1+c) \\ 0 \end{pmatrix}$$
$$= \frac{1}{2}(1+c)\left[1 + \frac{1}{2}\begin{pmatrix} -c & 1 \end{pmatrix}\begin{pmatrix} z \\ 0 \end{pmatrix}\frac{1}{z(z+c/2)}\right]$$
$$= \frac{1}{2}(1+c)\frac{z + c/2 - c/2}{z + c/2}$$
$$= \frac{1}{2}(1+c)\frac{z}{z + c/2},$$

which coincides with our previous finding. $\square$

11.6 Controller Design by Linear Quadratic Gaussian Theory

11.6.1 Introduction

Some aspects of the use of LQG theory for the design of a regulator are presented in this section.

It is seldom that the performance index is *a priori* given for the plant to be controlled. Instead, the weighting matrices Q_0, Q_1 and Q_2, as well as N, the horizon, which together shape up the performance index, should be regarded as *user design or tuning variables*. There are several design rules for the selection of these user variables.

- The absolute values of Q_0, Q_1 and Q_2 are not of importance, but their relative values are. In particular, the choices $(Q_0; Q_1; Q_2)$ and $(\alpha Q_0; \alpha Q_1; \alpha Q_2)$, with α an arbitrary scalar, give the same feedback vector L. As a consequence, one of the matrix elements is often set to a fixed value without causing any restriction.
- In most cases the asymptotic case of $N \to \infty$ is considered. Then Q_0 becomes obsolete, as the optimal gain L depends on the solution to the ARE (11.47).
- In the case where $N \to \infty$, the optimal regulator can be written as a linear time-invariant feedback. It can be interpreted as a form of *pole-placement algorithm*. The positions of the closed loop poles are affected by the choice of the weighting matrices Q_1 and Q_2.
- The weights Q_1 and Q_2 can be selected by a more pragmatic procedure. For ease of discussion, assume that the system has one input and that Q_1 is chosen diagonal. Then the stationary criterion can be written as

$$V = \sum_{i=1}^{n} Q_{1i,i} \mathbf{E}\, x_i^2(t) + Q_2 \mathbf{E}\, u^2(t) \,.$$

Initially, $Q_{1i,i}^{-1}$ and Q_2^{-1} should be chosen as the acceptable variances of $x_i(t)$ and $u(t)$ respectively. With the controller so designed, the variances $\mathbf{E}\, x_i^2(t)$ and $\mathbf{E}\, u^2(t)$ are evaluated for the closed loop system. If any variance is deemed to be too large, its corresponding weight in the criterion is increased and the design procedure is repeated.
- A special case of applying LQG is called generalized predictive control (GPC). One then has a finite N (often quite small) in the criterion. The feedback gain computed for the *initial* time is then used throughout; that is, the regulator will be

$$u(t) = -L(t_0)\hat{x}(t|t) \,, \qquad \text{all } t \geq t_0 \,,$$

where the gain $L(t_0)$ is computed from (11.6) and (11.7), and $\hat{x}(t|t)$ denotes the optimal state estimate given by (6.37). The GPC is often described in a slightly different fashion, with a receding horizon of the criterion. Assume that at time t_0 an LQG problem is solved for the interval $(t_0, t_0 + N)$. It results in an input sequence $u(t_0), \ldots u(t_0 + N - 1)$. Only $u(t_0)$ is applied. The procedure is repeated for the interval $(t_0 + 1, t_0 + N + 1)$, and $u(t_0 + 1)$ is determined. The procedure is repeated, shifting the interval one unit at a time. Assuming that the system, as well as the criterion, is time-invariant, this leads to the stated controller.

- The previous procedure can be extended to have different numbers of terms of $x^T Q_1 x$ and $u^T Q_2 u$ in the criterion. It is also possible to introduce constraints, for example requiring that the input $u(t)$ as well as the state variables are restricted to lie in specified intervals. The resulting controller will then become nonlinear, though. With such extensions the procedure is often called model predictive control (MPC).

As illustrations of the above design rules, three examples are offered. They are all based on simple dynamic models, but illustrate the impact of the weighting matrices.

Example 11.4 As a simple illustration of the effect of the weighting matrices Q_1 and Q_2, consider a sampled double integrator

$$x(t + 1) = \begin{pmatrix} 1 & h \\ 0 & 1 \end{pmatrix} x(t) + \begin{pmatrix} h^2/2 \\ h \end{pmatrix} u(t) + v(t) ,$$

$$y(t) = (1 \quad 0) x(t) + e(t) .$$

The process noise $v(t)$ is assumed to have the covariance matrix

$$R_1 = \mathbf{E}\, v(t) v^T(t) = r_v \begin{pmatrix} h^2/3 & h^2/2 \\ h^2/2 & h \end{pmatrix} .$$

The numerical values used are $h = 0.5$, $r_v = 10^{-3}$, and $r_e = \mathbf{E}\, e^2(t) = 10^{-3}$. The system is controlled by the regulator

$$u(t) = -L\hat{x}(t|t) + Mr(t) ,$$

where $r(t)$ is an external reference signal, L is the optimal feedback gain, and M is designed to give a unity static gain from $r(t)$ to $y(t)$. See Example 11.6 for details on the design of M.

Some different weighting matrices Q_1 and Q_2 are applied, and the closed loop system is simulated for $r(t)$ being a square wave. As initial values, $x(0) = (1 \quad 1)^T$, $\hat{x}(0|-1) = (0 \quad 0)^T$ are chosen. The results obtained are shown in Figure 11.1.

According to Lemma 11.4.2, the closed loop poles are determined by the eigenvalues of $F - GL$ ("the controller poles") and the eigenvalues of $F - KH$ ("the estimator poles"). The estimator poles do not vary with Q_1 and Q_2, and are, in this example, given by $z = 0.60 \pm i0.27$.

In case (a) in Figure 11.1, the weighting matrices are

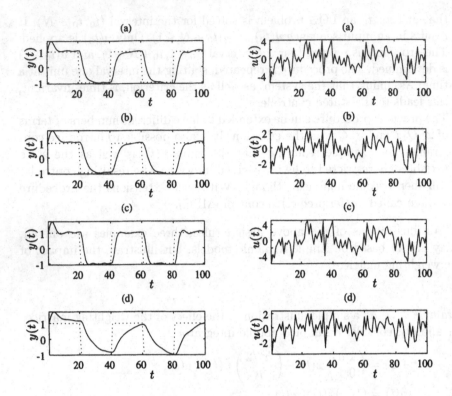

Fig. 11.1. Closed loop behaviour for some different cases, Example 11.4. *Left*: output signal (*solid lines*) and reference signal (*dotted lines*). *Right*: input signal

$$Q_1 = \begin{pmatrix} 1 & 0 \\ 0 & 1 \end{pmatrix}, \qquad Q_2 = 0.1,$$

giving controller poles in $z = 0.27$ and $z = 0.58$.

In case (b), the weighting of the input variable is increased and now

$$Q_1 = \begin{pmatrix} 1 & 0 \\ 0 & 1 \end{pmatrix}, \qquad Q_2 = 1,$$

giving controller poles in $z = 0.63 \pm i0.16$. Compared with case (a), the increased penalty on u has given a smaller control variable and a somewhat slower step response.

In case (c), the weighting of the first state (the output variable) is increased:

$$Q_1 = \begin{pmatrix} 10 & 0 \\ 0 & 1 \end{pmatrix}, \qquad Q_2 = 1,$$

leading to controller poles in $z = 0.43 \pm i0.28$. It can easily be seen in the diagram that the increased penalty on $y(t)$ has resulted in a faster step response. The closed loop poles have also moved closer to the origin.

Finally, in case (d), the weighting on the second state (the derivative of the output) has been increased compared with case (b):

$$Q_1 = \begin{pmatrix} 1 & 0 \\ 0 & 10 \end{pmatrix}, \qquad Q_2 = 1 ,$$

giving controller poles in $z = 0.24$ and $z = 0.85$. The behaviour of the system is, of course, not satisfactory in this case. The plots show clearly, however, that a large penalty of $\ddot{y}$ leads to outputs that change their trend (*i.e.* $\dot{y}$) very slowly.
$\qquad\qquad\qquad\qquad\qquad\qquad\qquad\qquad\qquad\qquad\qquad\qquad\qquad\qquad\qquad\qquad\quad$ □

Example 11.5 The purpose of this example is to illustrate the impact of the weighting matrices Q_1 and Q_2 on the variances of $u(t)$ and $y(t)$. Consider the system

$$x(t + 1) = \begin{pmatrix} 1.5 & 1 \\ -0.7 & 0 \end{pmatrix} x(t) + \begin{pmatrix} 1.0 \\ 0.5 \end{pmatrix} u(t) + v(t) ,$$
$$y(t) = (1 \ \ 0)x(t) ,$$

where

$$\mathbf{E}\, v(t)v^T(t) = \begin{pmatrix} 1 & 0 \\ 0 & 0 \end{pmatrix} .$$

The weighting matrices are chosen as

$$Q_1 = \begin{pmatrix} 1 & 0 \\ 0 & 0 \end{pmatrix}, \qquad Q_2 = \rho ,$$

so the criterion to be minimized is in fact

$$J = \mathbf{E}\,[y^2(t) + \rho u^2(t)] .$$

The relative weight ρ is varied. The specific value $\rho = 0$ corresponds to so-called minimal variance control, which was analyzed in Section 11.5.2. It can be shown that, in this case, $\mathbf{E}\,y^2(t) = 1$, $\mathbf{E}\,u^2(t) = 5.05$. The other extreme case, $\rho \to \infty$, corresponds to open loop operation, which in this example gives $\mathbf{E}\,y^2(t) = 8.85$, $\mathbf{E}\,u^2(t) = 0$.

The optimal regulator, $u(t) = -L\hat{x}(t|t)$, is computed for a number of ρ-values. The closed loop system has the form

$$\begin{pmatrix} x(t+1) \\ \tilde{x}(t+1) \end{pmatrix} = \begin{pmatrix} F - GL & GL(I - K_fH) \\ 0 & F - K_pH \end{pmatrix} \begin{pmatrix} x(t) \\ \tilde{x}(t) \end{pmatrix}$$
$$+ \begin{pmatrix} I \\ I \end{pmatrix} v(t) ,$$

$$\begin{pmatrix} y(t) \\ u(t) \end{pmatrix} = \begin{pmatrix} H & 0 \\ -L & L - LK_fH \end{pmatrix} \begin{pmatrix} x(t) \\ \tilde{x}(t) \end{pmatrix} ,$$

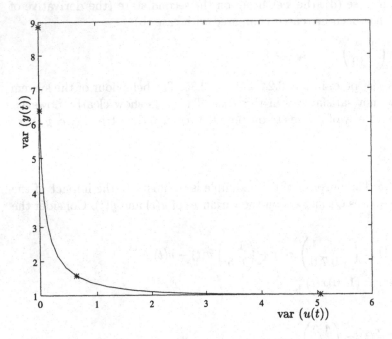

Fig. 11.2. Output variance versus input variance, Example 11.5, with ρ as a parameter. The values obtained for $\rho = 10^{-6}, 1, 10^2$ and 10^6 are marked

where K_f is the filter gain and $K_p = FK_f$ the predictor gain. The input and output variances can be found by first computing the covariance matrix of the extended state vector $(x^T(t) \quad \tilde{x}^T(t))^T$, solving a Lyapunov equation.

The closed loop system was simulated as well for some values of ρ. Some numerical results are shown in Figures 11.2 and 11.3.   □

Example 11.6 In practice, it is often important to include integral action in a regulator, at least if the process to be controlled is not integrating. This can be done in several ways. This example illustrates one way in which it can be used in order to avoid stationary errors when tracking a reference signal.

In order to simplify the illustration, a deterministic problem is treated, thus avoiding random effects. Consider a sampled representation of the harmonic oscillator given by the transfer function $G(s) = \omega_0^2/(s^2 + \omega_0^2)$. Using y and $\dot{y}$, scaled by ω_0^2, as state variables leads to the sampled-data model

$$x(t+1) = \begin{pmatrix} \cos \omega_0 h & (\sin(\omega_0 h))/\omega_0 \\ -\omega_0 \sin \omega_0 h & \cos \omega_0 h \end{pmatrix} x(t)$$
$$+ \begin{pmatrix} (1 - \cos \omega_0 h)/\omega_0^2 \\ (\sin(\omega_0 h))/\omega_0 \end{pmatrix} u(t),$$
$$y(t) = (\omega_0^2 \quad 0)x(t).$$

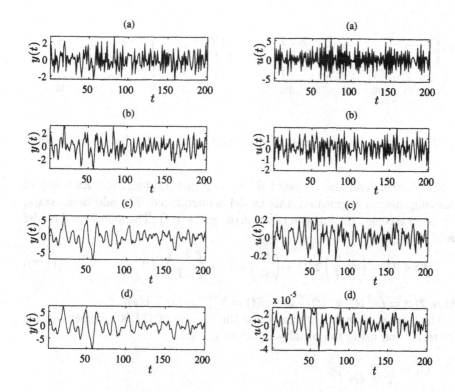

Fig. 11.3. Input and output realizations, Example 11.5, for some values of ρ: **(a)** $\rho = 10^{-6}$, **(b)** $\rho = 1$, **(c)** $\rho = 10^2$, **(d)** $\rho = 10^6$

For the numerical values $\omega_0 = 1$, $h = 0.5$, it is found that the penalty matrices $Q_1 = I$, $Q_2 = 1$ give a satisfactory result.

Consider the case when a reference signal $r(t)$ should be followed without any static error. Assume that the feedback law is set as

$$u(t) = -Lx(t) + Mr(t) .$$ (11.89)

The closed loop system becomes

$$x(t+1) = (F - GL)x(t) + GMr(t) ,$$
$$y(t) = Hx(t) .$$

The static gain from $r(t)$ to $y(t)$ is apparently

$$H(I - F + GL)^{-1}GM ,$$

and hence the appropriate value of M is

$$M = [H(I - F + GL)^{-1}G]^{-1} .$$

The input and output responses of the system using this feedback law are shown in Figure 11.4.

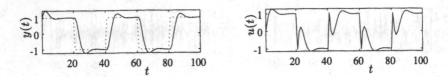

Fig. 11.4. Closed loop response, Example 11.6, using the feedback law (11.89)

Static errors can also be avoided by using integral feedback. As a way of achieving this, an augmented state model is introduced. The additional state, $z(t)$, is, in essence, the integral of the error $y(t) - r(t)$. The augmented model reads

$$\overline{x}(t+1) = \begin{pmatrix} F & 0 \\ H & I \end{pmatrix} \overline{x}(t) + \begin{pmatrix} G \\ 0 \end{pmatrix} u(t) + \begin{pmatrix} 0 \\ -I \end{pmatrix} r(t) , \qquad (11.90)$$

where $\overline{x}(t) = (x^T(t) \quad z^T(t))^T$ and $z(t) = \sum_{s=0}^{t-1} [y(s) - r(s)]$.

The design of an LQ regulator for the system of (11.90), neglecting the $r(t)$ term, and using the penalty matrices Q_2 and

$$\overline{Q_1} = \begin{pmatrix} Q_1 & 0 \\ 0 & Q_3 \end{pmatrix} ,$$

will result in a linear feedback law

$$u(t) = -\overline{L}\overline{x}(t)$$

$$\triangleq -L_1 x(t) - L_2 z(t)$$

$$= -L_1 x(t) - L_2 \sum_{s=0}^{t-1} [y(s) - r(s)] . \qquad (11.91)$$

The size of the penalty matrix Q_3 will affect the strength of the integrating term in the controller.

To obtain a faster servo response on changes of the reference signal, the feedback can be modified to include a direct term from $r(t)$ as well. It is reasonable to use the regulator

$$u(t) = -L_1 x(t) - L_2 \sum_{s=0}^{t-1} [y(s) - r(s)] + L_2 r(t) .$$

The behaviour of the system for some various values of Q_3 is displayed in Figure 11.5.

It can clearly be seen from the diagram that a small value of Q_3 results in a very slow system. When Q_3 is increased, it becomes increasingly "important" to minimize the stationary error, and the response becomes faster.

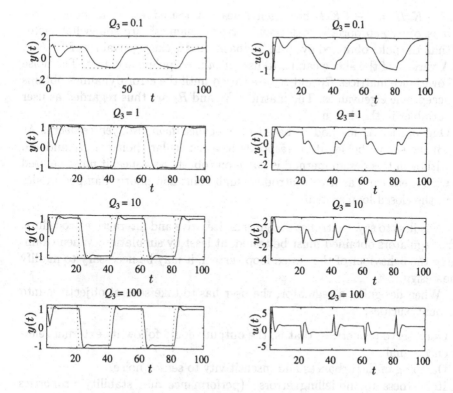

Fig. 11.5. Closed loop response, Example 11.6, using the state feedback with integral action (11.91). *Left*: output signal (*solid lines*) and reference signal (*dotted lines*). *Right*: input signal

When Q_3 is very large, the input tends to be quite spiky. Note that, in contrast to the regulator (11.89), no explicit information about the static gain is "given" to the controller in this case.   □

11.6.2 Choice of Observer Poles

In this section, some aspects of the choice of observer poles will be considered. As is known from Lemma 11.4, the observer or state estimator introduces additional states in the closed loop system. The associated dynamics is characterized by the eigenvalues of the matrix $F - K_p H$.

In many cases, the "true" noise covariance matrices R_1 and R_2 may be unknown. There are several ways to use the LQG theory in such cases for regulator design also, and still retain the general regulator structure described in Section 11.4.1.

- One possibility would be to substitute the Kalman filter approach with a direct pole-placement technique. Then the gain K_f is chosen such that

$F - K_p H = F - F K_f H$ has eigenvalues in a desired region. In loose terms, it is often appropriate that these "error dynamics" are somewhat faster than the poles obtained by state feedback, that is the eigenvalues of $F - GL$.

- A variant of the above can be to use an *indirect* pole placement. The noise covariance matrices R_1 and R_2 are varied until the error dynamics obtains acceptable eigenvalues. The matrices R_1 and R_2 are thus regarded as user variables in the design.

- One particular method is based on so-called *loop transfer recovery*. As will be shown below, it corresponds to a particular choice of R_1 and R_2. Although this design method is based on robustness issues, it must be used with great care, as it can introduce high gains and poorly damped modes in the closed loop system.

Needless to say, controller design is an iterative and interactive procedure. The regulators obtained must be tested, at least by simulation. When examining the behaviour of the closed loop system, it may be necessary to modify the design.

When designing a regulator, the user has to take several objectives into account, such as:

- Good servo properties; that is, the output should follow its external reference signal.
- Damping of disturbances and insensitivity to sensor noise.
- Robustness to modelling errors (performance and stability properties should be insensitive to variations in the underlying nominal model).

To be more specific, consider the control system of Figure 11.6. To simplify and focus on the system properties, we let $u(t)$ be a feedback of the control error, $u(t) = -G_R(q)[y(t) - r(t)]$, in Figure 11.6. For LQG design, one would rather have $u(t) = -G_R(q)y(t) - G_F(q)r(t)$. However, the simplified form of Figure 11.6 can be used when studying system properties such as those listed in points 1–4 below.

The *loop gain* is defined as

$$G_0(q) = G_S(q)G_R(q) . \tag{11.92}$$

In case the system is single-input, single-output the loop gain can also be written as

$$G_0(q) = G_R(q)G_S(q) . \tag{11.93}$$

In the multivariable case, equations (11.92) and (11.93) differ in general, but both definitions are useful. In what follows next the form of (11.92) will be used. The closed loop system is readily found to be

$$y(t) = [I + G_0(q)]^{-1}v(t) + [I + G_0(q)]^{-1}G_0(q)r(t) . \tag{11.94}$$

The control objectives can now be phrased in terms of the loop gain.

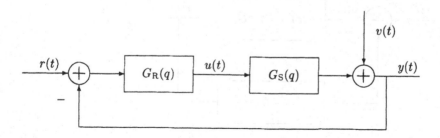

Fig. 11.6. Block diagram of a control system, $G_S(q)$ = system to be controlled, $G_R(q)$ = regulator; $u(t)$ = input signal, $y(t)$ = output signal, $v(t)$ = disturbance, $r(t)$ = reference signal

1. *Servo properties.* The transfer function from $r(t)$ to $y(t)$ should be close to I. This implies that

$$[I + G_0(q)]^{-1} G_0(q) - I = -[I + G_0(q)]^{-1}$$

must be small. This should be interpreted as

$$[I + G_0(e^{i\omega})]^{-1} \text{ small} \tag{11.95}$$

for the dominating frequency band of $r(t)$.
2. *Damping of disturbances.* The transfer function from $v(t)$ to $y(t)$ should be small. This again leads to the condition in (11.95), but now for the dominating frequency band of $v(t)$.
3. *Stability robustness.* For stability of the closed loop system, $G_0(z) + I$ must not have zeros outside the unit circle. This is often examined by drawing and inspecting the Nyquist curve; that is, plotting the loop gain $G_0(e^{i\omega})$ as a function of ω in the complex plane. Assuming that the open loop system is stable, the requirement is that the Nyquist curve must not encircle the point $z = -1$. (This is a simplified formulation; for unstable open loop systems the criterion is more involved. For multivariable systems, one considers $\det(I + G_0(e^{i\omega}))$.) As the open loop system may include some unmodelled dynamics, it is relevant to require that the Nyquist curve stay well away from $z = -1$. In continuous-time LQ problems it always holds that $|G_0(e^{i\omega}) + 1| \geq 1$. Such a condition is no longer guaranteed when an observer is included, or a discrete-time LQ(G) problem is treated. It illustrates, however, an attractive form of stability robustness.
4. *Sensitivity.* Consider a single-input, single-output system. The closed loop transfer function is

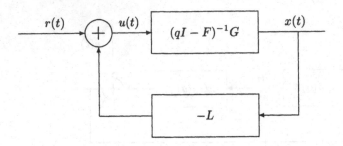

Fig. 11.7. State feedback of a deterministic system

$$G_c(q) = \frac{G_R(q)G_S(q)}{1 + G_R(q)G_S(q)} .$$

Assume that the open loop system $G_S(q)$ is uncertain or includes unmodelled dynamics. It is then relevant to require that the following sensitivity function is small:

$$S(q) = \frac{dG_c(q)/G_c(q)}{dG_S(q)/G_S(q)} = \frac{1}{1 + G_0(q)} . \tag{11.96}$$

Again, such a condition should be interpreted at the appropriate frequency band.

Objectives such as those above can be contradictory. It is good practice to characterize the behaviour of the closed loop system by the frequency properties of the loop gain $G_0(e^{i\omega})$. Roughly speaking, one normally requires the loop gain to be high for low frequencies in order to achieve insensitivity against modelling errors and good servo properties, and low for high frequencies to dampen noise and broadband disturbances, and to obtain stability robustness with respect to unmodelled high-frequency dynamics.

Consider now a deterministic linear system

$$x(t+1) = Fx(t) + Gu(t) , \tag{11.97}$$

controlled by a linear state feedback

$$u(t) = -Lx(t) + r(t) , \tag{11.98}$$

as depicted in Figure 11.7.

The feedback (11.98) is assumed to be chosen so that the closed loop system has good properties. The feedback vector L may be designed by using linear quadratic theory or applying a pole-placement technique. The loop gain (breaking the loop at u) is

$$G_0(q) = L(qI - F)^{-1}G .$$ (11.99)

In many circumstances, all state variables cannot be measured. If the state feedback (11.98) is replaced by feedback from reconstructed states, using an observer or LQG theory, the performance will in general become worse. The idea with *loop transfer recovery* (LTR) is to design the state estimator so that the loop gain G_0 is recovered to a large extent when exact states are not available. It turns out, as will be seen, that for discrete-time systems, *exact* loop transfer recovery is possible under some assumptions. The Kalman filter theory will be used as a method of designing a state estimator.

Let the output be given by

$$y(t) = Hx(t) ,$$ (11.100)

so the transfer function from u to y becomes

$$G_S(q) \overset{\triangle}{=} H(qI - F)^{-1}G .$$ (11.101)

Now make the following assumptions:

- The system has the same number of inputs and outputs, and the matrix HG is invertible. (The case of an additional number of outputs can be handled.)
- The system is minimum phase, meaning that $G_S^{-1}(q)$ is asymptotically stable.

Next, consider a regulator of the form

$$u(t) = -L\hat{x}(t|t) ,$$ (11.102)

where $\hat{x}(t|t)$ is the mean square optimal filter estimate designed as developed in Chapter 6. The regulator can now be described in state space form as (see (11.38)), K denoting the *filter* gain:

$$\hat{x}(t+1|t) = (F - GL)(I - KH)\hat{x}(t|t - 1) + (F - GL)Ky(t) ,$$
$$u(t) = -L(I - KH)\hat{x}(t|t - 1) - LKy(t) .$$ (11.103)

By using the matrix inversion lemma (Lemma 6.7), one can find the transfer function of the regulator. The regulator will be $u(t) = -G_R(q)y(t)$, where

$$\begin{aligned} G_R(q) &= LK + L(I - KH)[qI - (F - GL)(I - KH)]^{-1}(F - GL)K \\ &= L\{I + (I - KH)[qI - (F - GL)(I - KH)]^{-1}(F - GL)\}K \\ &= L\{I - (I - KH)q^{-1}(F - GL)\}^{-1}K \\ &= qL[qI - (I - KH)(F - GL)]^{-1}K . \end{aligned}$$ (11.104)

The loop gain for the system will now be $G_R(q)G_S(q)$ (see Figure 11.8).

Now make the following *specific choice* for designing the state estimator:

$$R_1 = GG^T , \qquad R_2 = 0 .$$ (11.105)

The ARE becomes

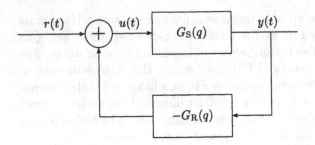

Fig. 11.8. Loop gain for the system with the regulator (11.104)

$$P = GG^T + F[P - PH^T(HPH^T)^{-1}HP]F^T ,$$ (11.106)

which, under the given assumptions, has the solution

$$P = GG^T .$$ (11.107)

The corresponding filter gain will be

$$
\begin{aligned}
K &= PH^T(HPH^T)^{-1} \\
&= GG^T H^T(HGG^T H^T)^{-1} \\
&= G(HG)^{-1} .
\end{aligned}
$$ (11.108)

Note that the assumption of HG being invertible is essential at this stage. The following result holds.

Lemma 11.9 *Consider the system of (11.97) with K given by (11.108). Under the given assumptions, there will be full loop transfer recovery in the sense that*

$$G_R(q)G_S(q) = G_0(q) ,$$ (11.109)

with $G_0(q)$ given by (11.99).

Proof First note that for the particular filter gain of (11.108)

$$(I - KH)G = G - G = 0 .$$ (11.110)

Hence

$$
\begin{aligned}
&G_R(q)G_S(q) - G_0(q) \\
&= qL[qI - (I - KH)F]^{-1}KH(qI - F)^{-1}G - L(qI - F)^{-1}G \\
&= L[qI - (I - KH)F]^{-1}\{qKH - [qI - (I - KH)F]\}(qI - F)^{-1}G \\
&= -L[qI - (I - KH)F]^{-1}\{(I - KH)(qI - F)\}(qI - F)^{-1}G \\
&= -L[qI - (I - KH)F]^{-1}(I - KH)G \\
&= 0 .
\end{aligned}
$$

In the second last equality, the assumption of a minimum phase system was used, and (11.110) was applied to establish the last equality.   ■

Example 11.7 As an illustration of LTR and alternative approaches for the design of the observer poles, consider control of the system

$$x(t+1) = \begin{pmatrix} \cos(\omega_0 h) & (\sin(\omega_0 h))/\omega_0 \\ -\omega_0 \sin(\omega_0 h) & \cos(\omega_0 h) \end{pmatrix} x(t)$$
$$+ \begin{pmatrix} (1 - \cos(\omega_0 h))/\omega_0^2 \\ (\sin(\omega_0 h))/\omega_0 \end{pmatrix} u(t) ,$$
$$y(t) = (\omega_0^2 \quad 0)x(t) ,$$

with the parameter values being $\omega_0 = 0.2$ and $h = 0.5$. The feedback gain L is designed using standard LQ theory. Using the parameters $Q_1 = I$, $Q_2 = 1$ gives

$$L = (0.651 \quad 1.314) .$$

The behaviour of the closed loop system for state feedback control is illustrated in Figure 11.9.

Next, LTR is applied. The behaviour of the closed loop system is not acceptable in this case. It turns out that the output follows the reference value almost as for a true state feedback. However, the input oscillates fiercely. It can be expected that a system so designed will be very sensitive to disturbances. It is instructive to compute the observer eigenvalues in this case. Set for simplicity

$$s = \sin(\omega_0 h) , \qquad c = \cos(\omega_0 h) .$$

Then one obtains

$$F - FKH = F[I - G(HG)^{-1}H]$$
$$= \begin{pmatrix} c & s/\omega_0 \\ -\omega_0 s & c \end{pmatrix} \begin{pmatrix} 0 & 0 \\ -\omega_0 s/(1-c) & 1 \end{pmatrix}$$
$$= \frac{1}{1-c} \begin{pmatrix} -s^2 & s(1-c)/\omega_0 \\ -\omega_0 sc & c(1-c) \end{pmatrix} ,$$

which has eigenvalues $\lambda_1 = 0$ and $\lambda_2 = -1$. It is the second eigenvalue that accounts for the severe input oscillations. In a more general setting, it can be shown that, for LTR, the observer dynamics have some poles in the origin and also at all zero positions of the open loop system. Compare with Exercise 11.8.

As a way of making the system less sensitive, a small R_2 matrix can be used instead of $R_2 = 0$. Case (c) in Figure 11.9 corresponds to the case where

$$R_1 = GG^T , \qquad R_2 = 10^{-3} .$$

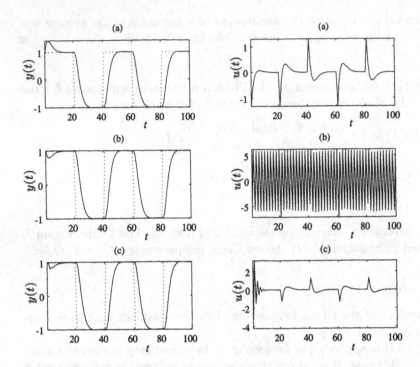

Fig. 11.9. Behaviour of the closed loop system, Example 11.7. *Left*: output signal (*solid lines*) and reference signal (*dotted lines*). *Right*: input signal. Cases: **(a)** state feedback, **(b)** LTR, **(c)** modified LTR

Compared with case (b), the oscillations are now damped while the output still follows the reference signal almost as previously. This idea, to use a modified LTR with R_2 being "a small matrix" instead of zero, is a general procedure when employing LTR for controller design.

For comparison, some other ways of choosing R_1 and R_2 were selected as well. In order to obtain a normalization, $R_1 = I$ was chosen fixed and the scalar R_2 was varied. Figure 11.10 displays the behaviour for the values 1, 10^2 and 10^4 of R_2. It can be seen, in these cases, that the output behaves similarly to that where state feedback is used, while the input signal has a much smoother behaviour than in the LTR cases. For very large values of R_2, the transient phase before the output reasonably follows the reference signal tends to be quite long.

The closed loop poles for the different cases are listed in Table 11.1. Note that the controller poles (the eigenvalues of $F - GL$) are the same in all cases.

As a further illustration, the Nyquist curves of the loop gain for the different cases are plotted in Figure 11.11.

It can be seen from the diagram that LTR gives no change in the loop gain. The regulator based on the modified LTR gives an insignificant change

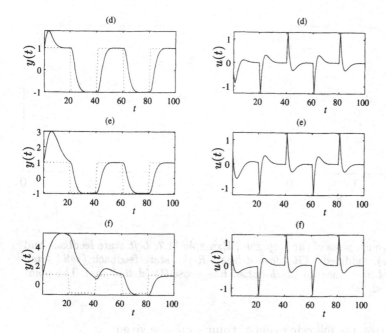

Fig. 11.10. Behaviour of the closed loop system, Example 11.7. *Left*: output signal (*solid lines*) and reference signal (*dotted lines*). *Right*: input signal. Cases: **(d)** $R_2 = 1$, **(e)** $R_2 = 10^2$, **(f)** $R_2 = 10^4$

Table 11.1. Closed loop poles, Example 11.7

Controller poles	$0.63 \pm i0.16$
Observer poles	
LTR	$-1, 0$
Approx LTR	$-0.57, -0.02$
$R_1 = I$, $R_2 = 1$	$0.50, 0.50$
$R_1 = I$, $R_2 = 100$	$0.84 \pm i0.13$
$R_1 = I$, $R_2 = 10000$	$0.95 \pm i0.05$

of the loop gain. Finally, cases (d) and (e) give a somewhat decreased stability margin, as the Nyquist curves in these cases move towards the point $z = -1$.   □

When applied in continuous-time, a possible drawback of the exact LTR approach is that the loop gain will be forced to have a "roll-off" as a pure integrator for high frequencies, as its pole excess will be one. If the open loop system has a larger pole excess than one, this implies that the regulator will have very high gain for large frequencies. Measurement errors at the output will then be amplified considerably at the input.

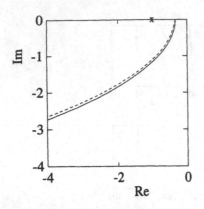
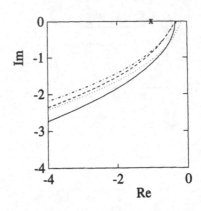

Fig. 11.11. Nyquist plots of the loop gain, Example 11.7. *Left:* state feedback and LTR (*solid line*), modified LTR (*dashed line*). *Right:* state feedback (*solid line*), case (d) (*dashed line*), case (e) (*dash-dotted line*), case (f) (*dotted line*). The point $s = -1$ is marked "x"

To summarize, the following rule of thumb may be given:

- First design an observer/state estimator by means of pole placement or the like. Check its robustness and sensitivity properties. If the Nyquist curve of the loop gain does not pass "close" to the point $z = -1$, then the design should be accepted from such a perspective.
- If the Nyquist curve of the loop gain passes close to $z = -1$, try a modified LTR design to improve the robustness and sensitivity properties.

11.A Appendix. Derivation of the Optimal Linear Quadratic Gaussian Feedback and the Riccati Equation from the Bellman Equation

In this section we will derive the optimal feedback result of Theorem 11.2 for complete state information, and therefore also the result of Theorem 11.1 for deterministic systems.

We will use the Bellman equation, see (10.53), which in this case reads

$$V(x(t), t) = \min_{u(t)} \mathbf{E} \left[x^T(t)Q_1 x(t) + x^T(t)Q_{12} u(t) + u^T(t)Q_{21} x(t) \right.$$

$$\left. + u^T(t)Q_2 u(t) + V(x(t+1), t+1)|x(t) \right]. \tag{11.111}$$

Considering the loss-to-go, $V(x(t), t)$, at time $t = N$ we find

$$V(x(N), N) = x^T(N)Q_0 x(N). \tag{11.112}$$

We next postulate that the general solution to the Bellman equation (11.111) will take the form

$$V(x(t),t) = x^T(t)S(t)x(t) + \sigma(t) , \tag{11.113}$$

where $S(t)$ is some nonnegative definite matrix, and $\sigma(t)$ is a nonnegative scalar. This is apparently true at least for $t = N$:

$$S(N) = Q_0 , \quad \sigma(N) = 0 , \tag{11.114}$$

and we will prove by induction (going backward in time) that it holds for any value of t. Further, the optimal feedback will be a linear time-varying one:

$$u(t) = -L(t)x(t) . \tag{11.115}$$

Assume now that the postulated solution (11.113) is true for time $t+1$. Inserting it into the Bellman equation (11.111) and making use of the system dynamics

$$x(t + 1) = Fx(t) + Gu(t) + v(t) \tag{11.116}$$

we get

$$\begin{aligned}
V(x(t),t) = \mathbf{E} &\left[\min_{u(t)} x^T(t)Q_1 x(t) + x^T(t)Q_{12}u(t) \right. \\
&+ u^T(t)Q_{21}x(t) + u^T(t)Q_2 u(t) \\
&\left. + (Fx(t) + Gu(t) + v(t))^T S(t+1)(Fx(t) + Gu(t) + v(t))|x(t) \right] \\
&+ \sigma(t+1) \\
= \min_{u(t)} \mathbf{E} &\left[x^T(t)(Q_1 + F^T S(t+1)F)x(t) \right. \\
&+ x^T(t)(F^T S(t+1)G + Q_{12})u(t) \\
&+ u^T(t)(G^T S(t+1)F + Q_{21})x(t) \\
&\left. + u^T(t)(Q_2 + G^T S(t+1)G)u(t) + \operatorname{tr} R_1 S(t+1) \right] \\
&+ \sigma(t+1) .
\end{aligned} \tag{11.117}$$

As the right-hand side is a quadratic function of $u(t)$, the optimal feedback is readily found to be

$$u(t) = -L(t)x(t) , \tag{11.118}$$

$$L(t) = \left(Q_2 + G^T S(t+1)G \right)^{-1} \left(G^T S(t+1)F + Q_{21} \right) . \tag{11.119}$$

Inserting the expression (11.118) for $u(t)$ into the Bellman equation (11.117) then verifies that the solution can indeed be written as in (11.113) with

$$\sigma(t) = \sigma(t+1) + \operatorname{tr}\left(S(t+1)R_1\right) \tag{11.120}$$

$$\begin{aligned}
S(t) = Q_1 + F^T S(t+1)F &- (F^T S(t+1)G + Q_{12}) \\
&\times (Q_2 + G^T S(t+1)G)^{-1}(G^T S(t+1)F + Q_{21}) .
\end{aligned} \tag{11.121}$$

Finally, the minimal loss, $\min \mathbf{E}\,\ell$, will be obtained by evaluating the expected value of $V(x(t_0), t_0)$:

$$\begin{aligned}
\min \mathbf{E}\,\ell &= \mathbf{E}\, V(x(t_0), t_0) \\
&= m^T S(t_0)m + \operatorname{tr}\left(S(t_0)R_0\right) + \sigma(t_0) \\
&= m^T S(t_0)m + \operatorname{tr}\left(S(t_0)R_0\right) + \sum_{t=t_0}^{N-1} \operatorname{tr}\left(S(t+1)R_1\right) .
\end{aligned} \tag{11.122}$$

In this calculation we used Lemma 2.1. The derived result (11.122) is precisely (11.14).

Exercises

Exercise 11.1 Consider the regulator

$$u(t) = -L\hat{x}(t|t-2) \,,$$

where $\hat{x}(t|t-2)$ is the stationary optimal two-step predictor. Rewrite the regulator in state space form, using $y(t)$ as the input to the regulator and $u(t)$ as the output.

Exercise 11.2 Verify that, for Example 11.5, $\mathbf{E}\,u^2(t) = 5.05$ when $\rho = 0$ and $\mathbf{E}\,y^2(t) = 8.85$ when $\rho \to \infty$.

Exercise 11.3 Consider the minimum phase system

$$A(q)y(t) = B(q)u(t) + C(q)e(t) \,,$$

where $\deg A - \deg B = k > 0$, and $C(z)$ has all zeros inside the unit circle. Suppose that the output variance is to be minimized but that the hardware to be used is slow compared with the sampling rate, so that $u(t)$ must be a function of $u(t-1), u(t-2), \ldots, y(t-\ell), y(t-\ell-1), \ldots$. Find the minimum variance regulator.

Exercise 11.4 Consider the system

$$y(t) - 0.8y(t-1) = 2.0u(t-3) + v(t) - 2v(t-1) \,,$$

where $v(t)$ is white noise of zero mean and unit variance. Find the regulator that minimizes the stationary output variance. What is that minimal variance?

Exercise 11.5 Consider the system

$$A(q)y(t) = B(q)u(t) + C(q)e(t)$$

with

$$A(q) = q^n + a_1 q^{n-1} + \ldots + a_n \,,$$
$$B(q) = b_0 q^m + \ldots + b_m \,, \qquad b_0 \neq 0, \ \ k = n - m > 0 \,,$$
$$C(q) = q^n + c_1 q^{n-1} + \ldots + c_n \,.$$

Assume that $C(z)$ has all zeros strictly inside the unit circle, so that the system is given in the innovations form. Find a regulator that, at time t, minimizes

$$V = \mathbf{E}\left[y^2(t+k) + \rho u^2(t)|Y^t\right] \,.$$

(a) Show that the regulator minimizing V is given by

$$[b_0qB(q)F(q) + \rho C(q)]u(t) = -b_0qL(q)y(t) .$$

Hint. Show first that the criterion can be written as

$$V = \mathbf{E}\left[F(q)e(t+1)\right]^2 + \left[\frac{qL(q)}{C(q)}y(t) + \frac{qB(q)F(q)}{C(q)}u(t)\right]^2 + \rho u^2(t) ,$$

where the usual predictor identity

$$z^{k-1}C(z) \equiv A(z)F(z) + L(z)$$

has been invoked.

(b) Show that the closed loop system satisfies

$$[b_0q^kB(q) + \rho A(q)]y(t) = [b_0qB(q)F(q) + \rho C(q)]e(t) .$$

Remark. Note that even if the open loop system is nonminimum phase (so that $B(z)$ has any zeros outside the unit circle) the closed loop system may become asymptotically stable by an appropriate choice of the control weighting ρ.   □

Exercise 11.6 Consider the unstable system

$$x(t+1) = 2x(t) + u(t) + v(t) ,$$
$$\mathbf{E}\,v(t)v(s) = \delta_{t,s} ,$$

and the criterion

$$V = \mathbf{E}\,\frac{1}{N}\left[q_0x^2(N) + \sum_{t=0}^{N-1} q_2u^2(t)\right] .$$

Determine the optimal control law explicitly. Assume the state to be exactly measurable. What is the limit $\bar{S} = \lim_{N\to\infty} S(t)$? What is the corresponding feedback gain? Also examine the closed loop poles and the set of solutions to the ARE.

Exercise 11.7 Consider the system

$$y(t) - 2y(t-1) = u(t-1) - 0.5u(t-2) + e(t) ,$$

and the criterion

$$V = \lim_{N\to\infty}\frac{1}{N}\mathbf{E}\sum_{t=1}^{N}[y^2(t) + \rho u^2(t)] .$$

Determine the optimal regulator when $u(t)$ depends on the available output data Y^t. Also determine the closed loop poles. What is the influence of the control weighting ρ? What happens if a further time delay is included in the system?

Exercise 11.8 Consider the observer dynamics using the LTR approach:

$$F_c = F - FG(HG)^{-1}H$$

(a) Show that F_c has at least one eigenvalue in the origin.
(b) Let $z \neq 0$ be a zero of the open loop transfer function (and assume that it is not also a pole). Show that z will also be an eigenvalue of the observer dynamics. Phrased differently, prove the implication

$$\det[H(zI - F)^{-1}G] = 0 \Rightarrow \det[zI - F_c] = 0.$$

Hint. In the matrix manipulations needed, Lemma 6.9 is useful.

Bibliography

For some other alternative texts on LQG control, see

Anderson, B.D.O., Moore, J.B., 1989. *Optimal Control. Linear Quadratic Methods.* Prentice Hall International, Hemel Hempstead, UK.

Åström. K.J., 1970. *Introduction to Stochastic Control.* Academic Press, New York.

Grimble, M.J., Johnson, M.A., 1988. *Optimal Control and Stochastic Estimation,* vol. 2. John Wiley & Sons, Chichester.

Kwakernaak, H., Sivan, R., 1982. *Linear Optimal Control Systems.* Wiley, New York.

Maybeck, P.S., 1982. *Stochastic Models, Estimation and Control,* vol. 3. Academic Press, New York.

Including some more terms in the generalized minimum variance criterion, treated in Exercise 11.5, and thus also penalizing future output deviations and control actions is often called generalized predictive control. See

Bitmead, R.R., Gevers. M., Wertz, V., 1990. *Adaptive Optimal Control. The Thinking Man's GPC.* Prentice Hall International, Hemel Hempstead, UK.

Clarke, D.W., Mohtadi, C., Tuffs, P.S., 1987. Generalized predictive control–Part I. The basic algorithm. *Automatica,* vol. 23, 137–148.

Clarke, D.W., Mohtadi, C., Tuffs, P.S., 1987. Generalized predictive control–Part II. Extensions and interpretations. *Automatica,* vol. 23, 149–160.

for this design method.
There are numerous books dedicated to control system design. The above texts on LQG design include many such aspects. Modern treatments of the design in a general setting include

Doyle, J.C., Francis, B.A., Tannenbaum, A.R., 1992. *Feedback Control Theory.* Macmillan, New York.

Glad, T. , Ljung, L., 2000. *Control Theory. Multivariable and Nonlinear Methods.* Taylor and Francis, London.

Goodwin, G.C., Graebe, S.F., Salgado, M.E., 2001. *Control System Design.* Prentice Hall, Upper Saddle River.

Maciejowski, J.M., 1989. *Multivariable Feedback Design.* Addison-Wesley, Wokingham, UK.

Skogestad, S., Postlethwaite, I., 1996. *Multivariable Feedback Control.* John Wiley, New York, NY.

The paper

Bitmead, R.R., Gevers. M., Wertz, V., 1989. Adaptation and robustness in predictive control. *Proceedings of 28th IEEE Conference on Decision and Control*, Tampa, FL.

summarizes many useful results on LTR, particularly in connection with discrete time LQG control.

Answers to Selected Exercises

2.1. (a) $\mathbf{E}\,\xi = 0$, $\mathbf{E}\,\xi^2 = a^2/3$, $\mathbf{E}\,\xi^3 = 0$, $\mathbf{E}\,\xi^4 = a^4/5$.

(b) $\mathbf{E}\,\xi = m$, $\mathbf{E}\,\xi^2 = m^2 + \sigma^2$, $\mathbf{E}\,\xi^3 = m^3 + 3m\sigma^2$,
$\mathbf{E}\,\xi^4 = m^4 + 6m^2\sigma^2 + 3\sigma^4$.

2.2. $\mathbf{E}\begin{pmatrix} \xi \\ \eta \end{pmatrix} = \frac{9}{14}\begin{pmatrix} 1 \\ 1 \end{pmatrix}$, $\quad \operatorname{cov}\begin{pmatrix} \xi \\ \eta \end{pmatrix} = \begin{pmatrix} 199/2940 & -5/588 \\ -5/588 & 199/2940 \end{pmatrix}$.

2.3. (a) $|\rho| < 1$.

(c) $p_{\eta|\xi=x}(y) \sim \mathbf{N}(2 - 2\rho + 2\rho y, 4 - 4\rho^2)$.

2.4. $B = R_{ve}R_e^{-1}$.

2.5. (a) $p_\zeta(z) = \int p_\xi(x) p_\eta(z - x)\, dx$.

(b) $\quad p_\zeta(z) = \begin{cases} 0 & z < 2a \\ (z - 2a)/(b - a)^2 & 2a \le z < a + b \\ (2b - z)/(b - a)^2 & a + b \le z < 2b \\ 0 & 2b \le z \end{cases}$.

2.7. (a) $\mathbf{E}\,\xi^2(t) = \mathbf{E}\,\eta^2(t) = (m^2 + \sigma^2)/2$, $\mathbf{E}\,\xi(t)\eta(t) = 0$.

(b) $\mathbf{E}\,\zeta = 0$, $\mathbf{E}\,\zeta^2 = 0$, $\mathbf{E}\,|\zeta|^2 = \sigma^2$, $\mathbf{E}\,|\zeta| = 2\sigma/\sqrt{2\pi}$.

3.1. $(q^3 - 2.3q^2 + 1.9q - 0.56)y(t) = (q^2 - 0.3q - 0.4)u(t)$
$+ (q^3 - 0.8q^2 - 0.35q + 0.49)e(t)$.

3.4. (a) $p(x(t+1)|x(t)) = \gamma(x(t+1); Fx(t), R_1)$,
$p(x(t+1)|x(t), y(t)) = \gamma(x(t+1); m, P)$,
$m = Fx(t) + R_{12}R_2^{-1}(y(t) - Hx(t))$, $P = R_1 - R_{12}R_2^{-1}R_{21}$.
The distributions coincide iff $R_{12} = 0$.

(b) For example:
$$\bar{F} = \begin{pmatrix} F - BH & B \\ H(F - BH) & HB \end{pmatrix},$$

$$\bar{R}_1 = \begin{pmatrix} I \\ H \end{pmatrix}(R_1 - R_{12}R_2^{-1}R_{21})(I \ \ H^T) + \begin{pmatrix} 0 & 0 \\ 0 & R_2 \end{pmatrix},$$

where $B = R_{12}R_2^{-1}$.

4.3. $y(t) = (q - 0.5)/(q - 0.8)\varepsilon(t)$, $\quad \mathbf{E}\,\varepsilon^2(t) = 182$.

4.4. $\mathbf{E}\,y^2(t) = r/(2a)$.

4.8. $d_0 > 0$, $|d_1| < d_0$.

4.11.

$$\begin{pmatrix} r(0) \\ r(1) \\ r(2) \end{pmatrix} = \frac{\lambda^2}{(1 - a_2)[(1 + a_2)^2 - a_1^2]} \begin{pmatrix} 1 + a_2 \\ -a_1 \\ a_1^2 - a_2(1 + a_2) \end{pmatrix}.$$

4.12.

$$r(k) = \frac{(-a)^k}{1 - ac}, \quad k \ge 0, \qquad r(-k) = \frac{(-c)^k}{1 - ac}, \quad k \ge 0.$$

4.13. $r(\tau) = (\alpha_1 e^{-\alpha_2 \tau} - \alpha_2 e^{-\alpha_1 \tau})/[2\alpha_1\alpha_2(\alpha_1^2 - \alpha_2^2)]$.

4.14. $f = e^{-ah}, \qquad \lambda^2 = \frac{r}{2a}(1 - e^{-2ah})$.

4.16. Set $A = A_R + iA_I$, $\quad P = P_R + iP_I$.

$$\tilde{A} = \begin{pmatrix} A_R & -A_I \\ A_I & A_R \end{pmatrix}, \qquad \tilde{P} = 0.5 \begin{pmatrix} P_R & -P_I \\ P_I & P_R \end{pmatrix}.$$

4.17.

$$y(t) = \frac{1}{q^2 - 0.5q + 0.5} \begin{pmatrix} q^2 - 0.5q & 0.25q \\ -0.5q + 0.25 & q^2 - 0.5q + 0.375 \end{pmatrix} \varepsilon(t),$$

$$\mathbf{E}\,\varepsilon(t)\varepsilon^T(t) = \begin{pmatrix} 2.5 & 1 \\ 1 & 2 \end{pmatrix}.$$

4.19. (a)

$$F^{(2)} = \begin{pmatrix} F^{(1)} & R_{12}^{(1)} R_2^{(1)-1} \\ 0 & 0 \end{pmatrix},$$

$$R_1^{(2)} = \begin{pmatrix} R_1^{(1)} - R_{12}^{(1)} R_2^{(1)-1} R_{21}^{(1)} & 0 \\ 0 & R_2^{(1)} \end{pmatrix},$$

$$H^{(2)} = \begin{pmatrix} H^{(1)} & 0 \end{pmatrix}, \quad R_2^{(2)} = 0.$$

(b) $\quad F = 0.8, \ H = 1, \ R_1 = 0.475, \ R_2 = 0.375$.

4.20. $r(0) > 0, \ -0.5(a+1) \le r(1)/r(0) \le 0.5(1-a)$.

4.21. There is a lack of uniqueness caused by an undetermined scaling factor.

4.23. (a) $R(-\mu, -\nu) = \beta_3 a^{2\nu-\mu}/(1 - a^3)$.

(c) $\text{var}(\hat{a} - a) = \frac{1}{N} \frac{1 - a^2}{a^{2(\mu-1)}}$, minimal for $\mu = 1$.

(d)

$$\text{var}(\hat{a} - a) = \frac{1}{N} \frac{\beta_2}{\beta_3^2} \frac{(1 - a^3)^2}{a^{2(2\nu-\mu-1)}}$$

$$\times \left[\frac{\beta_2^2}{(1 - a^2)^2}(1 + 2a^{2\nu-2\mu}) + \frac{\beta_4 - 3\beta_2^2}{1 - a^4} a^{2\nu-2\mu} \right],$$

minimal for $\mu = \nu = 1$.

5.1. $(x|y) \sim \mathbf{N}(\hat{x}, P), \quad \hat{x} = m_x + R_x C^T (CR_x C^T + R_e)^{-1}(y - Cm_x),$
$P = R_x - R_x C^T (CR_x C^T + R_e)^{-1} CR_x.$

5.2. (a) $(x|y) \sim \mathbf{N}(\hat{x}, P), \quad \hat{x} = m + R_x A^T (AR_x A^T + R_v)^{-1}(y - Am),$
$P = (R_x^{-1} + A^T R_v^{-1} A)^{-1}.$

(c) $\hat{x} \to (A^T R_v^{-1} A)^{-1} A^T R_v^{-1} y, \quad$ as $R_x^{-1} \to 0.$
$P \to (A^T R_v^{-1} A)^{-1}, \quad$ as $R_x^{-1} \to 0.$

(d) $\quad \mathbf{E}[b|X^t] = \dfrac{\sum u(t)[x(t+1) - ax(t)]}{\sum u^2(t)}, \qquad \text{var}[b|X^t] = \dfrac{\lambda^2}{\sum u^2(t)}.$

5.3. (a) $\hat{x}_{MS} = 0.5.$

(b) $\qquad$ $E[x|y] = \begin{cases} (y+r)/2, & -r < y \le r, \\ y, & r < y \le 1-r, \\ (1+y-r)/2, & 1-r < y \le 1+r. \end{cases}$

(c) $\hat{x}_{LLMS} = 0.5 + (y - 0.5)/(1 + 4r^2)$.

6.3. $\hat{y}(t + \tau|t) = Ce^{A\tau}x(t)$.

6.6. (a) $q_1 = 0$, $\qquad P_1 = \begin{pmatrix} 1 & c \\ c & c^2 \end{pmatrix}$ semidefinite.

$q_2 = c^2 - 1$, $P_2 = \begin{pmatrix} c^2 & c \\ c & c^2 \end{pmatrix}$ positive definite if $c^2 > 1$, indefinite if $c^2 < 1$.

(b) $q(t + 1) = c^2 q(t)/[1 + q(t)]$.

6.10. $\qquad \overline{K}_f = P \begin{pmatrix} I \\ (F - KH)^T \\ \vdots \\ (F - KH)^{T^m} \end{pmatrix} H^T (HPH^T + R_2)^{-1}$

where P is the solution to the standard ARE, and

$$K = FPH^T(HPH^T + R_2)^{-1},$$
$$P^{(0,i)} = (F - KH)^i P, \quad i \ge 0,$$
$$P^{(k,\ell)} = \left[I - \sum_{\mu=0}^{k-1} P(F - KH)^{T^\mu} H^T (HPH^T + R_2)^{-1} \right.$$
$$\left. H(F - KH)^\mu \right] (F - KH)^{\ell - k} P, \quad \ell \ge k \ge 1.$$

6.13. $\hat{x}(kh + \tau|kh) = e^{A\tau}\hat{x}(kh|kh)$, $\qquad 0 \le \tau < h$,

$P(kh + \tau|kh) = e^{A\tau}P(kh|kh)e^{A^T\tau} + \int_0^\tau e^{As}R_c e^{A^Ts}\, ds$.

6.14. $P(t + k|t) \ge P(t + k|t + 1)$ is always true.

$P(t + k|t) \ge P(t + k - 1|t)$ does not always hold.

6.15.
$$\hat{x}(t + 1|t) = \begin{pmatrix} 2y(t) - y(t - 1) \\ (y(t) - y(t - 1))/h \end{pmatrix},$$
$$\hat{x}(t|t) = \begin{pmatrix} y(t) \\ (y(t) - y(t - 1))/h \end{pmatrix}.$$

6.17. (a) SNR $= 1.889$.

(b)
$$\hat{x}(t + 1|t = 0.8\hat{x}(t|t - 1) + 0.4\tilde{y}(t)$$
$$y(t) = \hat{x}(t|t - 1) + \tilde{y}(t)$$

(c)
$$\begin{cases} h_0 = 0.5, \\ h_j = 0.4(0.8)^{j-1}, & j > 0 \\ h_j = 0.5(0.4)^{-j}, & -m \le j < 0. \end{cases}$$

6.20. (a) $\qquad F = \begin{pmatrix} 1 & h \\ 0 & 1 \end{pmatrix}$, $\qquad R_1 = \begin{pmatrix} h^3/3 & h^2/2 \\ h^2/2 & h \end{pmatrix}$,

(b)

$$\beta = (3 - \sqrt{3})/h \,,$$

$$P(t+h|t) = \frac{1}{6} \begin{pmatrix} h^3(2+\sqrt{3}) & h^2(3+\sqrt{3}) \\ h^2(3+\sqrt{3}) & h(6+\sqrt{3}) \end{pmatrix},$$

$$P(t|t) = \frac{1}{6} \begin{pmatrix} 0 & 0 \\ 0 & h\sqrt{3} \end{pmatrix}.$$

(c)

$$\hat{x}(t+\tau|t) = \frac{1}{q+(h\beta-1)} \begin{pmatrix} (1+\tau\beta)q + (h\beta-\tau\beta-1) \\ \beta(q-1) \end{pmatrix} y(t) \,,$$

$$P(t+\tau|t) = \frac{h}{\sqrt{12}} \begin{pmatrix} \tau^2 & \tau \\ \tau & 1 \end{pmatrix} + \begin{pmatrix} \tau^3/3 & \tau^2/2 \\ \tau^2/2 & \tau \end{pmatrix}.$$

6.25.

$$\overline{P} = \begin{pmatrix} P & (F-KH)P & (F-KH)^2P \\ P(F-KH)^T & P-PWP & \overline{P}_{23} \\ P\left((F-KH)^T\right)^2 & \overline{P}_{23}^T & \overline{P}_{33} \end{pmatrix},$$

where

$$W = H^T(HPH^T + R_2)^{-1}H \,,$$
$$\overline{P}_{23} = (I - PW)(F - KH)P \,,$$
$$\overline{P}_{33} = P - PWP - P(F-KH)^T W(F-KH)P \,.$$

The stationary filter gain is

$$\overline{K} = \begin{pmatrix} FPH^T(HPH^T + R_2)^{-1} \\ P(F-KH)^T H^T(HPH^T + R_2)^{-1} \\ P\left((F-KH)^T\right)^2 H^T(HPH^T + R_2)^{-1} \end{pmatrix}.$$

7.1.

$$\hat{y}(t+1|t) = \frac{1.3q}{q+0.5}y(t), \qquad \mathrm{E}\,\tilde{y}^2(t+1|t) = 4.$$

$$\hat{y}(t+2|t) = \frac{1.04q}{q+0.5}y(t), \qquad \mathrm{E}\,\tilde{y}^2(t+2|t) = 10.76.$$

7.2.

$$\phi_{\tilde{s}} = \frac{\lambda_v^2\lambda_e^2}{\lambda_\varepsilon^2}\frac{CC^*MM^*}{BB^*} + \frac{\lambda_v^4}{\lambda_\varepsilon^2}\frac{HH^*L_oL_o^*}{FF^*BB^*} \,.$$

7.3. (b) $G(q) = N_m(q)/D(q)$,
$D(q) = q^4 - 1.0738q^3 + 0.6832q^2 - 0.2128q + 0.0248$
$N_0(q) = 0.5631q^4 - 0.1905q^3$
$N_1(q) = 0.2721q^4 + 0.1549q^3$
$N_2(q) = 0.0321q^4 + 0.2240q^3 + 0.1774q^2$
$N_5(q) = 0.0012q^4 - 0.0312q^3 - 0.0029q^2 + 0.0833q + 0.1905 + 0.1774q^{-1}$
$N_{-1}(q) = 0.6542q^4 - 0.3942q^3$
$N_{-2}(q) = 0.5871q^4 - 0.4579q^3$
$N_{-5}(q) = 0.0387q^4 - 0.1562q^3$

7.9. (b)
$$\hat{y}(t+1|t) = \frac{(c-a)(1-ac)}{1+c^2-2ac}y(t),$$

$$\mathbf{E}\left[\hat{y}(t+1|t) - y(t+1)\right]^2 = \lambda^2\left[1 + \frac{c^2(c-a)^2}{1+c^2-2ac}\right].$$

(c) $n_1 \le n+1$, $n_2 = 0$.

8.3.

$$K(t) = \binom{1}{0}\frac{\lambda^2 c}{\lambda^2 + \alpha(t)},$$

$$c^2 < 1 \quad \alpha^{-1}(t) = \frac{1}{c^{2(t-t_0)}}\alpha^{-1}(t_0) + \frac{1}{\lambda^2 c^2}\frac{1-c^{-2(t-t_0)}}{1-c^{-2}},$$

$$c^2 = 1 \quad \alpha^{-1}(t) = \alpha^{-1}(t_0) + \frac{t-t_0}{\lambda^2}.$$

9.9. (b)

$$\hat{a}(t) = \hat{a}(t-1) + k(t)\varepsilon(t),$$

$$\varepsilon(t) = y(t) + \hat{a}(t-1)y(t-1),$$

$$k(t) = \frac{-y(t-1)r(t-1)}{\lambda^2 + y^2(t-1)r(t-1)},$$

$$r^{-1}(t+1) = r^{-1}(t) + \frac{y^2(t)}{\lambda^2}.$$

9.10. (a) $\hat{x}(t|t) = [r_0\sum_{s=1}^{t}y(s) + r_2\hat{x}(0|0)]/(r_2 + tr_0)$.

(b) $\text{var}[\frac{1}{t}\sum_{s=1}^{t}y(s)] = r_2/t$.

(c) $p(x(t)|Y^t)$ is uniformly distributed over
$(\max_{1\le s\le t}y(s) - a, \min_{1\le s\le t}y(s) + a)$.

(d) $\text{var}\left[\frac{1}{t}\sum_{s=1}^{t}y(s)\right] = \frac{a^2}{3t}$,

$\text{var}[\mathbf{E}\,(x(t)|Y^t)] = \frac{2a^2}{(t+1)(t+2)}$.

10.5. $u(t) = -(ab_0)/(r_b + b_0^2)x(t)$, with $b_0 = \mathbf{E}\,b$, $r_b = \text{var}(b)$.

11.1. For example,

$$\bar{x}(t+1) = \begin{pmatrix} F - KH & G \\ -LF & -LG \end{pmatrix}\bar{x}(t) + \binom{K}{0}y(t),$$

$$u(t) = (0\;\;I)\bar{x}(t).$$

11.3. $u(t) = -L(q)/[B(q)F(q)]y(t)$, where $z^{k+\ell+1}C(z) \equiv A(z)F(z) + L(z)$.

11.4. $u(t) = -0.096y(t) - 0.3u(t-1) - 0.24u(t-2)$.

11.6.

$$S(t) = \left[\frac{1}{4^{N-t}q_0} + \frac{4^{N-t}-1}{3q_2 4^{N-t}}\right]^{-1},$$

$$L(t) = \frac{2}{1 + q_2 S^{-1}(t+1)}.$$

$$\bar{S} = \lim_{N \to \infty} S(t) = 3q_2 , \qquad \bar{L} = 1.5 .$$

Closed loop pole in $z = 0.5$.

11.7. L is independent of ρ. $L = (2 \ 1)$, if an observable canonical state space representation is chosen.

Closed loop poles in $z = 0$ (multiple pole) and $z = 0.5$.

An increased delay introduces further poles in $z = 0$.

Index